Research Notes in Mathematics

Submission of proposals for consideration

Suggestions for publication, in the form of outlines and representative samples, are invited by the editorial board for assessment. Intending authors should contact either the main editor or another member of the editorial board, citing the relevant AMS subject classifications. Refereeing is by members of the board and other mathematical authorities in the topic concerned, located throughout the world.

Preparation of accepted manuscripts

On acceptance of a proposal, the publisher will supply full instructions for the preparation of manuscripts in a form suitable for direct photo-lithographic reproduction. Specially printed grid sheets are provided and a contribution is offered by the publisher towards the cost of typing.

Illustrations should be prepared by the authors, ready for direct reproduction without further improvement. The use of hand-drawn symbols should be avoided wherever possible, in order to maintain maximum clarity of the text.

The publisher will be pleased to give any guidance necessary during the preparation of a typescript, and will be happy to answer any queries.

Important note

In order to avoid later retyping, intending authors are strongly urged not to begin final preparation of a typescript before receiving the publisher's guidelines and special paper. In this way it is hoped to preserve the uniform appearance of the series.

Titles in this series

1 Improperly posed boundary value problems
A Carasso and A P Stone
2 Lie algebras generated by finite dimensional ideals
I N Stewart
3 Bifurcation problems in nonlinear elasticity
R W Dickey
4 Partial differential equations in the complex domain
D L Colton
5 Quasilinear hyperbolic systems and waves
A Jeffrey
6 Solution of boundary value problems by the method of integral operators
D L Colton
7 Taylor expansions and catastrophes
T Poston and I N Stewart
8 Function theoretic methods in differential equations
R P Gilbert and R J Weinacht
9 Differential topology with a view to applications
D R J Chillingworth
10 Characteristic classes of foliations
H V Pittie
11 Stochastic integration and generalized martingales
A U Kussmaul
12 Zeta-functions: An introduction to algebraic geometry
A D Thomas
13 Explicit *a priori* inequalities with applications to boundary value problems
V G Sigillito
14 Nonlinear diffusion
W E Fitzgibbon III and H F Walker
15 Unsolved problems concerning lattice points
J Hammer
16 Edge-colourings of graphs
S Fiorini and R J Wilson
17 Nonlinear analysis and mechanics: Heriot-Watt Symposium Volume I
R J Knops
18 Actions of fine abelian groups
C Kosniowski
19 Closed graph theorems and webbed spaces
M De Wilde
20 Singular perturbation techniques applied to integro-differential equations
H Grabmüller
21 Retarded functional differential equations: A global point of view
S E A Mohammed
22 Multiparameter spectral theory in Hilbert space
B D Sleeman
24 Mathematical modelling techniques
R Aris
25 Singular points of smooth mappings
C G Gibson
26 Nonlinear evolution equations solvable by the spectral transform
F Calogero
27 Nonlinear analysis and mechanics: Heriot-Watt Symposium Volume II
R J Knops
28 Constructive functional analysis
D S Bridges
29 Elongational flows: Aspects of the behaviour of model elasticoviscous fluids
C J S Petrie
30 Nonlinear analysis and mechanics: Heriot-Watt Symposium Volume III
R J Knops
31 Fractional calculus and integral transforms of generalized functions
A C McBride
32 Complex manifold techniques in theoretical physics
D E Lerner and P D Sommers
33 Hilbert's third problem: scissors congruence
C-H Sah
34 Graph theory and combinatorics
R J Wilson
35 The Tricomi equation with applications to the theory of plane transonic flow
A R Manwell
36 Abstract differential equations
S D Zaidman
37 Advances in twistor theory
L P Hughston and R S Ward
38 Operator theory and functional analysis
I Erdelyi
39 Nonlinear analysis and mechanics: Heriot-Watt Symposium Volume IV
R J Knops
40 Singular systems of differential equations
S L Campbell
41 N-dimensional crystallography
R L E Schwarzenberger
42 Nonlinear partial differential equations in physical problems
D Graffi
43 Shifts and periodicity for right invertible operators
D Przeworska-Rolewicz
44 Rings with chain conditions
A W Chatters and C R Hajarnavis
45 Moduli, deformations and classifications of compact complex manifolds
D Sundararaman
46 Nonlinear problems of analysis in geometry and mechanics
M Atteia, D Bancel and I Gumowski
47 Algorithmic methods in optimal control
W A Gruver and E Sachs
48 Abstract Cauchy problems and functional differential equations
F Kappel and W Schappacher
49 Sequence spaces
W H Ruckle
50 Recent contributions to nonlinear partial differential equations
H Berestycki and H Brezis
51 Subnormal operators
J B Conway
52 Wave propagation in viscoelastic media
F Mainardi
53 Nonlinear partial differential equations and their applications: Collège de France Seminar. Volume I
H Brezis and J L Lions
54 Geometry of Coxeter groups
H Hiller
55 Cusps of Gauss mappings
T Banchoff, T Gaffney and C McCrory

56 An approach to algebraic K-theory
A J Berrick
57 Convex analysis and optimization
J-P Aubin and R B Vintner
58 Convex analysis with applications in the differentiation of convex functions
J R Giles
59 Weak and variational methods for moving boundary problems
C M Elliott and J R Ockendon
60 Nonlinear partial differential equations and their applications: Collège de France Seminar. Volume II
H Brezis and J L Lions
61 Singular systems of differential equations II
S L Campbell
62 Rates of convergence in the central limit theorem
Peter Hall
63 Solution of differential equations by means of one-parameter groups
J M Hill
64 Hankel operators on Hilbert space
S C Power
65 Schrödinger-type operators with continuous spectra
M S P Eastham and H Kalf
66 Recent applications of generalized inverses
S L Campbell
67 Riesz and Fredholm theory in Banach algebra
B A Barnes, G J Murphy, M R F Smyth and T T West
68 Evolution equations and their applications
F Kappel and W Schappacher
69 Generalized solutions of Hamilton-Jacobi equations
P L Lions
70 Nonlinear partial differential equations and their applications: Collège de France Seminar. Volume III
H Brezis and J L Lions
71 Spectral theory and wave operators for the Schrödinger equation
A M Berthier
72 Approximation of Hilbert space operators I
D A Herrero
73 Vector valued Nevanlinna Theory
H J W Ziegler
74 Instability, nonexistence and weighted energy methods in fluid dynamics and related theories
B Straughan
75 Local bifurcation and symmetry
A Vanderbauwhede
76 Clifford analysis
F Brackx, R Delanghe and F Sommen
77 Nonlinear equivalence, reduction of PDEs to ODEs and fast convergent numerical methods
E E Rosinger
78 Free boundary problems, theory and applications. Volume I
A Fasano and M Primicerio
79 Free boundary problems, theory and applications. Volume II
A Fasano and M Primicerio
80 Symplectic geometry
A Crumeyrolle and J Grifone
81 An algorithmic analysis of a communication model with retransmission of flawed messages
D M Lucantoni
82 Geometric games and their applications
W H Ruckle
83 Additive groups of rings
S Feigelstock
84 Nonlinear partial differential equations and their applications: Collège de France Seminar. Volume IV
H Brezis and J L Lions
85 Multiplicative functionals on topological algebras
T Husain
86 Hamilton-Jacobi equations in Hilbert spaces
V Barbu and G Da Prato
87 Harmonic maps with symmetry, harmonic morphisms and deformations of metrics
P Baird
88 Similarity solutions of nonlinear partial differential equations
L Dresner
89 Contributions to nonlinear partial differential equations
C Bardos, A Damlamian, J I Díaz and J Hernández
90 Banach and Hilbert spaces of vector-valued functions
J Burbea and P Masani
91 Control and observation of neutral systems
D Salamon
92 Banach bundles, Banach modules and automorphisms of C*-algebras
M J Dupré and R M Gillette
93 Nonlinear partial differential equations and their applications: Collège de France Seminar. Volume V
H Brezis and J L Lions
94 Computer algebra in applied mathematics: an introduction to MACSYMA
R H Rand
95 Advances in nonlinear waves. Volume I
L Debnath
96 FC-groups
M J Tomkinson
97 Topics in relaxation and ellipsoidal methods
M Akgül
98 Analogue of the group algebra for topological semigroups
H Dzinotyiweyi
99 Stochastic functional differential equations
S E A Mohammed
100 Optimal control of variational inequalities
V Barbu
101 Partial differential equations and dynamical systems
W E Fitzgibbon III
102 Approximation of Hilbert space operators. Volume II
C Apostol, L A Fialkow, D A Herrero and D Voiculescu
103 Nondiscrete induction and iterative processes
V Ptak and F-A Potra
104 Analytic functions – growth aspects
O P Juneja and G P Kapoor
105 Theory of Tikhonov regularization for Fredholm equations of the first kind
C W Groetsch

106 Nonlinear partial differential equations and free boundaries
J I Díaz

107 Variational convergences for functions and operators
H Attouch

108 A layering method for viscous, incompressible L_p flows occupying R^n
A Douglis and E B Fabes

Partial differential equations and dynamical systems

W E Fitzgibbon III (Editor)
University of Houston

Partial differential equations and dynamical systems

Pitman Advanced Publishing Program
BOSTON · LONDON · MELBOURNE

PITMAN PUBLISHING LIMITED
128 Long Acre, London WC2E 9AN

PITMAN PUBLISHING INC
1020 Plain Street, Marshfield, Massachusetts 02050

Associated Companies
Pitman Publishing Pty Ltd, Melbourne
Pitman Publishing New Zealand Ltd, Wellington
Copp Clark Pitman, Toronto

First published 1984

AMS Subject Classifications: 3406, 3506, 5806

Library of Congress Cataloging in Publication Data
Main entry under title:

Partial differential equations and dynamical systems.

(Research notes in mathematics)
Includes bibliographical references.
1. Differential equations, Partial—addresses, essays, lectures. 2. Differentiable dynamical systems—addresses, essays, lectures. I. Fitzgibbon, W. E. (William Edward), 1945– . II. Series.
QA377.P295 1984 515.3′53 84-7769
ISBN 0-273-08644-8

British Library Cataloguing in Publication Data

Partial differential equations and dynamical systems.—(Research notes in mathematics; 101)
1. System analysis 2. Differential equations, Partial
I. Fitzgibbon, W. E. II. Series
515.3′53 QA402

ISBN 0-273-08644-8

Reproduced and printed by photolithography
in Great Britain by Biddles Ltd, Guildford

Contents

Preface

During the academic year, 1981-1982, the Mathematics Department of the University of Houston sponsored a "Year of Concentration in Partial Differential Equations and Dynamical Systems." This special year consisted of three major components: (i) a Distinguished Visitors Program; (ii) a Colloquium Program; and (iii) visiting appointments.

The Distinguished Visitors component involved one-week visits to the department by senior applied mathematicians. Each visitor gave the equivalent of three one-hour lectures. A list of the Distinguished Visitors follows:

Neal R. Amundson, University of Houston

Charles C. Conley, University of Wisconsin

Michael G. Crandall, University of Wisconsin

Constantine Dafermos, Brown University

Jim Douglas, Jr., University of Chicago

Morton E. Gurtin, Carnegie-Mellon University

Phillip J. Holmes, Cornell University

Lawrence E. Markus, University of Minnesota

Jerrold E. Marsden, University of California - Berkeley

Lawrence E. Payne, Cornell University

David G. Schaeffer, Duke University

Joel A. Smoller, University of Michigan

Hans F. Weinberger, University of Minnesota

The Colloquium Program was used to complement and supplement the

Distinguished Visitors Program. Twenty-six mathematicians visited the department through the component of the special year.

The remaining component consisted of visiting appointments for applied mathematicians whose research interest coincided with the mathematical focus of program and activities. The long term visitors to the department during the special year were: D.G. Aronson, University of Minnesota and J.W. Evans, University of California - San Diego, Visiting Professors Fall semester 1981; D.S. Levine, University of Pittsburgh, Visiting Assistant Professor, 1981-1982; and M. Mimura, Hiroshima University, Visiting Professor, Spring semester 1982.

This volume represents a partial survey of the mathematics which transpired during the year of concentration. The contributions contained herein fall into roughly three categories: general survey articles; self contained research papers; and shorter descriptions of research the details of which will appear elsewhere. Several of the articles suggest open problems. The nature of the contribution was left to the discretion of the contributor and the articles appear alphabetically by author.

The organization and sponsorship of the special year would not have been possible without financial support from a number of sources. The editor wishes to thank the College of Natural Sciences and Mathematics of the University of Houston and its former Dean, R.H. Walker; the University of Houston Energy Laboratory and its Director, A.C. Hildebrant; and the administration of the University Park Campus of the University of Houston. A grant from the Energy Laboratory was an essential part of the funding for the activities, and this support, together with financial support from the College of Natural Sciences and Mathematics and the Melrose Oil Royalty Fund, is gratefully acknowledged. Finally the program benefited from the personal generosity of Dr. Richard P. Kendall of Exxon Production Research Company.

Many members of the faculty and staff of the Mathematics Department contributed to the programs and activities of the Year of Concentration. The department also benefited from the encouragement and participation of members of the Departments of Chemical Engineering, Mechanical Engineering, and Physics of the University of Houston; members of the Departments of Mathematical Sciences and Mathematics of Rice University; and representatives from local industry.

If a dedication is appropriate for this volume, the dedication belongs to Professor Garret J. Etgen, Professor and Chairman of the Department of Mathematics. Not only did Professor Etgen procure the funding and do a great deal of the planning for the special year, but also his commitment of personal time and energy contributed greatly to its success.

Finally, special thanks for the diligent typing of this manuscript go to Angie Lang, Shun-Lan Lin and, in particular, to Leigh Ann Jacks who additionally assumed the task of the final collation of the manuscript.

The errors which most assuredly occur are the consequence of the carelessness and incompetence of the editor, and the editor herewith and henceforth apologizes.

William E. Fitzgibbon III

Houston, Texas

1984

N R AMUNDSON

Char combustion

1. INTRODUCTION.

While the combustion of carbon as a model for coal may be oversimplified, its modelling and resultant solution structure is very rich as an example of reaction and diffusion. A true model would involve the diffusion of oxygen and carbon dioxide through a boundary layer around the particle followed by diffusion of the same reactants through the interstices of the porous solid where heterogeneous reaction between the carbon dioxide and oxygen and the solid carbon obtains. Reaction products must then follow the same path from the particle. Heat is generated and/or absorbed during the course of the reactions and this must be conducted away. Since the rates at which the reactions take place are non-linear functions of the local state the resulting mathematical model can be a large coupled system of non-linear parabolic partical differential equations. It is further complicated by the fact that the transport of species may not be Fickian and and must be described by the Stefan-Maxwell equations.

2. THE MODEL.

Our object in this paper is not to describe the general problem above but rather to consider only what happens in the stagnant film boundary layer around the particle, assuming that the carbon is so impervious that all reaction between solid and gaseous species occurs at the surface of an infinite slab of carbon which we assume remains at a fixed position. In the more realistic case we will consider a sphere whose radius shrinks as reaction proceeds. We will consider first the quasi-steady state solutions for the following problem.

A stagnant gaseous film of prescribed thickness at the surface of an infinite slab has for one face, $x=0$, solid impervious carbon. At the other face, $x=\ell$, it is exposed to prescribed ambient conditions. In the ambient there will be oxygen, carbon dioxide, carbon monoxide, and nitrogen (derived from air). The oxygen must diffuse through this film to reach the surface where it reacts to form carbon monoxide

$$C+\tfrac{1}{2}O_2 \rightarrow CO \ , \ H_I \qquad \text{I}$$

where heat H_I is generated. The CO must then diffuse through the film, and, as it does, it reacts with O_2 to form carbon dioxide as

$$CO+\tfrac{1}{2}O_2 \rightarrow CO_2 \ , \ H_{III} \qquad \text{III}$$

where H_{III} is the heat generated. The CO_2 so formed may diffuse out through the film or it may diffuse to the surface of the carbon where it reacts as

$$CO_2+C \rightarrow 2CO \ , \ H_{II} \qquad \text{II}$$

to form more CO and at the same time absorbs heat H_{II}. Thus we must consider three diffusing species, the conduction of heat through the film, the reactions at the surface $x=0$ and the distributed reaction III in the film.

If we let g_1, g_2, g_3, g_4 stand for the mass fractions of CO_2, CO, O_2 and the temperature, respectively, then the rates of the reactions are given by

$$R_I = k_I g_3 e^{-\alpha_I/g_4}$$

$$R_{II} = k_{II} g_3 e^{-\alpha_{II}/g_4}$$

$$R_{III} = k_{III} g_3^{\frac{1}{2}} g_2 e^{-\alpha_{III}/g_4}$$

where the α_i and k_i are positive constants. The reaction-diffusion conservation equation for one dimension may then be written as

$$\frac{\partial}{\partial x}(\rho_f D_i \frac{\partial g_i}{\partial x}) + \nu_i M_i k_{III} g_2 g_3^{\frac{1}{2}} e^{-\alpha_{III}/g_4} = \rho_f \frac{\partial g_i}{\partial \theta} , \quad 0<x<\ell$$

where ρ_f is the fluid density, D_i the molecular diffusivity of species i , $D_4=K_f/c_f$ a thermal diffusivity, M_i the moledular weights, and $\nu_1=1$, $\nu_2=-1$, $\nu_3=-\frac{1}{2}$, and $\nu_4=H_{III}/c_f$. (We have written the unsteady state equation.)

At the particle surface there will be four boundary conditions which describe the relation between diffusion to or from the surface and the chemical reactions at the surface as well as the conduction of heat and its generation by the reactions

$$\rho_f D_3 \frac{\partial g_3}{\partial x} = \tfrac{1}{2} k_I g_3 e^{-\alpha_I/g_4}$$

$$-\rho_f D_2 \frac{\partial g_2}{\partial x} = k_I g_3 e^{-\alpha_I/g_4} + 2k_{II} g_1 e^{-\alpha_{II}/g_4}$$

$$\rho_f D_1 \frac{\partial g_1}{\partial x} = k_{II} g_1 e^{-\alpha_{II}/g_4}$$

$$-K_f \frac{\partial g_4}{\partial x} = k_I g_3 H_I e^{-\alpha_I/g_4} - k_{II} g_1 H_{II} e^{-\alpha_{II} g_4}$$

The first boundary condition (B.C.) merely reflects the fact that all of the oxygen which diffuses to the surface must react, the second says that all of the CO formed by the two reactions must diffuse away, the third says that all of the CO_2 which diffuses to the surface must react to form CO , and the last says that the net amount of heat generated at the surface must be conducted away. At the boundary $x=\ell$ we specify

$$g_1=g_{1b}=0\ ,\quad g_2=g_{2b}=0\ ,\quad g_3=g_{3b}\ ,\quad g_4=g_{4b}$$

where we have specified that there will be no CO_2 or CO in the ambient. (This only simplifies the analysis somewhat.) If we were to solve the unsteady state problem initial conditions of the g_i would necessarily be specified.

3. LIMITING SOLUTIONS.

The general numerical solution of the problem as stated turns out to be difficult and it is instructive and informative to consider first quasi-steady state solutions. These are solutions in which we assume that the processes which take place are so slow that at each moment the solutions are in the steady state--all time derivations are zero. For the slab this would mean that the boundary at $x=0$ must be moved toward the right since it is being consumed by the reaction. For the problem of the sphere we would have to make some assumption about the thickness of the boundary layer and its relation to the radius since the latter will be diminishing. Also for the sphere its temperature will change with time so that the quasi-steady state model is less accurate but in this case only two transient equations are necessary--one for the sphere temperature and the other for particle size. More about this later. For the moment, however, let us consider the problem with all time derivations for the slab geometry zero. This we refer to as the quasi-steady state. This problem is still not a trivial problem to solve numerically and to obtain some bounds on the solution we consider some limiting solutions.

A. We assume that there is no reaction in the boundary layer; i.e., the CO formed at the surface diffuses out of the boundary layer so $k_{III}=0$, no CO_2 is present, and the only surface reaction is I. Thus for the slab within the boundary layer we have

$$\frac{d^2 g_i}{dx_i} = 0 \ , \quad 0<x<\ell$$

and the species and temperature profiles are straight lines. We assume ρ_f is constant and $D_2=D_3$. The B.C. at $x=0+$ are then

$$\rho_f D_3 \frac{dg_3}{dx} = \tfrac{1}{2} k_I g_3 e^{-\alpha_I/g_4}$$

$$-\rho_f D_3 \frac{dg_2}{dx} = k_I g_3 e^{-\alpha_I/g_4}$$

$$K_f \frac{dg_4}{dx} = -k_I g_3 H_I e^{-\alpha_I/g_4}$$

The solution is purely algebraic and geometric since the slope of the g_3 straight line must be half of that of the g_2 line at $x=0+$ and since g_{3b} and g_{4b} are specified with $g_{2b}=0$ the whole solution may be drawn as shown in Figure 1. We observe that the values used in the boundary conditions are the limiting values of g_i as the surface is approached $(x=0+)$, which values we call g_{is}. Note also that we assume always that g_{4s} is the temperature of the solid surface. No heat is conducted into the solid in this quasi-steady state slab problem. It is interesting now to determine how the set of straight line solution profiles change as the specified ambient temperature g_{4b} is varied. In order to illustrate this we vary g_{4b}

and determine how g_{4s} will vary obtaining thereby a solution locus of g_{4s} vs. g_{4b}. This solution locus is shown as Curve A in Figure 4, such a locus having asymptotes for very low and very large g_{4b}. At low g_{4b}, $g_{4b}=g_{4s}$. For very high g_{4b}, g_{4s} approaches g_{4b} plus the adiabatic temperature rise for reaction R_I. In Figure 4 the curve A may be monotonic increasing for some values of the parameters.

B. This solution is obtained by assuming as in A that there is no reaction in the boundary layer, $k_{III}=0$, but in this case only CO_2 is produced at the reaction surface by the appropriate combination of all <u>three</u> reactions ($k_{III}\neq 0$ at x=0+). The appropriate boundary conditions are

$$\left.\begin{aligned}
\rho_f D_3 \frac{dg_3}{dx} &= k_I g_3 e^{-\alpha_I/g_4} + k_{II} g_1 e^{-\alpha_{II}/g_4} \\
\rho_f D_1 \frac{dg_1}{dx} &= k_{II} g_1 e^{-\alpha_{II}/g_4} + k_I g_3 e^{-\alpha_I/g_4} \\
K_f \frac{dg_4}{dx} &= \tfrac{1}{2} k_I g_3 e^{-\alpha_I/g_4} H_I - k_{II} g_1 e^{-\alpha_{II}/g_4} H_{II} \\
&\quad + (k_I g_3 e^{-\alpha_I/g_4} + k_{II} g_1 e^{-\alpha_{II}/g_4}) H_{II}
\end{aligned}\right\} , \quad x=0+$$

In obtaining these boundary conditions it is assumed that reaction R_I proceeds and that the CO formed must be consuded <u>instantly</u> by its oxidation as in R_{III}. The CO_2 formed at the surface can also react to form CO . Thus all three reactions occur at the boundary x=0+ . The third boundary condition reflects the generation of heat from R_I, the absorption of heat by R_{II} and the generation of heat by burning of CO at the surface. Since in $0<x<\ell$

$$\frac{d^2 g_i}{dx^2} = 0$$

the solution profiles are again straight lines and the problem is an algebraic-geometric one and is given in Figure 2. If as in A we allow g_{4b} to vary and calculate g_{4s} to obtain a locus of quasi-steady state solutions we obtain Curve B in Figure 4 which becomes asymptotic to Curve A for low values of g_{4b} and stops at the intersection with the next limiting solution.

C. This case is somewhat more complicated than the previous two and involves a different assumption for reaction R_{III}. We assume that R_{III} is infinitely fast and that the reaction takes place at a point where the flux of CO outwards is exactly equal to one half the flux of oxygen inwards, i.e., where the reaction stoichiometry, $CO+\frac{1}{2}O_2=CO_2$, is exactly met. This This in reality is the standard flame front assumption. Let $x=\xi$ be the position of the sharp flame front, then

$$\left.\begin{aligned}
&\frac{d^2 g_i}{dx^2} = 0 \ , \quad 0<x<\ell \quad , \quad x\neq\xi \\
&-\rho_f D_2 \frac{dg_2}{dx} = 2k_{II} g_1 e^{-\alpha_{II}/g_4} \\
&\rho_f D_1 \frac{dg_1}{dx} = k_{II} g_1 e^{-\alpha_{II}/g_4} \\
&K_f \frac{dg_4}{dx} = k_{II} g_1 e^{-\alpha_{II}/g_4} \ H_{II} \\
&\left.\rho_f D_2 \frac{dg_2}{dx}\right)_{\xi-} = \left. 2\rho_f D_3 \frac{dg_3}{dx}\right)_{x=\xi+}
\end{aligned}\right\} \quad x=0+$$

$$g_2=0 \ , \quad \xi<x\ell$$

$$g_3=0 \ , \quad 0<x \ \xi$$

$$K_f\frac{dg_4}{dx})_{\xi-} - K_f\frac{dg_4}{dx})_{\xi+} + \rho_f D_2\frac{dg_2}{dx})_{\xi-} H_{III} = 0$$

$$g_4(\xi-) = g_3(\xi+)$$

$$g_1=g_{1b}=g_2=g_{2b}=0 \ , \quad g_3=g_{3b} \ , \quad g_4=g_{4b} \ , \quad x=\ell-$$

While this looks more complicated its solution now is in terms of two piece-wise straight lines and therefore is algebraic-geometric. In this case there is only heat absorption at the surface $x=0$ so there is a maximum in the temperature and CO_2 profiles at $x=\xi$. The profiles are shown in Figure 3 and the locus of solutions obtained by allowing g_{4b} to vary is shown as Curve C in Figure 4. This terminates where it intersects with Curve B and becomes asymptotic to Curve A for large g_{4b}. We should say here that the position of the flame front varies for different values of g_{4b} and for low values of g_{4b} the flame front approaches the surface and this is why Curves C and B intersect.

4. THE FEASIBILITY REGION.

The quasi-steady state equations for the slab are given by

$$\frac{d}{dx}(\rho_f D_i\frac{dg_i}{dx}) + \nu_i M_i k_{III} g_2 g_3^{\frac{1}{2}} e^{-\alpha_{III}/g_4} = 0 \qquad (1)$$

for $0<x<\ell$ and at $x=0+$

$$\rho_f D_3 \frac{dg_3}{dx} = \tfrac{1}{2} k_I g_3 e^{-\alpha_I/g_4} \tag{2}$$

$$-\rho_f D_2 \frac{dg_2}{dx} = k_I g_3 e^{-\alpha_I/g_4} + 2k_{II} g_1 e^{-\alpha_{II}/g_4} \tag{3}$$

$$\rho_f D_1 \frac{dg_1}{dx} = k_{II} g_1 e^{-\alpha_{II}/g_4} \tag{4}$$

$$-K_f \frac{dg_4}{dx} = k_I g_3 e^{-\alpha_I/g_4} H_I - k_{II} g_1 e^{-\alpha_{II}/g_4} H_{II} \tag{5}$$

and at $x=\ell-$

$$g_1=0 \ , \ g_2=0 \ , \ g_3=g_{3b} \ , \ g_4=g_{4b}$$

We suppose that all but g_1, g_2, g_3, g_4 are known in these equations and all those things that look like constants are positive. The set of equations (1) - (5) is certainly a set which is amenable to numerical calcuation and the natural mode is to assume values of g_1, g_2, g_3, g_4 at $x=0+$ from which the derivatives can be computed at $x=0+$ and march to $x=\ell-$ using standard iteration procedures to correct the guesses on the g_i . It develops that this almost never works and one must be more clever. Examination of equations (2) - (5) reveals that these four equations can be combined into three equations by considering them in pairs and eliminating the source terms to give

$$\alpha_i \frac{d^2 g_i}{dx^2} = \alpha_j \frac{d^2 g_i}{dx^2}$$

where

$$\alpha_i = \nu_i M_i$$

These may be integrated twice to give

$$\alpha_i(g_j-g_{js})-\alpha_j(g_i-g_{is}) = A_{ij}x$$

with g_{is} being $\lim_{x\to 0+} g_i = g_{is}$ and where the A_{ij} depend and can be determined by the boundary conditions at $x=0+$. The system we have now consists of one non-linear differential equation, three algebraic equations and one boundary condition at $x=0+$ and one at $x=\ell-$ since

$$\alpha_i(g_{jb}-g_{js})-\alpha_j(g_{ib}-g_{is}) = A_{ij}\ell$$

Suppose now we choose g_1 as a key component so that the system is

$$\frac{d}{dx}(\rho_f D_1 \frac{dg_1}{dx})+\nu_1 M_1 k_{III} e^{-\alpha_{III}/g_4} g_2 g_3^{\frac{1}{2}} = 0 \,, \quad 0 < x < \ell$$

$$\rho_f D_1 \frac{dg_1}{dx} = k_{II} e^{-\alpha_{II}/g_4} g_1 \,, \quad x = 0+$$

$$\alpha_1(g_j-g_{js})-\alpha_j(g_1-g_{1s}) = A_{1j}x$$

$$\alpha_1(g_{jb}-g_{js})-\alpha_j(g_{1b}-g_{1s}) = A_{1j}\ell$$

$$g_1 = g_{1b} \,, \quad g_4 = g_{4b} \,, \quad x = \ell-$$

Now in order to circumvent the previous numerical difficulties we consider the set of three algebraic equations

$$\alpha_1(g_{jb}-g_{js})-\alpha_j(g_{1b}-g_{1s}) = A_{1j}\ell$$

The unknowns in these equations are g_{is} , i=1 to 4 . We know that physically g_{is} can never be negative. If we set successively $g_{1s}=0$, $g_{2s}=0$, $g_{3s}=0$, then we will obtain three sets of three equations each which can be solved as

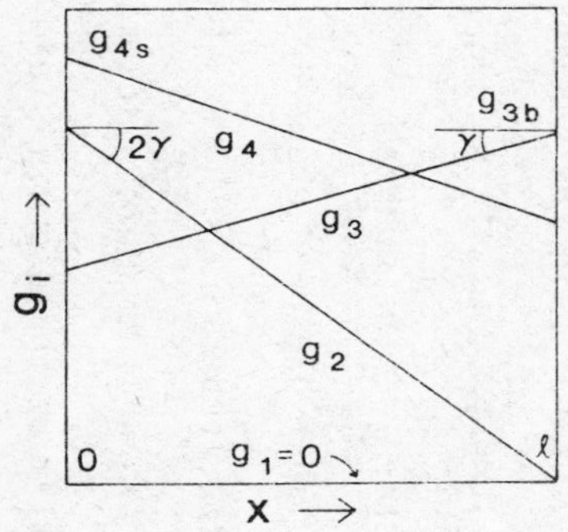

Figure 1

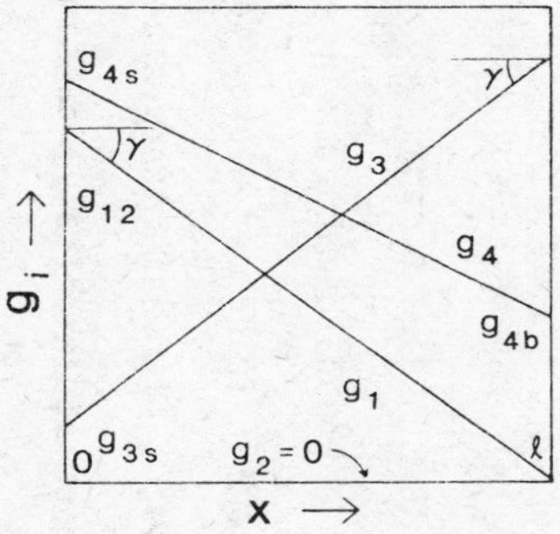

Figure 2

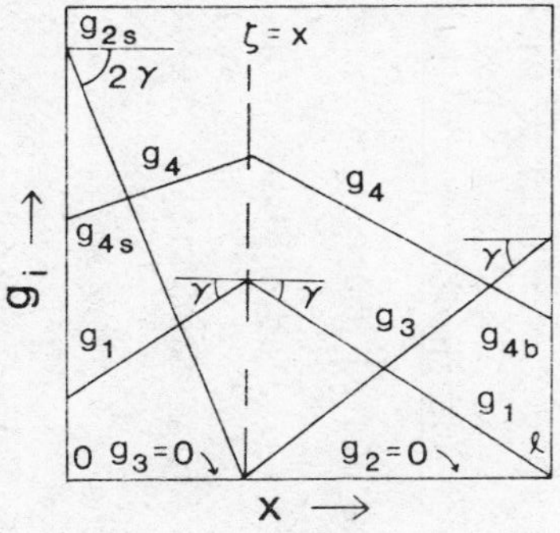

Figure 3

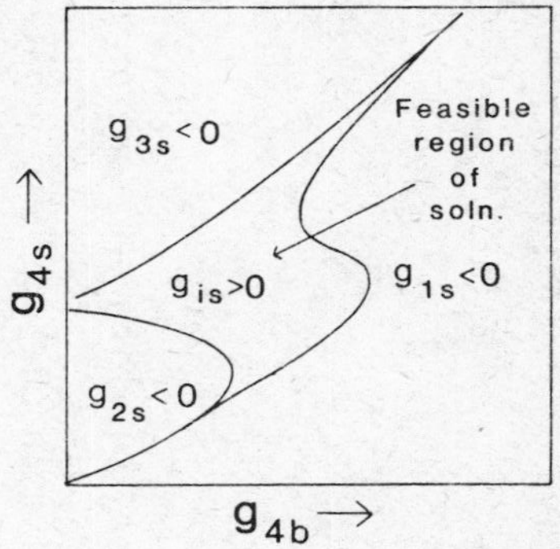

Figure 4

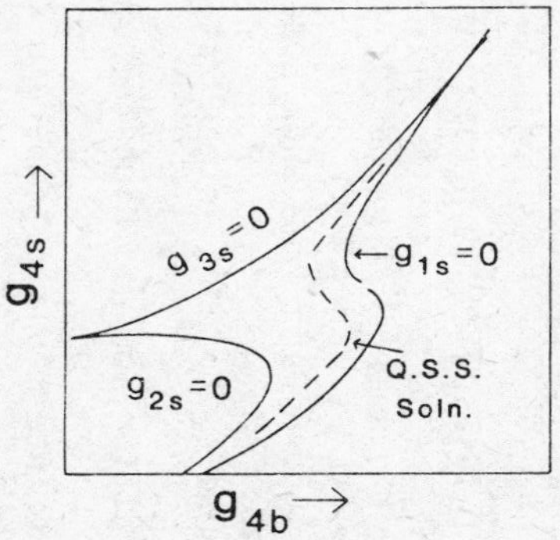

Figure 5

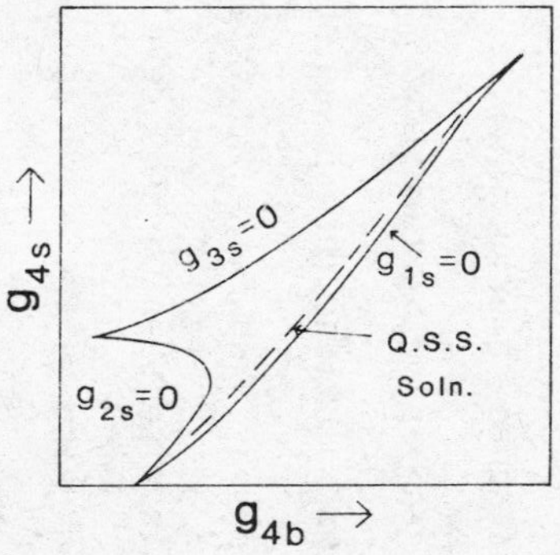

Figure 6

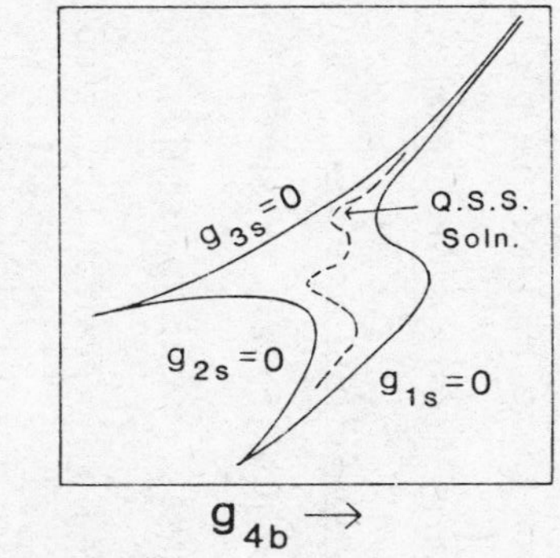

Figure 7

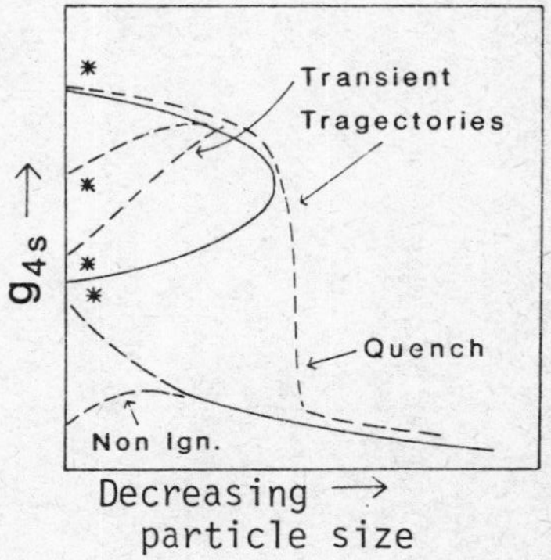

Figure 8

$$g_{4b} = F_1(g_{4s}) \; , \; g_{1s} = 0$$

$$g_{4b} = F_2(g_{4s}) \; , \; g_{2s} = 0$$

$$g_{4b} = F_3(g_{4s}) \; , \; g_{3s} = 0$$

In the plane of (g_{4b}, g_{4s}) these are three loci on which each one of the g_{is} , i=1 to 3 , is equal to zero and so positive on one side and negative on the other. These loci may be plotted and it is not difficult to show that these three loci are identical with the loci obtained from the three limiting solutions A, B, and C. In Figure 4 the region where $g_{is}>0$ for all i is then where all of the quasi-steady state solutions must lie. For low temperatures, i.e., g_{4b} small, the solution g_{4s} must lie in the lower asymptotic region while for high temperatures the solution g_{4s} must lie in the upper asymptotic region. Note that once g_{4s} is known g_{1s}, g_{2s}, g_{3s} are known and hence $g_1(x)$, $g_2(x)$, $g_3(x)$, and $g_4(x)$ can be calculated. We observe in these asymptotic regions g_{4s} is determined as soon as g_{4b} is specified, the solutions then being obtained in a standard manner. Once these solutions are known then it is a simple matter to obtain nearby solutions in a stepwise fashion thus developing a locus of quasi-steady state solutions beginning in the lower asymptotic region and eventually passing out the upper asymptotic regions. How this locus behaves in the feasibility region itself is a point of great interest. This behavior will be determined in detail by the numerical values of the parameters of the problem.

In Figures 5, 6, and 7 we show schematically what has been observed thus far. In Figure 5 the quasi-steady state solution locus is the normal one and shows that for low and high temperatures there will be unique solutions while at some intermediate temperatures there will be three quasi-steady states, although depending upon the shape of the feasibility region there may be a unique steady state as shown in Figure 6. In Figure 7 the quasi-steady state locus shows five solutions for some range of g_{4b} . In other cases the quasi-steady state locus may be identical to the limiting solution locus ($g_{1s}=0$) . The locus almost never follows the $g_{2s}=0$ limiting solution but often at high temperatures becomes identical with the upper limiting solution $g_{3s}=0$, the case where there is a flame and no oxygen reaches the surface. The general pathology of these solutions has not been examined in detail. Routine stability analyses and even some numerical solutions show that in the case of these quasi-steady state solutions the intermediate ones are unstable and in the case of five solutions the second and fourth are unstable. This whole question should be examined in more detail.

5. THE SPHERICAL PARTICLE.

For the spherical particle we will consider the simplest non-trivial transient model which may be written as

$$\frac{1}{r^2}\frac{\partial}{\partial r}(r^2\rho_f D_i\frac{\partial g_i}{\partial r})+\nu_i M_i k_{III} g_2 g_3^{\frac{1}{2}} e^{-\alpha_{III}/g_4} = \frac{\partial g_i}{f\,\partial\theta}\ , \quad a < r < b$$

$$\left.\begin{aligned}
&\rho_f D_3\frac{\partial g_3}{\partial r} = \tfrac{1}{2}k_I g_3 e^{-\alpha_I/g_4} \\
&-\rho_f D_2\frac{\partial g_2}{\partial r} = k_I g_3 e^{-\alpha_I/g_4}+2k_{II} g_1 e^{-\alpha_{II}/g_4} \\
&\rho_f D\frac{\partial g_1}{\partial r} = k_{II} g_1 e^{-\alpha_{II}/g_4} \\
&\frac{a}{3}c_s\rho_s\frac{\partial g_{4s}}{\partial\theta} = K_f\frac{\partial g_4}{\partial x} + k_I g_3 H_I e^{-\alpha_I/g_4}-k_{II} g_1 H_{II} e^{-\alpha_{II}/g_4} \\
&\frac{\partial a}{\partial\theta} = -\frac{M_S}{\rho_s}k_I g_3 e^{-\alpha_I/g_4}+k_{II} g_1 e^{-\alpha_{II}/g_4}
\end{aligned}\right\}\ r=a+$$

We should observe that for a problem of this complexity there are a hierarchy of models. In this one we have assumed that the particle temperature is uniform but changing--an assumption that is fairly realistic since at high temperatures the conductivity of carbon is high.

With the spherical particle we can compute quasi-steady state solutions with no more difficulty than for slab and we can also obtain the limiting solutions for a fixed particle size. With the sphere it is interesting to obtain the locus of quasi-steady state solutions with g_{4s} plotted against particle size, all other parameters the same. The general features obtained are shown schematically in Figure 8 where we see that for a given particle size there may be three steady state solutions. We then must ask what will happen if we introduce a particle in the environment shown by * on the ordinate. It will begin to diminish in size, and if the quasi-steady state solutions have any significance its trajectory with time must seek out those states. The dotted lines on Figure 8 show that indeed it does, following the upper state as long as it can and then must quench to the lower steady state. It can not follow the intermediate state since it is unstable and so it rises to the upper state or follows to the lower state. Thus we see here the phenomenon of quenching as well as lack of ignition. One of the surprising results is that the time of flight calculation reveals that it may take longer for small particles to burn than larger particles, a result of the fact that the larger particle remains on the upper branch for a longer time.

More complicated models show that the solution structure is changed in detail but that the gross behavior is not substantially altered.

REFERENCES

1. Hugo S. Caram and Neal R. Amundson, Diffusion and Reaction in a Stagnant Boundary Layer, Ind. Eng. Chem. Fund., 16 (1977), 171-181.

2. Christos Georgakis, John Congalidis, and Yam Yee Lee, Physical Interpretation of the Feasibility Region in the Combustion of Char by Use of the Single and Double Film Theories, Ind. Eng. Chem. Fund., 19 (1980), 98-103.

3. Eduardo Mon and Neal R. Amundson, Diffusion and Reaction in a Stagnant Boundary Layer, 2. Ind. Eng. Chem. Fund., 17 (1978), 313-321; 3. Ind. Eng. Chem. Fund., 18 (1979), 162-168; 4. Ind. Eng. Chem. Fund., 19 (1980), 243-250.

4. S. Sundaresan and Neal R. Amundson, Diffusion and Reaction in a Stagnant Boundary Layer, 5. Ind. Eng. Chem. Fund., 19 (1980), 344-351; 6. Ind. Eng. Chem. Fund., 19 (1980), 351-357; 7. Am. Inst. Chem. E. J., 27 (1981), 679-686.

Neal R. Amundson
University of Houston

D G ARONSON

Bifurcation phenomena associated with nonlinear diffusion mechanisms

The topic to be discussed is appropriate for these proceedings since it concerns problems in partial differential equations which are attacked, at least in part, from a dynamical systems point of view. To begin with we shall give an introduction to nonlinear diffusion mechanisms and then discuss some diffusion-interaction problems which arise in population dynamics and theoretical ecology.

1. NONLINEAR DIFFUSION MECHANISMS.

Perhaps the simplest example of a nonlinear diffusion mechanism occurs in the study of isentropic flow of an ideal gas in a homogeneous porous medium. If $u = u(x,t)$ represents the density of the gas, then its evolution is governed by the equation

$$\frac{\partial u}{\partial t} = \Delta(u^m) \tag{1}$$

where $\Delta = \sum_{j=1}^{d} \frac{\partial^2}{\partial x_j^2}$ and $m > 1$ is a constant. Although equation (1) is often called the porous medium equation it also arises in other contexts, for example, plasma physics and population dynamics.

Since u is interpreted as the density of the gas it is natural to assume that $u \geq 0$. If the equation is written in traditional Fickian diffusion form

$$\frac{\partial u}{\partial t} = \text{div } (mv^{m-1} \text{ grad } u)$$

it is apparent that the diffusivity is mu^{m-1} . Therefore, the diffusivity vanishes with the density. The most striking consequence of this nonlinearly degenerate diffusivity is the finite speed of propagation of disturbances from rest. This is in stark contrast to the classical linear diffusion case, $m = 1$, where, as is well known, there is an infinite speed of propagation.

To be more specific about the finite speed of propagation consider the initial value problem

$$\begin{aligned} &\frac{\partial u}{\partial t} = \Delta(u^m) \qquad \text{in} \quad \mathbb{R}^d \times \mathbb{R}^+ \\ &u(\cdot,0) = u_0 \qquad \text{in} \quad \mathbb{R}^d \end{aligned} \tag{2}$$

where u_0 is a given nonnegative function. For the sake of the simplest illustration of finite speed assume

$$\text{supp} \quad u_0 \quad \text{is homoemorphic to} \quad \overline{B}_1(0) . \tag{3}$$

Under this assumption, problem (2) does not have a solution in the classical sense. However, under reasonable circumstances, there is a perfectly well defined weak solution which will be discussed with no further qualification. The key fact here is that under the hypothesis (3), the solution $u(x,t)$ of (2) exists and $\text{supp}\, u(\cdot,t)$ is homeomorphic to $\overline{B}_1(0)$ for all $t \in \mathbb{R}^+$. The set $\{(x,t)\colon x \in \partial\text{supp}\, u(\cdot,t),\ t \in \mathbb{R}^+\}$ is called the interface. It is much studied for $d = 1$, but as yet very little is known about it for $d > 1$.

2. SOME POPULATION DYNAMICS.

Gurtin & MacCamy [3] have modeled the dynamics of spatially distributed biological populations whose tendency to migrate is influenced by the local population density. Their model has the general form

$$\frac{\partial u}{\partial t} = \Delta\varphi(u) + f(u) \tag{4}$$

Here $\varphi(u)$ describes the dispersal or migration mechanism while $f(u)$ describes the underlying population dynamics. In particular, if the population in question likes to avoid crowds then $f(0) = 0$, $f'(0) = 0$, $f > 0$, and $f' > 0$ on $\mathbb{R}^+$.

The model problem which motivates the theory to be subsequently described in these talks comes form considering (4) for $x \in \mathbb{R}$ with specific choices for φ and f; in particular,

$$\varphi(u) = u^m$$

for some constant $m > 1$, and

$$f(u) = u(1-u)(u-a)$$

for some constant $a \in (0,1)$.

3. MODEL PROBLEM.

The work discussed here is joint work with M. C. Crandall & L. A. Peletier [1]. Consider the initial-boundary value problem

$$\begin{aligned} &u_t = (u^m)_{xx} + f(u) \quad \text{in} \quad (-L,L) \times \mathbb{R}^+ \\ &u(\pm L,\cdot) = 0 \quad \text{in} \quad \mathbb{R}^+ \\ &u(\cdot,0) = u_0 \quad \text{in } [-L,L] \end{aligned} \tag{5}$$

where u_0 is a given nonnegative function on $[-L,L]$ with $u_0(\pm L) = 0$. A solution of problem (5) will be interpreted as a trajectory in some function space and will therefore be written as $u(t;u_0)$. The problem to be discussed is:

$$\lim_{t\uparrow\infty} u(t;u_0) = ?$$

In order to begin to answer this question it is necessary to know something about the structure of the set of equilibrium solutions of problem (5). For each $L \varepsilon \mathbb{R}^+$ let $\mathcal{E}(L)$ denote the collection of nonnegative solutions of the problem

$$\begin{aligned}(v^m)'' &= f(v) = 0 \quad \text{in} \quad (-L,L)\\ v(\pm L) &= 0 \;.\end{aligned} \tag{6}$$

Clearly $\{0\} \subset \mathcal{E}(L)$ regardless of the value of L. Next we look for positive solutions, that is, solutions v such that $v > 0$ on $(-L,L)$. It is easy to verify that we must have $v \in (0,1)$. We now proceed by means of the method of first integrals, that is we multiply both sides of equation (6) by $(v^m)'$ and integrate. The result is

$$\frac{1}{2}(v^m)'^2 + F(v) = \text{const.}$$

where

$$F(v) = \int_0^v \eta^{m-1} f(\eta) \;.$$

Define α by $F < 0$ on $(0,\alpha)$, $F(\alpha) = 0$. Note that $\alpha \in (0,1)$ if and only if $a \in (0, \frac{m+1}{m+3})$.

Proposition. Suppose $a \in (0, \frac{m+1}{m+3})$. Then v is a positive solution of problem (6) if and only if

$$\sqrt{\frac{m}{2}}\int_{v(x)}^{\mu} \frac{\eta^{m-1}\, d\eta}{\sqrt{F(\mu) - F(\eta)}} = |x|$$

<u>for all</u> $|x| \le L$ <u>where</u> $\mu \in [\alpha,1)$ <u>and</u> $L \in \mathbb{R}^+$ <u>are related by the equation</u>

$$\lambda(\mu) \equiv \sqrt{\frac{m}{2}}\int_{0}^{\mu} \frac{\eta^{m-1}\, d\eta}{\sqrt{F(\mu) - F(\eta)}} = L .$$

Given L there exists a solution of (6)

$$v = v(\cdot;\mu(L))$$

if L is in the range of $\lambda(\mu)$ and $\mu(L)$ is such that $\lambda(\mu(L))$.

<u>Proposition</u>. The two branches of $\mu_{\pm}(L)$ are as shown in the figure.

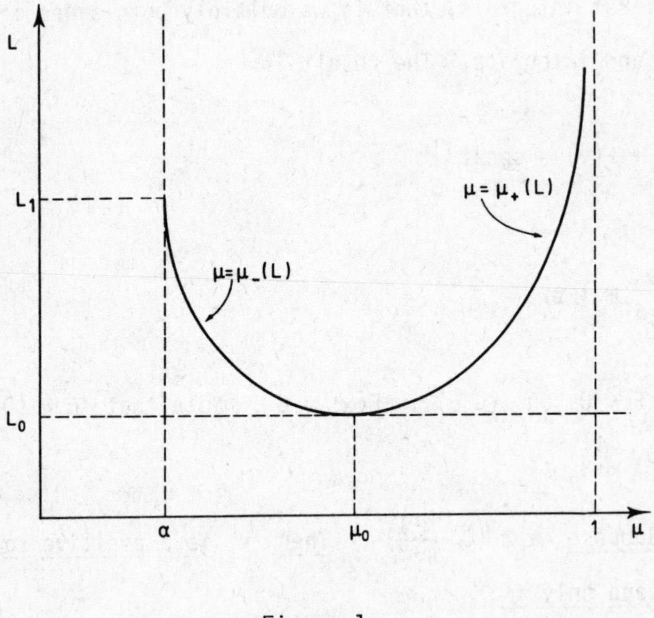

Figure 1

In particular $\mu_+(L) \uparrow 1$ as $L \to \infty$ and there exists an $L_1 = L_1(m) \in \mathbb{R}^+$ such that $\mu_-(L) \uparrow \alpha$ as $L \uparrow L_1$. Moreover $L_1(m) \to \infty$ as $m \downarrow 1$.

We introduce the notation $p(\cdot;L) \equiv v(\cdot;\mu_-(L))$ and $q(\cdot,L) = v(\cdot;\mu_+(L))$. For $L \varepsilon (L_0, L_1)$ the picture is:

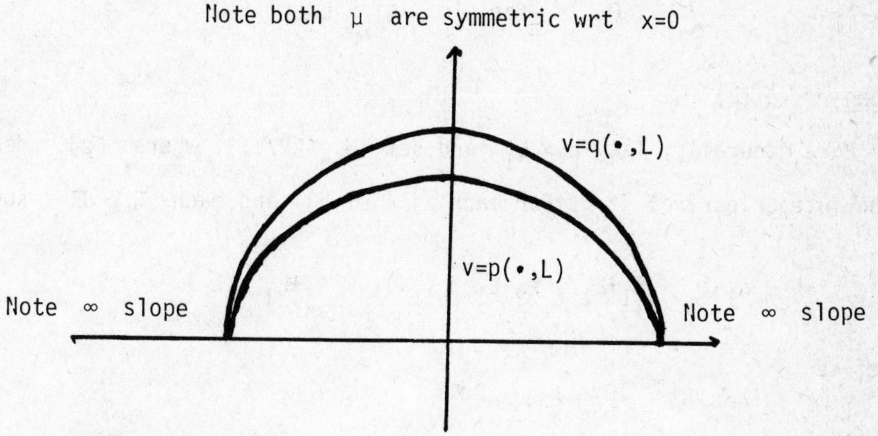

So far we know that

$$\mathscr{E}(L) = \begin{cases} \{0\} & \text{for } 0 < L < L_0 \\ \{0\} \cup \{p\} \cup \{q\} & \text{for } L_0 \le L \le L_1 \end{cases}$$

We wish to know what happens for $L > L_1$.

Observe that

$$(v^m)'(\pm L;\mu_-(L)) \quad \begin{cases} \neq 0 & \text{for } L \in [L_0,L_1) \\ = 0 & \text{for } L = L_1 . \end{cases}$$

Therefore we can extend $p(\cdot;L_1) = v(\cdot;\mu_-(L))$ as 0 outside $(-L_1,L_1)$. The resulting function is a nonnegative (weak) equilibrium on $(-L,L)$ for any $L > L_1$. More formally we have the following result.

Proposition. Suppose $L > L_1$. For every $\xi \in (-L,L)$ such that

$$-L \le \xi - L_1 < \xi + L_1 \le L$$

define

$$r(x) = \begin{cases} p(x-\xi;L_1) & \text{for } |x-\xi| \le L_1 \\ 0 & \text{for } |x-\xi| > L_1 . \end{cases}$$

Then $r \in \mathcal{E}(L)$.

More generally, let $L > L_1$ and set $N = [L/L_1]$, where $[z]$ denotes the integer part of z . For each $j \in [1,N]$ and each $\xi \in \mathbb{R}^j$ such that

$$-L \le \xi_1 - L_1 < \xi_1 + L_1 \le \xi_2 - L_1 < \ldots.. < \xi_j + L_1 \le L \tag{7j}$$

define

$$r(x;\underset{\sim}{\xi}) = \begin{cases} p(x-\xi_\ell;L_1) & \text{if } |x-\xi_\ell| \le L_1 \text{ for } \ell = 1,2,\ldots, j \\ 0 & \text{otherwise} \end{cases}$$

Finally let

$$\mathcal{P}(L) = \left\{ \bigcup_{j=1}^{N} \{r(\cdot;\underset{\sim}{\xi}) : \underset{\sim}{\xi} \text{ satisfies (7j)}\right\} .$$

Proposition. *If* $L \ge L_1$ *then* $\mathcal{E}(L) = \{0\} \cup \mathcal{P}(L) \cup \{q\}$.

We return to the original question: $u(t;u_0) \to ?$ Formally there is a Liapunov function for problem (5):

$$V(\zeta) = \int_{-L}^{L} \{\tfrac{1}{2} \zeta^{m/2} - m F(y)\} \, dx$$

Therefore we expect $\lim_{t\uparrow\infty} u(t;u_0) \in \mathcal{E}(L)$. There are two problems with this:

(i) It is hard to get enough regularity (compactness) to actually use the Liapunov function.

(ii) $\mathcal{E}(L)$ contains a continuum for $L > L_1$.

4. SOME GENERAL THEORY.

Consider the initial-boundary value problem:

$$\begin{aligned} &u_t = (u^m)_{xx} + f(u) \quad \text{in} \quad (-L,L) \times \mathbb{R}^+ \\ &u(\pm L,\cdot) = 0 \quad \text{in } \mathbb{R}^+ \\ &u(\cdot,0) = u_0 \quad \text{in} \quad [-L,L] \end{aligned} \tag{8}$$

where now we simply assume $f \in \text{Lip}\,(\mathbb{R})$ and $L > 0$. We shall use the notation

$$\Omega = (-L,L) \quad \text{and} \quad Q_T = \Omega \times (0,T) .$$

Definition. u is a weak solution of problem (8) on $[0,T]$ if

$$u \in C([0,T]: L^1(\Omega)) \cap L^\infty(Q_T)$$

and

$$\int_\Omega u\varphi|_t - \int_{Q_t}\int (u\varphi_t + u^m\varphi_{xx}) = \int_\Omega u_0\varphi(x,0) + \int_{Q_t}\int f\varphi$$

for all test functions $Q_t \in C^{2,1}$ with $\varphi|_{x=\pm L} = 0$, and $t \in [0,T]$.

Theorem. Suppose $f(0) = f(1) = 0$ and $u_0 \in [0,1]$. Then there exists a unique weak solution u of problem (8). Moreover, $u \in [0,1]$ and $(u^m)_x \in L^2(\Omega)$ for each $t \in (0,T]$.

In addition to the existence and uniqueness theory we have the following absolutely indispensable comparison principle.

Definition. u is a subsolution (supersolution) of problem (8) on $[0,T]$

if

$$u \in C([0,T] : L^1(\Omega)) \cap L^\infty(Q_T)$$

and

$$\int_\Omega u\varphi|_t - \int_{Q_t}\int (u\varphi_t + u^m\varphi_{xx}) \le (\ge) \int_\Omega u_0\varphi(0) + \int_{Q_t}\int f\varphi$$

for all $\varphi \in C^{2,1}(\overline{Q}_T)$ with $\varphi \ge 0$ and $\varphi(\pm L) = 0$.

Theorem. Let u be a subsolution of (8) with initial datum u_0 and $\hat{u}$ be a supersolution of (8) with initial datum $\hat{u}_0$. If $u_0 \le \hat{u}_0$ then

$$u(\cdot;u_0) \le u(\cdot;\hat{u}_0) .$$

Let $\mathcal{E}(L)$ denote the collection of all nonnegative equilibrium solutions of problem (8) and

$$X = \{ u \in L^\infty(\Omega): u \in [0,1] , (u^m)_x \in L^2(\Omega) \} .$$

X is a complete metric space with

$$d(u,v) \equiv ||u-v||_{L^1(\Omega)} + ||(u^m-v^m)_x||_{L^2(\Omega)} .$$

For $u_0 \in X$ define the ω-limit set

$$\omega(u_0) = \{ w \in X : \exists \{t_n\} \text{ with } t_n \to \infty \text{ and } u(t_n;u_0) \to w \} .$$

Theorem. $\omega(u\ )$ is nonempty and connected in X . Moreover, for all $u_0 \in X$,

$$\omega(u\ ) \subset \mathcal{E}(L) .$$

The first assertion is standard. The main point of the theorem is the second assertion. The proof involves some rather technical regularity theory. The details can be found in [1]. This result is most useful when $\mathcal{E}(L)$ is a singleton or consists only of isolated elements. In the application to the model problem $\mathcal{E}(L)$ sometimes contains a continuum. The following observation will be helpful.

Corollary. Let $\mathcal{S} \subset X$ be such that

> $\mathcal{S}$ is positively invariant (that is, $u_0 \in \mathcal{S}$ implies $u(t;u_0) \in \mathcal{S}$, for all $t \in \mathbb{R}^+$)

and

$$\mathcal{S} \cap \mathcal{E}(L) = \{w\} .$$

Then $u_0 \in \mathcal{S}$ implies that $u(t;u_0) \to w$.

This corollary provides the link between the dynamical systems theory and the comparison theory for our partial differential equation problem. Using it together with what we know about the structure of $\mathcal{E}(L)$, we will be able to say a good deal about the stability of various equilibria of our model problem.

5. THE MODEL PROBLEM REVISITED.

We now apply the general theory to the model problem discussed in 3.

(i) $0 < L < L_0$: In this case $\mathcal{E}(L) = \{0\}$ and $\lim_{t\uparrow\infty} u(t;u_0) = 0$ for all $u_0 \in X$.

(ii) $L_0 < L \le L_1$. In this case $\mathcal{E}(L) = \{0\} \cup \{p(\cdot;L)\} \cup \{q(\cdot;L)\}$.

Choose ℓ and ζ such that $-L \le \zeta-\ell < \zeta+\ell \le L$ and define

$$\tilde{P}(x) = \begin{cases} p(x-\zeta;\ell) & \text{for } |x-\zeta| \le \ell \\ 0 & \text{for } |x-\zeta| > \ell . \end{cases}$$

It is easy to verify that $v = \tilde{p}$ is a subsolution and $v \equiv 1$ is a supersolution for the model problem on $(-L,L) \times \mathbb{R}^+$. Thus the set

$$\mathcal{S} \equiv \{u_0 \in X : \tilde{p} \le u_0 \le 1\}$$

is positively invariant. Moreover $\mathcal{S} \cap \mathcal{E}(L) = \{q(\cdot;L)\}$. Therefore $u_0 \in \mathcal{S}$ implies $\lim_{t\uparrow\infty} u(t;u_0) = q(\cdot;L)$.

Figure 3

(a) Domain of attraction for $q(\cdot,L)$ where $L \in (L_0,L_1)$

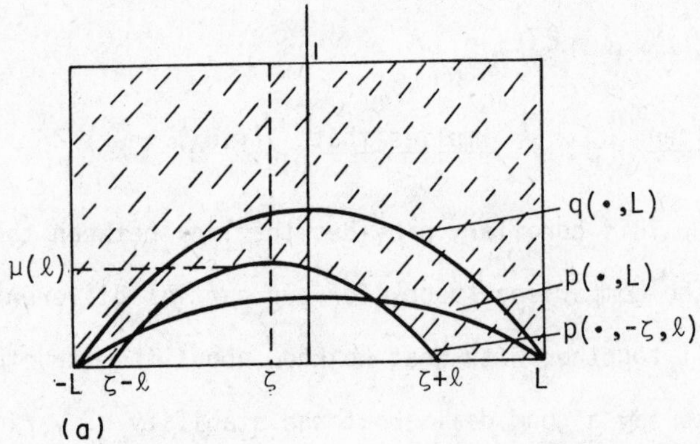

(b) Domain of attraction for 0 where $L \in (L_0,L_1)$

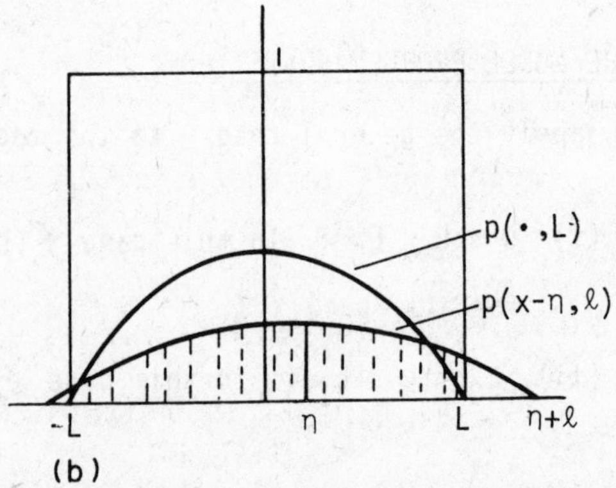

Now choose η and ℓ such that $\eta-\ell \le -L < L \le \eta+\ell$ and define

$$\mathcal{S} \equiv \{u_0 \in X : \ 0 \le u_0 \le p(\cdot;\ell)\}$$

Then $\mathcal{S}$ is positively invariant and $\mathcal{S} \cap \mathcal{E}(L) = \{0\}$. Thus $u_0 \in \mathcal{S}$ implies that $\lim_{t\uparrow\infty} u(t;u_0) = 0$.

(iii) $L > L_1$. In this case $\mathcal{E}(L) = \{0\} \cup p_{[L/L_1]}(L) \cup \{q(\cdot;L)\}$. Choose $\zeta_1, \zeta_2 \in (-L,L)$ such that $\zeta_1 < \zeta_2$ and $-L \le \zeta_1-L_1 < \zeta_2+L_1 \le L$. The function

$$\hat{p}(x) = \begin{cases} \max\{p(x-\zeta_1;L_1),\ p(x-\zeta_2;L_1)\} & \text{for } \zeta_1-L_1 \le x \le \zeta_2+L_1 \\ 0 \ \text{ otherwise} & \end{cases}$$

is a subsolution for the model problem on $(-L,L) \times \mathbb{R}^+$. Thus the set

$$\mathcal{S} \equiv \{u_0 \in X : \hat{p} \le u_0 \le 1\} .$$

is positively invariant. Moreover, since $\mathcal{S} \cap \mathcal{E}(L) = \{q(\cdot;L)\}$ it follows $u_0 \in \mathcal{S}$ implies $\lim_{t\uparrow\infty} u(t;u_0) = q(\cdot;L)$

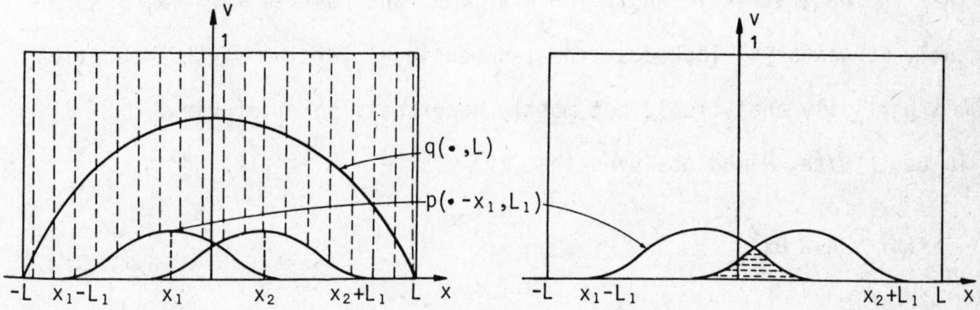

Figure 4

An isolating set for $u \equiv 0$ is given by $\mathcal{S} = \{u_0 \in X|: 0 \le u_0 \le \bar{p}\}$ where

$$\bar{p}(x) = \begin{cases} \min p(x-\zeta_1;L_1), p(x-\zeta_2;L_1)\} & \text{for } \zeta_1-L_1 \le x \le \zeta_2+L_1 \\ 0 & \text{otherwise} \end{cases}$$

6. ANOTHER PROBLEM.

In a study of natural regulation of animal populations, Gurney & Nisbet [2] introduced the model

$$\frac{\partial u}{\partial t} = \Delta(u^2) + f(|x|)u$$

to describe the dynamics of a population of crowd avoiders whose intrinsic growth rate depends on the position within the habitat. Of particular interest is the case when

$$f(\eta) \begin{cases} > 0 & \text{for } \eta < R \\ < 0 & \text{for } \eta > R \end{cases}$$

so that the habitat is friendly for $|x| < R$ and hostile for $|x| > R$. Recently T. Namba [4] looked at the 1-dimensional version of this equation from a partially analytical, but mostly numerical, point of view.

To begin with, Namba observed that if

$$f(x) = A - Bx^2$$

then the steady state equation

$$(v^2)'' + f(x)v = 0$$

has a solution

$$\bar{v}(x) = \begin{cases} \frac{B}{48}\left(x - \sqrt{\frac{7A}{B}}\right)^2\left(x + \sqrt{\frac{7A}{B}}\right)^2 & \text{for } |x| \le \sqrt{\frac{7A}{B}} \\ 0 & \text{for } |x| > \sqrt{\frac{7A}{B}} . \end{cases}$$

Note that $\bar{v}$ satisfies both zero Dirichlet and zero Neuman conditions at the end points of every interval $(-L,L)$ with $L \ge \sqrt{\frac{7A}{B}}$.

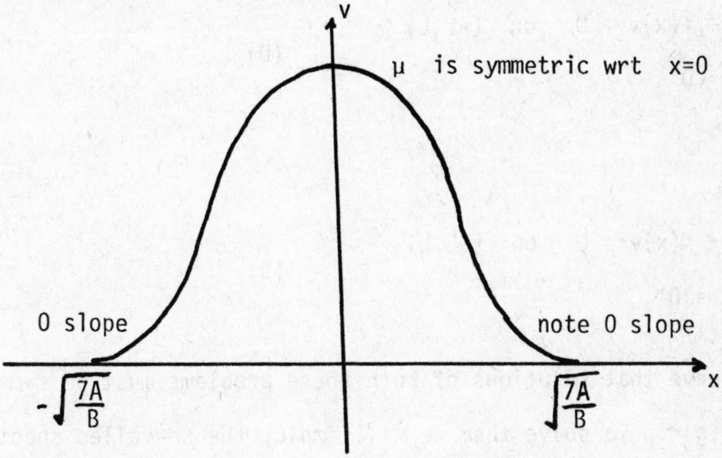

Namba calls $\bar{v}$ a singular solution and produces some numerical evidence for its stability. Moreover he has numerical evidence for existence of singular solutions for more general f's . Actually it is not difficult to prove existence of a singular solution under fairly general hypothesis on f (Peletier and the author have recently shown this by a fairly routine shooting argument.). However, so far, the uniqueness of the singular solution remains intractable.* To get some feeling for the overall problem, I have investigated a special case which can be done almost explicitly, that is, by manipulating awful elliptic integrals.

Consider the 1-dimensional problem with

*Added, Jan '83. Since these lectures were given, Joel Spruck has shown me how to prove uniqueness for a very reasonable class of growth functions f .

$$f(x) = \begin{cases} 1 & \text{for } |x| < 1 \\ -1 & \text{for } |x| > 1 . \end{cases}$$

We are interested in finding positive solutions to the following two equilibrium problems:

$$\begin{aligned} &(v^m)'' + f(x)v + 0 \quad \text{on } (-L,L) \\ &v(\pm L) = 0 \end{aligned} \qquad \text{(D)}$$

and

$$\begin{aligned} &(v^m)'' + f(x)v = 0 \quad \text{on } (-L,L) \\ &v'(\pm L) = 0 . \end{aligned} \qquad \text{(N)}$$

First observe that solutions of both these problems must be symmetric about the origin. To solve them we will imploy the so-called shooting method. First replace v by $w = v^m$ so that the equation becomes

$$w'' + f(x)w^{1/m} = 0 .$$

For $\alpha \in \mathbb{R}^+$ we look at initial value problem

$$\begin{aligned} &w'' + f(x)w^{1/m} = 0 \\ &w(0) = \alpha , \quad w'(0) = 0 \end{aligned}$$

and see what the solution does as x increases.

To begin with it is useful to rescale. Introduce new dependent and independent variables

$$W(z) = \frac{1}{\alpha} w(x) \quad \text{and} \quad x = \alpha^{\frac{2-\mu}{2}} z \quad \text{with} \quad \mu = \frac{1}{m} + 1 .$$

Then

$$\ddot{W} + h_\zeta(z)W^{\frac{1}{m}} = 0 \quad W(0) = 1 \,, \quad \dot{W}(0) = 0$$

where

$$h_\zeta(z) = \begin{cases} +1 & \text{for } |z| < \alpha^{\frac{-2}{2}} \equiv \zeta \\ -1 & \text{for } |z| > \zeta \,. \end{cases}$$

For $z \in (0,\zeta)$, $\ddot{W} + W^{\mu-1} = 0$ implies

$$\dot{W}^2(\zeta) = \frac{2}{\mu}\{1-W^\mu(\zeta)\} \,.$$

For $z > \zeta$, $\ddot{W} - W^{\mu-1} = 0$ implies

$$\dot{W}^2 = \frac{2}{\mu}\{W^\mu - (2W^\mu(\zeta) - 1)\} \,.$$

Suppose ζ^* is such that $2W^\mu(\zeta^*) - 1 = 0$. Then for $z > \zeta^*$

1. $\dot{W}^2 = \frac{2}{\mu} W^\mu$, $W^\mu(\zeta^*) = \frac{1}{2}$

and we get the singular solution.

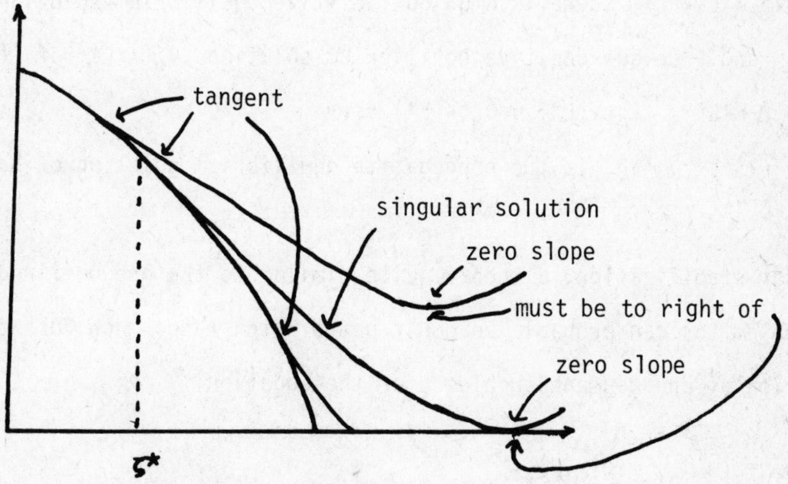

2. If $\zeta \in (0,\zeta^*)$ we get the solution of the Neumann problem (N)

3. If $\zeta > \zeta^*$ we get the solution of the Dirichlet problem (D)

If we translate all this back to original variables, it can be shown that one gets the following picture:

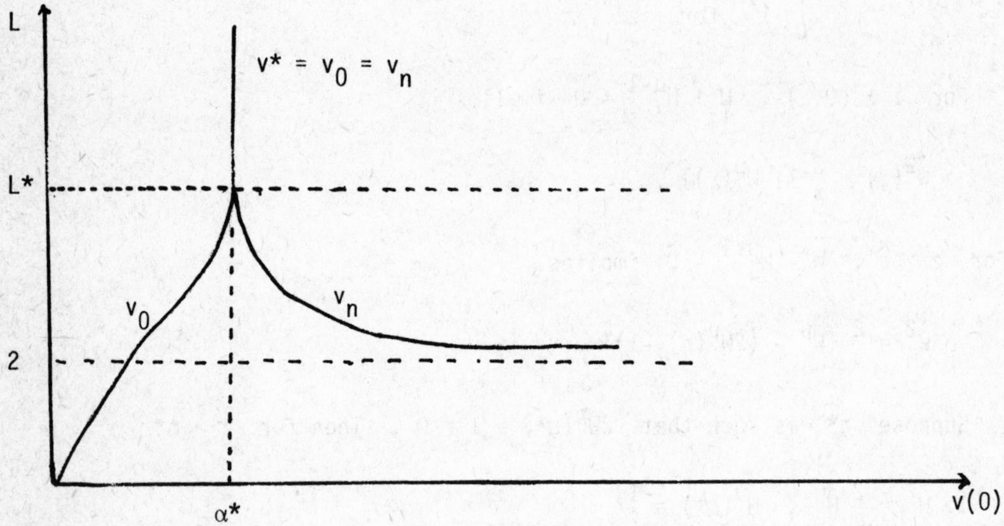

For $0 < L \leq 2$, there is a unique positive equilibrium solution $v_D(\cdot;L)$ of (D) and no nontrivial solution of (N).

For $2 < L < L^*$ there is a unique positive equilibrium solution $v_D(\cdot;L)$ of (D) and a unique positive equilibrium solution $v_N(\cdot;L)$ of (N).

For $L = L^*$, $v_D(\cdot;L^*) = v_N(\cdot;L^*) = v^*$

For $L \geq L^*$, v^* is the nonnegative equilibrium solution of both (D) and (N).

As for stabilization, a theory quite similar to the one outlined above for problem (5) can probably be constructed without too much fuss for both the Dirichlet and Neumann problems for the equation

$$u_t = (u^m)_{xx} + g(x,u) ,$$

where g is Lipschitz continuous in u and, say, bounded and measurable in x. Moreover, one can use the bifurcation diagram to construct isolating invariant sets to prove the stability of the nontrivial solutions.

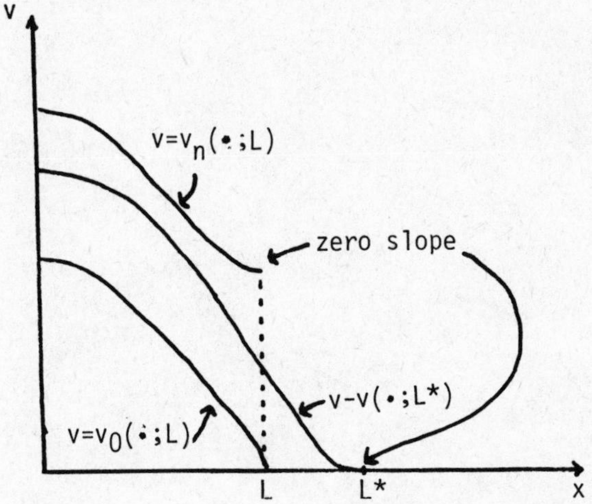

REFERENCES

1. Donald Aronson, Michael G. Crandall and L.A. Peletier, Stabilization of Solutions of a Degenerate Nonlinear Diffusion Problem, Nonlinear Analysis, TMA, 6 (1982) p. 1001-1022.

2. W.S.C. Gurney and R.M. Nisbet, The Regulation of Inhomogeneous Populations, J. Theor. Biol., 52 (1975), p. 441-457.

3. M.E. Gurtin and R.C. MacCamy, On the Diffusion of Biological Populations, Math. Biosci., 33 (1977), 35-49.

4. T. Namba, Density-dependent Dispersal and Spatial Distribution of a Population, J. Theor. Biol., 86 (1980), p. 351-363.

Donald G. Aronson
University of Minnesota

J F G AUCHMUTY*

Bifurcation analysis of reaction-diffusion equations v. rotating waves on a disc

1. INTRODUCTION

One of the fascinating aspects of reaction-diffusion equations is the variety of special types of solutions they exhibit. In this paper we shall develop a bifurcation analysis of rotating-wave solutions for such equations on a disc.

This work has been motivated to a large extent by the numerical simulations of Erneux and Herschkowitz-Kaufman (8) and by many of the ideas developed in Winfree (14). In particular the numerical work in (8) indicates that rotating waves are possible, asymptotically stable, solutions of reaction-diffusion equations.

A number of others, including Cohen, Neu and Rosales (6), Greenberg (10) and (11) and Kopell and Howard (12) have studied rotating-spriral-wave solutions; generally for λ-ω systems. The equations to be studied in this paper are considerably more general and the methods used will be quite different. The work here is much closer in spirit to that described in (5). Unfortunately, the analysis described there does not apply to discs; the only two dimensional domains allowed were annuli. The extension to discs requires considerable changes in the analysis to allow for the existence of a "center".

In the next section the problem will be described and in section 3 the associated linear problems will be analyzed. Section 4 describes the local

*This research was partially supported by NSF grant MCS 8201889.

bifurcation analysis and section 5 describes the global theory. The last section is devoted to the particular example of the Brusselator whose numerical simulation was described in (8) and which appears to exhibit these solutions as asymtotically stable solutions.

2. ROTATING WAVES ON A DISC.

Let Ω be the open unit ball in $\mathbb{R}^2$ and consider the problem of finding solutions of the weakly coupled parabolic system

$$\frac{\partial u_j}{\partial t} = D_j \Delta u_j + f_j(u_1, u_2, \ldots, u_m; \mu) \quad \text{on} \quad \Omega \times (0,T) \tag{2.1}$$

$$\text{subject to} \quad \frac{\partial u_j}{\partial r} = 0 \quad \text{on } \partial\Omega \times (0,T), \quad 1 \le j \le m. \tag{2.2}$$

Here $D_1, D_2, \ldots, D_m$ are positive constants, Δ represents the Laplacian on Ω and a point x in Ω has Cartesian coordinates (x_1, x_2) and polar coordinates (r, θ).

Let Λ be an open interval in $\mathbb{R}$ and $f: \mathbb{R}^m \times \Lambda \to \mathbb{R}^m$ be a continuous function whose components f_i obey

(F1): $f_i(0; \mu) = 0$ for all i, μ

(F2): $f_i(u, \mu)$ is C^2 on $\mathbb{R}^m \times \Lambda$.

A function $u: \overline{\Omega} \times (0,T) \to \mathbb{R}^m$ will be said to be a solution of (2.1)-(2.2) if it is continuous and obeys the equations and the boundary conditions in a distributional sense.

A solution u of (2.1)-(2.2) is said to be a rotating wave if there are functions v_j and a wave-speed c $(\neq 0)$ with

$$u_j(r, \theta, t) = v_j(r, \theta - ct) \quad \text{for} \quad 1 \le j \le m \tag{2.3}$$

and all $(r, \theta, t) \in \Omega \times (0,\infty)$.

Let $\xi = \theta - ct$, then v may be considered as a function from $\overline{\Omega}$ to $\mathbb{R}^m$ and it obeys the system of equations

$$D_j \Delta v_j + c \frac{\partial v_j}{\partial \xi} + f_j(v_1, \ldots, v_m; \mu) = 0 \quad \text{on} \quad \Omega \tag{2.4}$$

$$\text{subject to} \quad \frac{\partial v_j}{\partial r} = 0 \quad \text{on} \quad r = 1, \quad 1 \leq j \leq m . \tag{2.5}$$

Here we are using polar coordinates (r, ξ) on Ω and the functions $v_j(r, \xi)$ will be periodic of period 2π in ξ for each $0 < r < 1$. The Laplacian is given by

$$\Delta = \frac{1}{r} \frac{\partial}{\partial r} \left(r \frac{\partial}{\partial r}\right) + \frac{1}{r^2} \frac{\partial^2}{\partial \xi^2} .$$

Our interest is in showing that there are non-trivial solutions of (2.4) and (2.5) and, in particular, that one may construct such solutions by bifurcation-theoretic methods. The wave-speed c is not known a priori and plays a role analogous to that of an eigenvalue.

To study these questions we shall formulate (2.4)-(2.5) as non-linear compact operator equations in an appropriate Banach space. Let $X_0 = C(\overline{\Omega}; \mathbb{R}^m)$ be the space of all continuous $\mathbb{R}^m$-valued functions defined on $\overline{\Omega}$. It is a Banach space under the norm

$$||v||_0 = \sum_{k=1}^{m} \sup_{x \in \overline{\Omega}} |v_k(x)| .$$

Let X_1 be the subspace of X_0 of all functions which are continuously differentiable on Ω and have uniformly bounded derivatives there. It is a Banach space under the norm $||v||_1 = ||v||_0 + ||\partial_1 v||_0 + ||\partial_2 v||_0$ where $\partial_i v_j = \frac{\partial v_j}{\partial x_i}(x) \quad 1 \leq i \leq 2, \; 1 \leq j \leq m.$

In the rest of this paper, the topology will be these norm topologies unless otherwise indicated.

Consider the problem of solving

$$-\Delta u + u = g \quad \text{on} \quad \Omega \tag{2.6}$$

$$\text{subject to} \quad \frac{\partial u}{\partial r} = 0 \quad \text{on} \quad \partial\Omega \tag{2.7}.$$

It is well-known that if $g \in L^p(\Omega)$ with $1 < p < \infty$, there is a unique solution u of (2.6)-(2.7). This solution lies in $W^{2,p}(\Omega)$ and is given by

$$u(x) = \int_\Omega G_N(x, y)\, g(y) dy \tag{2.8}$$

where $G_N: \overline{\Omega} \times \overline{\Omega} \to [0, \infty]$ is an L^2-function which obeys

(i) $G_N(x, y) = G_N(y, x)$ for all (x, y)

and (ii) $G_N(x, y) \geq 0$ for all $(x, y) \in \Omega \times \Omega$.

Equation (2.4) may be written

$$-D_j(\Delta v_j - v_j) = c\frac{\partial v_j}{\partial \xi} + D_j v_j + f_j(v_1,\ldots,v_m; \mu).$$

Thus using the equivalence of (2.6)-(2.7) and (2.8) one has that $v \in X_0$ will be a solution of (2.4)-(2.5) iff v is a solution of the non-linear system of integro-differential equations

$$D_j v_j(x) = \int_\Omega G_N(x, y)\left[c\frac{\partial v_j}{\partial \xi}(y) + \tilde{f}_j(v(y); \mu)\right] dy \tag{2.9}$$

Here $\tilde{f}_j(v; \mu) = D_j v_j + f_j(v_1, v_2,\ldots, v_m; \mu)$ and $1 \leq j \leq m$.

This system of equations may be considered as an operator equation on X_1. Define the maps $F: X_1 \times \Lambda \to X_0$, $A: X_1 \to X_0$ and $G: X_0 \to X_1$ by

$$F_j(v, \mu)(x) = \tilde{f}_j(v(x); \mu)$$

$$A_j v(x) = \frac{\partial v_j}{\partial \xi}(r, \xi) = x_1 \frac{\partial v_j}{\partial x_2}(x) - x_2 \frac{\partial v_j}{\partial x_1}(x)$$

$$\text{and} \quad G_j v(x) = D_j^{-1} \int_\Omega G_N(x, y) v_j(y) dy$$

and $1 \le j \le m$. The fact that these operators have the indicated ranges is obvious for F and A. For G it follows from the Sobolev theory for (2.6)-(2.7). Here F_j, A_j and G_j each represent the j-th component of the respective maps F, A and G.

Equation (2.9) may now be rewritten as a fixed point problem. Define $\mathbb{A}: X_1 \to X_1$, $\mathcal{F}: X_1 \times \Lambda \to X_1$ and $\mathbb{G}: X_1 \times \mathbb{R} \times \Lambda \to X_1$ by

$$\mathbb{A}v = GAv\,, \quad \mathcal{F}(v, \mu) = GF(v, \mu) \quad \text{and}$$

$$\mathbb{G}(v, c, \mu) = c\mathbb{A}v + \mathcal{F}(v, \mu)$$

Then (2.9) can be written as the operator equation

$$v = c\mathbb{A}v + \mathcal{F}(v, \mu) = \mathbb{G}(v, c, \mu) \tag{2.10}$$

One is interested in finding the values of (c, μ) for which there is a non-trivial solution in X_1 of this fixed point problem.

<u>Lemma 2.1.</u> $\mathbb{A}$, $\mathcal{F}(\cdot, \mu)$ are continuous and compact maps.

<u>Proof.</u> From the definition of X_1, X_0, one observes that A is a bounded, linear map of X_1 to X_0, hence it is continuous.

When $v_j \in C^\circ(\overline{\Omega})$, the solution u of (2.6)-(2.7) lies in $W^{2,p}(\Omega)$ for all $1 < p < \infty$. The Sobolev imbedding theorem then implies that $G_N v_j \in C^1(\Omega)$ and that $v_j \mapsto G_N v_j$ is compact for each j. Hence each component of $\mathbb{A}$ is a compact map of $C(\overline{\Omega})$ to $C^1(\Omega)$ and thus $\mathbb{A}$ is compact, being the product of compact maps.

Since $\tilde{f}: \mathbb{R}^m \times \Lambda \to \mathbb{R}^m$ is continuous, one has the map $F(\cdot, \mu): X_1 \to X_0$ is continuous. As indicated in the previous paragraph $G: X_0 \to X_1$ is a compact linear map hence $\mathcal{F}(\cdot, \mu) = GF(\cdot, \mu)$ is a compact and continuous map.

Here, and henceforth, we are using compact in the sense of mapping bounded sets into sets whose closure is compact.

<u>Corollary</u>. $\mathfrak{G}(\cdot, c, \mu)$ is a continuous and compact map of X_1 into itself for each (c, μ) in $\mathbb{R} \times \Lambda$.

3. BIFURCATION POINTS AND LINEAR THEORY.

We would like to show that there are non-trivial solutions of (2.4)-(2.5) by showing there are such solutions of (2.10). The parameter μ will be adjustable and we shall look for non-zero functions v and non-zero wave-speeds c obeying (2.10). The condition $c \neq 0$ implies that one has a non-stationary, or time-dependent, solution of the original equations.

A function v in X_1 is said to be a non-trivial solution of (2.10) if v is not the zero function in X_1. Let

$$\mathcal{S} = \{(v, c, \mu) \in X_1 \times \mathbb{R} \times \Lambda : \ v = \mathfrak{G}(v, c, \mu)\} ,$$

$$\mathcal{S}_0 = \{(0, c, \mu) : (c, \mu) \in \mathbb{R} \times \Lambda\}.$$

$\mathcal{S}$ is the set of all solutions of (2.10) and $\mathcal{S}_0$ is the set of trivial solutions of (2.10).

A point $(0, \hat{c}, \hat{\mu})$ is called a bifurcation point for a rotating wave if, in each open neighborhood U of $(0, \hat{c}, \hat{\mu})$ in $X_1 \times \mathbb{R} \times \Lambda$ there is a solution (v, c, μ) of (2.10) with $cv \neq 0$.

A mapping $N : X_1 \to X_1$ is said to be Fréchet differentiable at a point u in X_1 if there is a continuous linear operator $L : X_1 \to X_1$ obeying

$$\lim_{||h||_1 \to 0} ||h||_1^{-1} \, ||N(u + h) - N(u) - Lh||_1 = 0$$

When this holds we shall write $L = D_u N(u) \in \mathcal{L}(X_1)$ with $\mathcal{L}(X_1)$ being the space of all continuous linear maps of X_1 to itself endowed with the operator-norm topology.

The mapping N is continuously Frechet differentiable on a subset U of X_1 if it is Fréchet differentiable at each point of U and the map $DN: U \to \mathcal{L}(X_1)$ is continuous.

<u>Lemma 3.1</u>. For each (c, μ) in $\mathbb{R} \times \Lambda$, $\mathfrak{G}(\cdot, c, \mu)$ is Fréchet differentiable on X_1 and its derivative at a point v in X_1 is given by

$$D_v \mathfrak{G}(v, c, \mu)h(x) = c\mathbf{A}h(x) + GB(\mu)h(x) \tag{3.1}$$

where

$$B_i(\mu)h(x) = \sum_{j=1}^{m} b_{ij}(x, \mu)h_j(x) \quad \text{for} \quad 1 \le i \le m \tag{3.2}$$

and $b_{ij}(x, \mu) = \dfrac{\partial \tilde{f}_i}{\partial v_j}(v(x); \mu)$. $D_v \mathfrak{G}(v, c, \mu)$ is a compact linear map of X_1 to itself.

<u>Proof</u>. Since $\mathbf{A}$ is a compact linear map, its derivative is itself. Let h be a arbitrary function in X_1 and consider

$$\mathcal{F}(v+h, \mu) - \mathcal{F}(v, \mu) - GB(\mu)h$$

$$= G[F(v+h, \mu) - F(v, \mu) - B(\mu)h] .$$

Its j-th component

$$= D_j^{-1} \int_\Omega G_N(x, y)[\tilde{f}_j(v(y) + h(y); \mu) - \tilde{f}_j(v(y); \mu)$$

$$- \sum_{j=1}^{m} b_{ij}(y, \mu)h_j(y)]dy$$

If $||h||_1 \leq \delta$ one has there exists a constant C such that

$$\frac{||F(v+h, \mu) - F(v, \mu) - B(\mu)h||_0}{||h||_1} \leq C||h||_0 ,$$

since $\tilde{f}_j$ is twice continuously differentiable. Since G is a continuous linear map from X_0 to X_1, this implies

$$\lim_{||h||_1 \to 0^+} ||\mathcal{F}(v+h, \mu) - \mathcal{F}(v, \mu) - GB(\mu)h||_1 ||h||_1^{-1} = 0$$

and consequently $\mathcal{F}(\cdot, \mu)$ is Fréchet differentiable at v in X_1. Since $B(\mu)$ and A are continuous linear maps of X_1 to X_0 and G is compact from X_0 to X_1, the composition $G(cA + B(\mu))$ is compact from X_1 to itself. Thus $D_v\mathfrak{G}(v, c, \mu)$ is compact. □

Equation (2.10) can be written as

$$v - \mathfrak{G}(v, c, \mu) = 0 \tag{3.3}$$

From the implicit function theorem, one knows that this equation has a unique curve of solutions in a neighborhood of the solution $(0, c_0, \mu_0)$ provided the linear operator $L_0: X_1 \to X_1$ defined by

$$L_0 h = h - D_v\mathfrak{G}(0, C_0, \mu_0)h$$

$$= h - C_0 Ah - GB(\mu_0)h \tag{3.4}$$

is continuously invertible.

The fact that $D_v\mathfrak{G}(0, c_0, \mu_0)$ is compact implies that L_0 is continuously invertible if and only if 1 is not an eigenvalue of $D_v\mathfrak{G}(0, c_0, \mu_0)$. Thus the possible bifurcation points of (2.10) occur when one has non-trivial solutions h in X_1 of

$$h = c_0 GAh + GB(\mu_0)h \tag{3.5}$$

Using the definitions of A, G and $B(\mu_0)$ one observes that h is a solution of (3.5) if and only if $h \in X_1$ and it is a distributional solution of the system

$$M_{0j}h = D_j \Delta h_j + c_0 \frac{\partial h_j}{\partial \xi} + \sum_{k=1}^{m} b_{jk}(\mu_0) h_k = 0 \quad \text{in } \Omega \quad 1 \le j \le m \tag{3.6}$$

$$\text{subject to} \quad \frac{\partial h_j}{\partial r}(1, \xi) = 0 \quad \text{for} \quad 0 \le \xi \le 2\pi . \tag{3.7}$$

Here we have written $b_{jk}(\mu_0)$ for $\frac{\partial f_j}{\partial v_k}(0; \mu_0)$.

Equations (3.6)-(3.7) constitute a linear boundary value problem and we would like to find the values of c_0 for which it has a non-trivial solution.

Consider the standard eigenvalue problem

$$- \Delta u = \lambda u \quad \text{in } \Omega \tag{3.8}$$

$$\frac{\partial u}{\partial r} = 0 \quad \text{in } \partial\Omega \tag{3.9}$$

From classical analysis, the eigenvalues are $\lambda_{\ell n} = (j'_{\ell n})^2$ where $j'_{\ell n}$ is the n-th zero of $J'_\ell(r)$ and $J_\ell(r)$ is the usual Bessel function. Here ℓ is a non-negative integer, $n \ge 1$ and we are using the notation of Abramowitz and Stegun (1) Chapter 9.

When $\ell = 0$, the corresponding eigenfunction is $J_0(\sqrt{\lambda}_{\ell n} r)$. When $\ell \geq 1$ there are two linearly independent real eigenfunctions corresponding to each ℓ, n namely

$$J_\ell(\sqrt{\lambda}_{\ell n} r)\cos\ell\xi \quad \text{and} \quad J(\sqrt{\lambda}_{\ell n} r)\sin\ell\xi \; .$$

For notational convenience, we shall use the complex, normalized eigenfunctions

$$e_{\ell n}(r, \xi) = c_{\ell n} \, J_{|\ell|}(\sqrt{\lambda}_{\ell n} r)e^{i\ell\xi} \tag{3.10}$$

with ℓ any integer, $n \geq 1$, $\lambda_{-\ell n} = \lambda_{\ell n}$ for $\ell > 0$ and

$$c_{\ell n}^{-1/2} = 2\pi \int_0^1 r \, J_\ell(\sqrt{\lambda_{\ell n}} r)^2 dr \tag{3.11}$$

With this notation, when $\ell \geq 1$, $e_{\ell n}$ and $e_{-\ell n}$ will be two linearly independent complex eigenfunctions corresponding to the eigenvalue $\lambda_{\ell n}$.

Let $L^2(\Omega)$ be the usual complex Hilbert space of (equivalence classes of) Lebesgue measurable functions on Ω with the inner product

$$\langle f, g \rangle = \int_\Omega f(x)\overline{g(x)} \, dx$$

$L^2(\Omega; \mathbb{C}^m)$ will be the space of all functions $f\colon \Omega \to \mathbb{C}^m$ each of whose components f_i is in $L^2(\Omega)$. It is a Hilbert space under the inner product

$$[f, g] = \sum_{i=1}^{m} \langle f_i, g_i \rangle \; .$$

The family $\{e_{\ell n}\colon -\infty < \ell < \infty, n \geq 1\}$ defined by (3.10) is a complete orthonormal basis in $L^2(\Omega)$ from standard results in spectral theory and the normalization (3.11).

Let $B(\mu) = (b_{jk}(\mu))$ and $\mathcal{D} = \text{diag}(D_1, D_2, \ldots, D_m)$ be the $m \times m$ matrices whose entries are the coefficients in (3.6). The problem of finding non-trivial solutions of (3.6)-(3.7) may be reduced to a finite dimensional eigenvalue problem by the following result.

Theorem 3.2. (3.6)-(3.7) has a non-trivial solution if and only if there is an eigenvalue $\lambda_{\ell n}$ of (3.8)-(3.9) such that $\pm i\ell c_0$ are eigenvalues of $\lambda_{\ell n}\mathcal{D} - B(\mu_0)$.

Proof. Suppose h is a non-trivial solution of (3.6)-(3.7) then each h_j has a representation (in the L^2-sense) of the form.

$$h_j(r, \xi) = \sum_{p=-\infty}^{\infty} \sum_{q=1}^{\infty} a_{jpq} e_{pq}(r, \xi) \tag{3.12}$$

Take the inner product of $M_{oj}h$ with $e_{\ell n}$, then

$$(- \lambda_{\ell n} D_j + i\ell c_0) a_{j\ell n} + \sum_{k=1}^{m} b_{jk}(\mu_0) a_{k\ell n} = 0$$

for each j, ℓ, n. This is a matrix equation for the column m-vector $a^{(\ell n)}$ whose entries are $a_{j\ell n}$. It may be rewritten as

$$[\lambda_{\ell n}\mathcal{D} - B(\mu_0)] a^{(\ell n)} = i\ell c_0 a^{(\ell n)} \tag{3.13}$$

When h is non-trivial, there must be some non-trivial coefficients $a_{j\ell n}$ in (3.12) and, hence, there will be a non-trivial solution of (3.13) for some ℓ, n.

Conversely, if (3.13) has a non-trivial solution $a^{(\ell n)}$ then the functions

$$h(r, \xi) = a^{(\ell n)} e_{\ell n}(r, \xi)$$

will be non-trivial solutions of (3.6)-(3.7) lying in X_1.

Since $\lambda_{\ell n}$, $\mathfrak{D}$ and $B(\mu_0)$ are real matrices when $i\ell c_0$ is an eigenvalue of $\lambda_{\ell n}\mathfrak{D} - B(\mu_0)$ then $-i\ell c_0$ is also an eigenvalue. □

Corollary 1. The system (3.6)-(3.7) has a linearly independent pair of non-trivial real solutions $h^{(1)}$, $h^{(2)}$ given by

$$h_j^{(1)}(r, \xi) = c_j J_\ell(\nu r)\cos(\ell\xi + \psi_j) \qquad 1 \le j \le m \tag{3.14}$$

$$h_j^{(2)}(r, \xi) = c_j J_\ell(\nu r)\sin(\ell\xi + \psi_j)$$

with ℓ a positive integer and $(c_1, c_2, \ldots, c_m) \neq 0$ in $\mathbb{R}^m$

provided (i) ν is a positive zero of $J'_\ell(r)$

and (ii) $\nu^2\mathfrak{D} - B(\mu_0)$ has a pair of complex eigenvalues $\pm i\zeta$.

When (i) - (ii) holds, then $c_0 = \ell^{-1}\zeta$ in (3.6).

Proof. When (3.14) is a solution, the boundary condition (3.7) implies that ν is a zero of $J'_\ell(r)$. When $\nu = 0$, $\ell \ge 1$, the expression (3.14) becomes trivial, hence $\nu > 0$.

From the proof of the theorem; one sees that there will be a non-trivial complex solution of (3.6)-(3.7) of the form $ae_{\ell n}(r, \xi)$ with $a \in \mathbb{C}^m$ if and only if $\lambda_{\ell n}\mathfrak{D} - B(\mu_0)$ has $\pm i\ell c_0$ as eigenvalues. Identifying ν^2 and $\lambda_{\ell n}$ and $\zeta = \ell c_0$, one can take $h^{(1)}$ and $h^{(2)}$ to be the real and complex parts of $ae_{\ell n}(r, \xi)$. Thus $a_j = c_j e^{i\psi_j}$ for $1 \le j \le m$ with $c_j = |a_j|$ being real.

Remark In (3.14), one can normalize the ψ_j's so that $\psi_1 = 0$ as any complex multiple of $ae_{\ell n}(r, \xi)$ will also be a complex eigenfunction of (3.6)-(3.7).

One would now like to relate the solutions of (3.6)-(3.7) to those of (3.5). Note first that $\lambda_{\ell n} \ge \lambda_{\ell 1} \ge \ell^2$ for all $\ell \ge 1$ from (9.5.2) in

(1) and $\lambda_{\ell n} \to +\infty$ as either ℓ or n increases to infinity.

Thus for fixed μ_0, there could only be a finite number of matrices of the form $\lambda_{\ell n}\mathfrak{D} - B(\mu_0)$ which could have pure imaginary eigenvalues. Suppose that there are M of these matrices which have pure imaginary eigenvalues, one would like to know the dimension of the null space of L_0.

Lemma 3.3 Suppose $\lambda_{\ell n}\mathfrak{D} - B(\mu_0)$ has $\{\pm i\zeta_1, \pm i\zeta_2, \ldots, \pm i\zeta_k\}$ as distinct, non-zero, purely imaginary eigenvalues. Suppose that $i\zeta_j$ is an eigenvalue of geometric multiplicity m_j of $\lambda_{\ell n}\mathfrak{D} - B(\mu_0)$. Then (3.6)-(3.7) has exactly $2m_j$ real, linearly independent, solutions of the form (3.14) with $\nu = \sqrt{\lambda_{\ell n}}$ and $c_0 = \ell^{-1}|\zeta_j|$.

Proof. Follows from the proof of corollary 1 to theorem 3.2 as each linearly independent complex solution of (3.13) yields exactly 2 real linearly independent solutions of (3.6)-(3.7).

One observes that if $i\zeta_1$ and $i\zeta_2$ with $|\zeta_1| \neq |\zeta_2|$ are eigenvalues of $(\lambda_{\ell n}\mathfrak{D} - B(\mu_0))$ then the corresponding bifurcating waves-speeds will be distinct.

Theorem 3.4. Suppose $\ell \geq 1$, $c_0 \neq 0$ and $\pm i\ell c_0$ are eigenvalues of $\lambda_{\ell n}\mathfrak{D} - B(\mu_0)$ of geometric multiplicity $m_{\ell n}$. Then (3.6)-(3.7) has $2(\sum_{n=1}^{\infty} m_{\ell n})$ linearly independent real solutions of the form (3.14).

Proof. From lemma 3.3, for each n, one has $2m_{\ell n}$ linearly independent real solutions of (3.6)-(3.7). Summing over n one has the desired result since eigenfunctions corresponding to distinct eigenvalues are linearly independent. Note that there will only be finitely many non-zero values of $m_{\ell n}$ in this sum.

Corollary. Suppose $c_0 \neq 0$. Then there are $2(\sum_{\ell=1}^{\infty} \sum_{n=1}^{\infty} m_{\ell n})$ real, linearly independent solutions of (3.5) where $m_{\ell n}$ is defined as above.

In particular, the real dimension of the null space of L_0 is always even and finite when $c_0 \neq 0$.

Thus to check whether, when $\mu = \mu_0$, there might be a non-trivial solution of (3.5) for some $c_0 \in \mathbb{R}$ one must see if any of the matrices $\lambda_{\ell n}\mathcal{D} - B(\mu_0)$ with ℓ, n positive integers have pure imaginary, non-zero eigenvalues. To see if there is a non-trivial solution for a specific $c_0 (\neq 0)$ one must check whether $\pm i\ell c_0$ is an eigenvalues of $\lambda_{\ell n}\mathcal{D} - B(\mu_0)$. These are the basic bifurcation criteria. In the next section we shall discuss the actual existence of bifurcating waves.

4. BIFURCATION OF ROTATING WAVES.

To construct a new family of non-trivial solutions of (2.10) near $(0, c_0, \mu_0)$ we shall use an implicit function argument as in (5), theorem 5. Just as there we shall assume the interaction term is linear in the parameter μ, so that

$$f_j(v; \mu) = f_j^{(1)}(v) + \mu f_j^{(2)}(v). \tag{4.1}$$

for v in $\mathbb{R}^m$. This is true for the specific examples to be considered and the following results can often be generalized to cover the case where it doesn't hold.

When (4.1) holds one may write, in an obvious notation,

$$\mathcal{F}(v, \mu) = \mathcal{F}^{(1)}(v) + \mu\mathcal{F}^{(2)}(v)$$

and $\quad B(\mu_0)h = B^{(1)}h + \mu_0 B^{(2)}h$.

Equation (2.10) may be rewritten as

$$L_0 v = [I - c_0 \mathbb{A} - GB(\mu_0)]v = (c - c_0)\mathbb{A}v + (\mu - \mu_0)GB^{(2)}v \qquad (4.2)$$

$$+ N(v, \mu)$$

where $N(v, \mu) = \mathcal{F}(v, \mu) - GB(\mu)v$.

When L is a linear map, $N(L)$ and $R(L)$ will denote its kernel and range respectively. If Y is a subspace of X_1 then dim Y is the algebraic dimension of Y and codim Y is the dimension of the complement of Y is X_1 (when it exists).

<u>Theorem 4.1</u>. Suppose $\mathbb{G}: X_1 \times \mathbb{R} \times \Lambda \to X_1$ is continuously Fréchet differentiable on an open neighborhood of $(0, c_0, \mu_0)$. Let $L_0 = I - D_v\mathbb{G}(0, c_0, \mu_0)$ and assume that dim $N(L_0)$ = codim $R(L_0) = 2$. Suppose $h^{(1)}$, $h^{(2)}$ span $N(L_0)$ and $k^{(1)}$, $k^{(2)}$ span the complement Z, of $R(L_0)$ in X_1. Assume further that there is an h in $N(L_0)$ obeying

$$\text{der}\begin{pmatrix} [\mathbb{A}h, k^{(1)}] & [GB^{(2)}h, k^{(1)}] \\ [\mathbb{A}h, k^{(2)}] & [GB^{(2)}h, k^{(2)}] \end{pmatrix} \neq 0 \qquad (4.3)$$

Then there is an open neighborhood I of 0 in $\mathbb{R}^1$ and continuous functions $v: I \to X_1$, $c: I \to \mathbb{R}$ and $\mu: I \to \Lambda$ obeying $v(0) = 0$, $c(0) = c_0$ and $\mu(0) = \mu_0$ such that $(v(\varepsilon), c(\varepsilon), \mu(\varepsilon))$ is a non-trivial solution of (2.10) for $\varepsilon(\neq 0)$ in I.

<u>Proof.</u> Write $X_1 = Y_0 \oplus Y_1$ where $Y_0 = N(L_0)$ and Y_1 is the complement of Y_0 in X_1. Choose $h(\neq 0)$ in Y_0 and look for solutions of (4.1) of the form

$$v(\varepsilon) = \varepsilon[h + w(\varepsilon)] \qquad (4.4)$$

with $w(\varepsilon)$ in Y_1. Define $H: Y_1 \times \mathbb{R} \times \Lambda \times \mathbb{R} \to X_1$ by

$$H(w, c, \mu, \varepsilon) = L_0 w - (c - c_0)\mathbb{A}(h + w) - (\mu - \mu_0)GB^{(2)}(h + w)$$
$$- \varepsilon^{-1}N(\varepsilon(h + w), \mu).$$

When $v(\varepsilon)$ is a solution of (4.1), then $w(\varepsilon)$ will be a zero of H for given c, μ. Moreover from the assumptions on $\mathbb{G}$, H is continuously Fréchet differentiable near $(0, c_0, \mu_0, 0)$ and

$$H(0, c_0, \mu_0, 0) = 0$$

The derivatives of H are

$$D_w H(0, c_0, \mu_0, 0) = L_0, \quad D_c H(0, c_0, \mu_0, 0) = - \mathbb{A}h$$

$$\text{and} \quad D_\mu H(0, c_0, \mu_0, 0) = - GB^{(2)}h$$

We now use a version of the implicit function theorem described in Fife (9) theorem 1. The only requirment that remains to be verified is that if $P: X_1 \to Z_1$ is the canonical projection then $PD_\lambda H(0, c_0, \mu_0, 0)$ is one-to-one and onto where $\lambda = (c, \mu)$.

From the above expressions for $D_c H$ and $D_\mu H$, one sees that this will be true iff (4.3) holds with $k^{(1)}, k^{(2)}$ being a basis for Z_1.

Corollary. Suppose $\mathbb{G}$ is k-times continuously Fréchet differentiable (resp. real analytic) in an open neighborhood of $(0, c_0, \mu_0)$ and the other assumptions of the theorem hold. Then the nontrivial solutions $(v(\varepsilon), c(\varepsilon), \mu(\varepsilon))$ of (2.10) are (k-1)-times continuously Fréchet differentiable (resp. real analytic) on I.

Proof. This follows from the natural extension of Fife's result. See Crandall and Rabinowitz (7) for an analogous theorem in the case of Hopf bifurcation.

The condition (4.3) may be regarded as a transversality condition, and as will be seen in the example it enables one to solve recursively for the power series expansions of the solution when the equation is real analytic.

From (3.14) and (4.4), one sees that under the assumptions of the theorem, the bifurcation wave near a bifurcation point has the form

$$v_j(r, \xi; \varepsilon) = \varepsilon c_j J_\ell(\sqrt{\lambda_{\ell n}}\, r)\cos(\ell\xi + \psi_j + \theta) + O(\varepsilon)$$

for some $\ell \geq 1$ and some $0 \leq \theta \leq 2\pi$. The choice of h in $N(L_0)$ corresponds to a particular choice of phase θ. Thus, to a first approximation, the bifurcating wave will be characterized by the two positive integers ℓ, n. ℓ characterizes the azimuthal structure and n characterizes the radial behavior of the solution. For a particular case, we shall do a more detailed analysis in section 6.

It is often useful to have a version of (4.3) which does not require the computation of the Green's function G. Equation (2.4) can be rewritten, when f has the form (4.1), as

$$D_j \Delta v_j + c_0 \frac{\partial v_j}{\partial \xi} + \sum_{k=1}^{m} b_{jk}(\mu_0) v_k$$

$$= (c_0 - c)\frac{\partial v_j}{\partial \xi} + (\mu_0 - \mu) \sum_{k=1}^{m} b_{jk}^{(2)}(\mu_0) v_k + r_j(v, \mu) \tag{4.5}$$

where $b_{jk}^{(2)}(\mu_0) = \dfrac{\partial f_j^{(2)}}{\partial v_k}(\mu_0)$ and $r_j(v, \mu) = f_j(v; \mu) - \sum_{k=1}^{m} b_{jk}(\mu) v_k$

The left hand side of (4.5) is precisely the operator M_{oj} defined in (3.6). The adjoint operator to M_{oj} is

$$M^*_{0j}w = D_j \Delta w_j - c_0 \frac{\partial w_j}{\partial \xi} + \sum_{k=1}^{m} b_{kj}(\mu_0) w_k \tag{4.6}$$

subject to $\frac{\partial w_j}{\partial r}(1, \xi) = 0$ for $0 \le \xi \le 2\pi$ (4.7).

The criterion (4.3) is equivalent to requiring that for any non-trivial solution h of (3.6)-(3.7), one has

$$\det \begin{pmatrix} [\frac{\partial h}{\partial \xi}, k^{(1)}] & [B^{(2)}(\mu_0)h, k^{(1)}] \\ [\frac{\partial h}{\partial \xi}, k^{(2)}] & [B^{(2)}(\mu_0)h, k^{(2)}] \end{pmatrix} \neq 0 \tag{4.8}$$

whenever $K^{(1)}$, $k^{(2)}$ are two real, linearly independent solutions of $M^*_{0j}w = 0$ and $\dim N(M^*_0) = 2$.

5. GLOBAL BRANCHING OF WAVES

Theorem 4.1 is a local result in that it only quarantees a curve of non-trivial solutions close to the bifurcation point. By using topological methods one can prove a non-local result. These results are analogous to those of Alexander-Auchmuty (2) and depend on an abstract result of Alexander and Fitzpatrick (3).

One says that global bifurcation occurs at $(0, c_0, \mu_0)$ in $\mathcal{S}_0$ if there is a non-empty connected subset $\mathcal{B}_0$ of $\mathcal{S}-\mathcal{S}_0$ with $(0, c_0, \mu_0)$ in the closure $\overline{\mathcal{B}}_0$ and at least one of the following is true

(i) $\mathcal{B}_0$ is unbounded in the natural induced metric from $\mathcal{S}$.

(ii) there is a $(\overline{v}, \overline{c}, \overline{\mu})$ in $\overline{\mathcal{B}}_0$ with $\overline{\mu} \notin \Lambda$.

(iii) there is a $(c_1, \mu_1) \neq (c_0, \mu_0)$ with $(0, c_1, \mu_1) \in \overline{\mathcal{B}}_0$.

(i) says that either the norm of the solution in X_1, or the wavespeed c or the parameter μ becomes arbitrarily large (in absolute value) in $\mathcal{B}_0$.

Condition (ii) says that μ could go to the boundary of Λ, while (iii) says that in the closure of $\mathcal{B}_0$ there is another bifurcation point $(0, c_1, \mu_1)$.

Let $L(c, \mu): X_1 \to X_1$ be defined by

$$L(c, \mu) = I - c\mathbb{A} - GB(\mu) \tag{5.1}$$

(c.f. (3.4)). The global bifurcation theorem requires a condition on $L(c, \mu)$ rather than the transversality condition (4.3). As before let $L_0 = L(c_0, \mu_0)$.

Theorem 5.1. Suppose f obeys (F1) - (F2) and

(i) there is a neighborhood U of (c_0, μ_0) in $\mathbb{R} \times \Lambda$ such that 0 is not in the spectrum of $L(c, \mu)$ for (c, μ) in $U - \{(c_0, \mu_0)\}$,

(ii) there is an $\varepsilon > 0$ and some $\delta > 0$ such that when $\max(|c - c_0|, |\mu - \mu_0|) < \delta$ and $w \in N(L_0)$ then $||L(c, \mu)w|| \geq \varepsilon \max(|c - c_0|, |\mu - \mu_0|)||w||$,

(iii) $\dim N(L_0) = 2r$ with r odd,

then global bifurcation occurs at $(0, c_0, \mu_0)$.

Proof. This follows from the theorem in section 3 of (2). The condition (P1) of that theorem holds in our case from lemma 2.1. Condition (P2) is proved in lemma 3.1 and (P3) follows from the properties of $B(\mu)$ as a function of μ. Condition (P4) has been incorporated as condition (i) above.

6. ROTATING WAVES FOR THE BRUSSELATOR.

In this section, we will explicitly construct certain analytic approximations to the rotating wave solutions of the Brusselator on the unit disc. For other analyses of these equations see Nicolis and Prigogine (13) Chapter

7 or Auchmuty and Nicolis (4).

Here we shall write $u_1(r, \theta, t) = X(r, \theta, t) - A$

$$u_2(r, \theta, t) = Y(r, \theta, t) - B/A ,$$

then the equations for u_1, u_2 are

$$\frac{\partial}{\partial t}\begin{pmatrix} u_1 \\ u_2 \end{pmatrix} = \begin{pmatrix} D_1\Delta u_1 + (\mu - 1)u_1 + A^2u_2 + r(u_1, u_2) \\ D_2\Delta u_2 + \mu u_1 - A^2u_2 - r(u_1, u_2) \end{pmatrix} \tag{6.1}$$

with $r(u_1, u_2) = \mu A^{-1}u_1^2 + u_1u_2(2A + u_1)$.

Here μ replaces B in the usual form of these equations and we want to find rotating wave-solutions subject to the boundary conditions (2.2).

$$\text{Let} \quad u_j(r, \theta, t) = v_j(r, \theta - ct) \tag{6.2}$$

for $j = 1, 2$, $0 \leq r < 1$ and $0 < \theta < 2\pi$. If $\xi = \theta - ct$, then

$$D_i\Delta v_i + c\,\frac{\partial v_i}{\partial \xi} + f_i(v_1, v_2; \mu) = 0 \tag{6.3}$$

where

$$f_1(v_1, v_2; \mu) = (\mu - 1)v_1 + A^2v_2 + r(v_1, v_2)$$

and (6.4)

$$f_2(v_1, v_2; \mu) = -\mu v_1 - A^2v_2 - r(v_1, v_2).$$

We would like to find solutions v in X_1 which obey

$$\frac{\partial v_i}{\partial r}(1, \xi) = 0 \quad \text{for } 0 \leq \xi \leq 2\pi \quad i = 1, 2 \tag{6.5}$$

The equations (3.6) become

$$M_{01}h = D_1\Delta h_1 + c_0 \frac{\partial h_1}{\partial \xi} + (\mu_0 - 1)h_1 + A^2 h_2 = 0$$

$$M_{02}h = D_2\Delta h_2 + c_0 \frac{\partial h_2}{\partial \xi} - \mu_0 h_1 - A^2 h_2 = 0 \tag{6.6}$$

subject to (6.5) with h in place of v.

Now $B(\mu) = \begin{pmatrix} \mu-1 & A^2 \\ -\mu & -A^2 \end{pmatrix}$

and

$$\lambda\mathcal{D} - B(\mu) = \begin{pmatrix} \lambda D_1 - \mu + 1 & -A^2 \\ \mu & \lambda D_2 + A^2 \end{pmatrix}$$

$\lambda\mathcal{D} - B(\mu)$ will have a pair of pure imaginary eigenvalues $\pm i\zeta$ iff (i) its trace is zero and (ii) its determinant $\Delta(\lambda)$ is positive. When these conditions hold $\zeta = \sqrt{\Delta(\lambda)}$.

(i) will hold with $\lambda = \lambda_{\ell n}$ iff

$$\mu = \mu_{\ell n} = 1 + A^2 + \lambda_{\ell n}(D_1 + D_2). \tag{6.7}$$

when $\mu = \mu_{\ell n}$, then

$$\Delta(\lambda_{\ell n}) = A^2 + \lambda_{\ell n}A^2(D_1 - D_2) - \lambda_{\ell n}^2 D_n^2 . \tag{6.8}$$

There are only a finite number of eigenvalues $\lambda_{\ell n}$ of (3.8) - (3.9) for which $\Delta(\lambda_{\ell n}) > 0$. More specifically, if $\hat{\lambda}(A, D_1, D_2)$ is the unique positive solution of

$$D_2^2 \lambda^2 - \lambda A^2(D_1 - D_2) - A^2 = 0 \tag{6.9}$$

then $\Delta(\lambda_{\ell n}) > 0$ iff $0 \le \lambda_{\ell n} < \hat{\lambda}(A, D_1, D_2)$. Thus if there are bifurcating waves for the system (6.3) - (6.5) they only arise from the lowest eigenvalues of the Neumann problem (3.8) - (3.9).

Approximations to these bifurcating waves can be obtained near a bifurcation point by the familiar methods of analytic branching theory; see (4) for the computation of Hopf bifurcations in this manner for this system. We would like to approximate non-trivial solutions of (6.3) - (6.5) near $(0, c_{\ell n}, \mu_{\ell n})$ where $\mu_{\ell n}$ is given by (6.7) and $C_{\ell n} = \ell^{-1}\sqrt{\Delta(\lambda_{\ell n})}$ with $\ell \ge 1$ and $\Delta(\lambda_{\ell n}) > 0$.

Henceforth we shall assume these conditions and replace the subscripts ℓn by 0 for notational convenience. Then (6.3) - (6.4) can be written

$$M_0 v = (c_0 - c)\frac{\partial v}{\partial \xi} + (\mu_0 - \mu)\begin{pmatrix} v_1 \\ -v_1 \end{pmatrix} - \begin{pmatrix} r(v_1, v_2) \\ -r(v_1, v_2) \end{pmatrix} \tag{6.10}$$

where M_0 is the operator defined by (6.5) - (6.6) and $v = \begin{pmatrix} v_1 \\ v_2 \end{pmatrix} \in X_1$.

When $c_0 = c_{\ell n}$, $\mu_0 = \mu_{\ell n}$, the complex solutions of (6.6) are complex multiples of

$$h(r, \xi) = \begin{pmatrix} 1 \\ -r_{\ell n}\, e^{i\phi_{\ell n}} \end{pmatrix} J_\ell(\sqrt{\lambda_{\ell n}}\, r)\, e^{i\ell\xi} \tag{6.11}$$

where

$$r_{\ell n}^2 = A^{-2}(A^2 + \lambda_{\ell n}(D_1 + D_2)) = A^{-2}(\mu_{\ell n} - 1)$$

and $0 < \phi_{\ell n} < \pi/2$ is the solution of

$$\tan \phi_{\ell n} = \frac{\sqrt{\Delta(\lambda_{\ell n})}}{A^2 + \lambda_{\ell n} D_2} .$$

The functions $h^{(1)}$, $h^{(2)}$ corresponding to (3.14) are the real and complex parts of (6.11).

In a similar manner, the real eigenfunctions of $M_0^* w = 0$ are $k^{(1)}$, $k^{(2)}$ where these are the real and complex parts of

$$K(r, \xi) = \begin{pmatrix} r_{\ell n} e^{i\phi_{\ell n}} \\ 1 \end{pmatrix} J_\ell(\sqrt{\lambda_{\ell n}}\, r)\, e^{i\ell\xi} . \tag{6.12}$$

$k^{(1)}$, $k^{(2)}$ span the complement of $R(M_0)$ in X_1 from the Fredholm alternative.

Suppose that near $(0, c_0, \mu_0)$, the non-trivial solutions of (6.10) have power series expansions of the form

$$c(\varepsilon) = \sum_{m=0}^{\infty} c_m \varepsilon^m \quad , \quad \mu(\varepsilon) = \sum_{m=0}^{\infty} \mu_m \varepsilon^m$$

and

$$v(\varepsilon) = \sum_{m=1}^{\infty} \varepsilon^m v^{(m)} \tag{6.13}$$

with $v^{(m)}$ orthogonal to $\ker M_0$ for $m \geq 2$.

Substitute these expressions in (6.10) and equate equal powers of ε. One finds that $M_0 v^{(1)} = 0$ and

$$-M_0 v^{(m)} = \sum_{j=1}^{m-1} [c_j \frac{\partial v^{(m-j)}}{\partial \xi} + \mu_j \begin{pmatrix} v_1^{(m-j)} \\ -v_1^{(m-j)} \end{pmatrix}] + \begin{pmatrix} r^{(m)} \\ -r^{(m)} \end{pmatrix} \quad m \geq 2 \tag{6.14}$$

where each $r^{(m)}$ is a function of $A, \mu_0, \ldots, \mu_{m-2}$, and the components of $v^{(1)}, \ldots, v^{(m-1)}$.

In particular

$$r^{(2)} = \mu_0 A^{-1} v_1^{(1)^2} + 2A\, v_1^{(1)}\, v_2^{(1)} ,$$

$$\text{and}\quad r^{(3)} = A^{-1}[2\mu_0 v_1^{(1)} v_1^{(2)} + \mu_1 v_1^{(1)^2}]$$

$$+ 2A[v_1^{(1)} v_2^{(2)} + v_1^{(2)} v_2^{(1)}] + v_1^{(1)^2} v_2^{(1)} ,$$

are the first two such expressions.

Take $v^{(1)} = h^{(1)}$ to be the real part of (6.11). This choice is a normalization of ε and of a phase. Thus

$$v^{(1)}(r, \xi) = \begin{pmatrix} \cos \ell\xi \\ -r_{\ell n} \cos(\ell\xi + \phi_n) \end{pmatrix} J_\ell(\sqrt{\lambda_{\ell n}}\, r). \qquad (6.15)$$

and

$$r^{(2)}(r, \xi) = \left[\frac{\mu_0}{A} \cos^2 \ell\xi - 2A\, r_{\ell n} \cos\ell\xi \cos(\ell\xi + \phi_n)\right] J_\ell^2(\sqrt{\lambda_{\ell n}}\, r). \qquad (6.16)$$

The equation (6.14) with $m = 2$ is a linear, inhomogeneous equation for $v^{(2)}$. It has a unique solution $v^{(2)}$ which is orthogonal to $N(M_0)$ in $L^2(\Omega; \mathbb{R}^2)$ provided the right-hand side is orthogonal to both $k^{(1)}$ and $K^{(2)}$. That is

$$c_1 < \frac{\partial h^{(1)}}{\partial \xi}, k^{(j)} > + \mu_1 < \begin{pmatrix} h^{(1)} \\ -h^{(1)} \end{pmatrix}, k^{(j)} > = < \begin{pmatrix} -r^{(2)} \\ r^{(2)} \end{pmatrix}, k^{(j)} >$$

for $j = 1, 2$. This is a pair of equation for c_1, μ_1.

Doing all the substitutions ans evaluations one finds $c_1 = \mu_1 = 0$. For general m, one finds that (6.14) has a solution for $v^{(m)}$ provided

$$\begin{pmatrix} < \frac{\partial h^{(1)}}{\partial \xi}, k^{(1)}> & < \begin{pmatrix} h_1^{(1)} \\ -h_1^{(1)} \end{pmatrix}, k^{(1)}> \\ < \frac{\partial h^{(1)}}{\partial \xi}, k^{(2)}> & < \begin{pmatrix} h_1^{(1)} \\ -h_1^{(1)} \end{pmatrix}, k^{(2)}> \end{pmatrix} \begin{pmatrix} c_{m-1} \\ \mu_{m-1} \end{pmatrix} = \begin{pmatrix} d_1^{(m-1)} \\ d_2^{(m-1)} \end{pmatrix} \tag{6.17}$$

where $d_1^{(m-1)}$ is a vector which can be computed. The matrix on the left hand side is precisely the matrix (4.8) as, in this case, $B^{(2)}(\mu_0) = \begin{pmatrix} 1 & 0 \\ -1 & 0 \end{pmatrix}$.

A computation shows that this matrix

$$= \pi r_{\ell n} I_{\ell n} \begin{pmatrix} 2\ell \sin \phi_{\ell n} & \cos \phi_{\ell n} - r_{\ell n}^{-1} \\ 0 & \sin \phi_{\ell n} \end{pmatrix}$$

where $I_{\ell n} = \int_0^1 r J_\ell^2(\sqrt{\lambda_{\ell n}}\, r)\, dr > 0.$

This matrix is always non-singular as $0 < \phi_{\ell n} < \pi/2$ so the transversality conditions (4.3) or (4.8) hold at every possible bifurcation point for this system.

Having found that $c_1 = \mu_1 = 0$, one can then construct the solution $v^{(2)}$ of

$$-M_0 v^{(2)} = \begin{pmatrix} r^{(2)} \\ -r^{(2)} \end{pmatrix} \tag{6.18}$$

which is orthogonal to $N(M_0)$ by using eigenfunction expansions. Since $r^{(2)}$ is given by (6.16) and is non-zero, $v^{(2)}$ is non-trivial. In particular one find that

$$\int_0^{2\pi}\int_\Omega^1 rv^{(2)}(r, \xi)drd\xi \neq 0$$

so that $v^{(2)}$ has a non-trivial constant component in its eigenfunction expansion.

One can continue and compute c_2, μ_2. My computations indicated that μ_2 was always negative but c_2 did not seem to have any obvious sign conditions for all A, D_1, D_2 positive. Thus these bifurcating waves all appear to bifurcate subcritically, with

$$\mu = \mu_{\ell n} + \mu_2\varepsilon^2 \qquad c = c_{\ell n} + c_2\varepsilon^2$$

$$\text{and} \quad v(\varepsilon) = \varepsilon\begin{pmatrix} \cos\ell\xi \\ \\ -r_{\ell n}\cos(\ell\xi + \phi_{\ell n}) \end{pmatrix} J_\ell(\sqrt{\lambda_{\ell n}}\, r) + \varepsilon^2 v^{(2)}(r, \xi)$$

with μ_2 negative and c_2 non-zero in general.

To see if this bifurcation is global one must verify certain assumptions on the operator $L(c, \mu)$ defined by (5.1).

Theorem 6.1. Suppose $\mu_0 = \mu_{\ell n}$ is defined by (6.7) with ℓ, n positive integers and $\Delta(\lambda_{\ell n}) > 0$. Then there is global bifurcation of non-trivial rotating-wave solutions of (6.1) - (6.2) at $(0, c_0, \mu_0)$ with $c_0 = \ell^{-1}\sqrt{\Delta(\lambda_{\ell n})}$.

Proof. To prove this theorem one must verify the conditions of theorem 5.1. For each pair of positive integers (ℓ, n), there is a corresponding eigenvalue $\lambda_{\ell n}$ of (3.8)-(3.9). These eigenvalues are isolated. Hence each critical value $\mu_{\ell n}$ of μ is isolated from the expression (6.7) so condition (i) holds.

When $\mu = \mu_{\ell n}$, $\pm i\ell c_0 = \sqrt{\Delta(\lambda_{\ell n})}$ are eigenvalues of multiplicity 1 of $\lambda_n \mathcal{D} - B(\mu_{\ell n})$ and there is only one such value of λ at which $\lambda\mathcal{D} - B(\mu_{\ell n})$ has a pair of pure imaginary eigenvalues. Hence from the corollary to theorem 3.4, there are exactly 2 linearly independent real solutions of (3.5) and thus $\dim N(L_0) = 2$.

Finally $L(c, \mu)h^{(j)} = (c - c_0)GAh^{(j)} + (\mu - \mu_0)G\begin{pmatrix} h_1^{(j)} \\ -h_1^{(j)} \end{pmatrix}$, using the fact that $L(c_0, \mu_0)h^{(j)} = 0$ for $j = 1, 2$.

Now $Ah^{(j)}(r, \xi) = \begin{pmatrix} -\ell\sin\ell\xi \\ +\ell r_{\ell n} \sin(\ell\xi + \phi_n) \end{pmatrix} J_1(\sqrt{\lambda_{\ell n}}\, r)$.

so $||Ah^{(j)}||_1 = \ell||h^{(j)}||_1$, and, moreover, $||h_2^{(j)}||_1 = r_{\ell n}||h_1^{(j)}||_1$ and $r_{\ell n} > 1$. (where we are just using the X_1-norm on a single component of $h^{(j)}$). Since $GAh^{(j)}$ and $G\begin{pmatrix} h_1^{(j)} \\ -h_1^{(j)} \end{pmatrix}$ have components which are proportional to $\lambda_{\ell n}^{-1}Ah^{(j)}$ and $\lambda_{\ell n}^{-1}\begin{pmatrix} h_1^{(j)} \\ -h_1^{(j)} \end{pmatrix}$ one verifies that condition (ii) of theorem 5.1 holds and thus the result follows. □

REFERENCES.

1. M. Abramowitz, and I.A. Stegun, (eds) Handbook of Mathematical Functions, Nat. Bureau of Standards A.M.S. Vol 55 (1964).

2. J.C. Alexander, and J.F.G. Auchmuty, Global Bifurcation of Waves, Manuscripta Math 27, (1979) pp 159-166.

3. J.C. Alexander, and P.M. Fitzpatrick, Bifurcation of Fixed Points of parametrized condensing operaters, J. Func. Anal.

4. J.F.G. Auchmuty, and G. Nicolis, Bifurcation Analysis of reaction-diffusion equations III, Chemical oscillations, Bull. Math Bio 38, (1976) pp 325-350.

5. J.F.G. Auchmuty, Bifurcating waves, Annals N.Y. Acad. of Sci., 316, (1979) pp 263-278.

6. D.S. Cohen, J.C. Neu and R.R. Rosales, Rotating Spiral Wave Solutions of Reaction-diffusion equations, SIAM J. Ap. Math 35(1978) pp 536-547.

7. M.G. Crandall, and P.H. Rabinowitz, The Hopf Bifurcation theorem, MRC Report #1604, (1976) Univ. of Wisconsin.

8. T. Erneux, and M. Herschkowitz-Kaufman, Rotating Waves as asymptotic solutions of a model chemical reaction, J. Chem. Phys. 66 (1977) pp 248-253.

9. P.C. Fife, Branching Phenomena in Fluid Dynamics and chemical reaction-diffusion theory, in Eigenvalues of non-linear problems, CIME lectures (1974) pp 25-83. Cremonese, Rome.

10. J.M. Greenberg, Periodic Solutions to reaction-diffusion equations, SIAM. J. Appl. Math 30 (1976) pp 199-205.

11. J.M. Greenberg, Spiral Waves for λ-ω Systems, SIAM J. Appl. Math 39 (1980) pp 301-309.

12. N. Kopell, and L.N. Howard, Target Patterns and horseshoes from a perturbed central force problem, Stud. Appl. Math 64 (1981) pp 1-56.

13. G. Nicolis, and I. Prigogine, Self-organization in nonequilibrium systems, John Wiley, New York (1977).

14. A.T. Winfree, The Geometry of biological time, Springer Verlag New York (1980).

Giles Auchmuty
University of Houston

P N BROWN

A nonlinear diffusion model in population biology

§1. INTRODUCTION

We consider here the nonlinear-diffusion interaction equation

$$u_t = [(\alpha u + \beta)u]_{xx} + u(1 - u) , \quad -L < x < L , \; t > 0 , \tag{1.1}$$

$$u(\pm L,t) \equiv 0 , \; t \geq 0 , \tag{1.2}$$

$$u(x,0) = u^o(x) \geq 0 , \quad -L \leq x \leq L . \tag{1.3}$$

This is a model for a species which is undergoing a "logistic" growth rate and dispersing throughout a spatial habitat $(-L,L)$ touching a completely hostile environment at its boundary. Here, $u(x,t)$ represents the species density, $\alpha \geq 0$ and $\beta > 0$ are constants, and the dispersal rate is nonlinear and "density-dependent" in that it depends upon the species density. Thus, the species u disperses not only randomly, but also away from high concentrations of itself. Density-dependent diffusion models have been studied by several authors, e.g. Mimura [8], and Okubo [10] (see Gurtin & MacCamy [5] or Aronson [1] for the derivation of a model of this type). We emphasize that in the following β will always be a positive constant. This has the effect of making (1.1) nondegenerate in the sense that disturbances still propagate with infinite speed (see §2), as opposed to degenerate nonlinear-diffusive terms (e.g. $\beta = 0$ above) considered by Aronson [1] and others, where there is a finite speed of propagation.

When $\alpha = 0$ and $\beta = 1$, equation (1.1) reduces to <u>Fisher's equation</u>, for which the large-time behavior of positive solutions is well-known (cf. [6]). Briefly, if $L < \pi/2$, then all solutions of (1.1) - (1.3) (with $\alpha = 0$ and $\beta = 1$) decay to zero as $t \to \infty$. However, for each $L > \pi/2$, there exists a unique positive steady-state solution which is a global attractor. Our interest here is to investigate the effect upon the large-time bahavior of solutions that occurs when one replaces the linear diffusive term in Fisher's equation by a nonlinear mechanism. Several authors have recently considered two-species competition models with non-linear diffusive terms (cf. [8] and [9]). There, the effect of these terms involves the destablizing of a spatially uniform steady-state solution with both species coexisting, which then gives rise to a bifurcating family of spatially non-uniform steady-state solutions with both species segregated in space. For two-species competition with linear diffusive terms, no destablization and bifurcation occurs (cf. [2]). We show below that the inclusion of a nonlinear-diffusive term in Fisher's equation does not have the same effect. Namely, the behavior of solutions of (1.1) - (1.3) is completley analogous with that of Fisher's equation. These preliminary results seem to indicate that in one-species models, nondegenerate nonlinear diffusion plays essentially the same role as linear diffusion.

The rest of the paper is organized as follows. In §2, we prove a maximum principle and comparison theorem. In §3, we discuss the existence of positive steady-state solutions of (1.1) - (1.2), and finally in §4, discuss their asymptotic stability.

§2. COMPARISON RESULTS

In this section we prove a comparison theorem for a more general version of (1.1) - (1.3). We begin with two lemmas. First, let

$$\phi(s) = (\alpha s + \beta)s \quad \text{with} \quad \beta > 0 \quad \text{and} \quad \alpha \geq 0.$$

We note that $\phi(s) > 0$ for $s > 0$ and $\phi(s) < 0$ iff $\frac{-\beta}{\alpha} < s < 0$, with $\phi'(s) = 0$ only when $s = \frac{-\beta}{2\alpha}$. Also, $\phi(\frac{-\beta}{2\alpha})$ is a global minimum. Next, for any $T > 0$ let

$$\Omega_T = \{(x,t): -L < x < L \ , \ 0 < t \leq T\} \ .$$

<u>Lemma 2.1</u>: Let $h(x,t) \leq 0$ in Ω_T and let $u(x,t) \in C^0(\bar{\Omega}_T) \cap C^2(\Omega_T)$ satisfy

$$\phi(u)_{xx} - u_t + h(x,t)u < 0 \quad \text{in} \quad \Omega_T \ , \tag{2.1}$$

with

$$u(x,0) \geq 0 \ , \quad -L \leq x \leq L \ , \quad \text{and} \tag{2.2}$$

$$u(\pm L,t) \geq 0 \ , \quad 0 \leq t \leq T \ . \tag{2.3}$$

Then $u(x,t) > 0$ in Ω_T.

<u>Proof</u>: Let $w(x,t) \equiv \phi(u(x,t))$ and suppose that w has a nonpositive minimum at a point $(x_0,t_0) \in \Omega_T$. Then

$$w_{xx}(x_0,t_0) \geq 0 \ , \quad w_x(x_0,t_0) = 0 \ , \quad \text{and} \quad w_t(x_0,t_0) \leq 0 \ .$$

Also,

$$w_t = \phi'(u)u_t \ , \quad w_x = \phi'(u)u_x \ , \quad \text{and}$$

$$w_{xx} = \phi''(u)u_x^{\ 2} + \phi'(u)u_{xx} \ , \tag{2.4}$$

and from the definition of $\phi(s)$ it follows that if $w(x_0,t_0) = \phi(u(x_0,t_0)) \le 0$, then $u(x_0,t_0) \le 0$.

If $\phi'(u(x_0,t_0)) > 0$, then (2.4) implies that $u_t(x_0,t_0) \le 0$. Thus, using (2.1)

$$0 > w_{xx}(x_0,t_0) - u_t(x_0,t_0) + h(x_0,t_0)u(x_0,t_0) \ge 0 \qquad (2.5)$$

which is a contradiction.

If $\phi'(u(x_0,t_0)) \le 0$, then (2.2) and the fact that $\phi'(s) > 0$ for $s \ge 0$ imply that there exists a smallest time $t_0' > 0$ such that

$$\phi'(u(x_0,t_0') = 0$$

and

$$\phi'(u(x_0,t)) > 0 \quad \text{for} \quad 0 \le t < t_0' \quad . \qquad (2.6)$$

From the definition of $\phi(s)$ we clearly must then have $u(x_0,t_0') = \frac{-\beta}{2\alpha}$ and $\phi(\frac{-\beta}{2\alpha})$ is the global minimum of $\phi(s)$, $s \in \mathbb{R}$. Now,

$$w_{xx}(x_0,t_0') = 2\alpha u_x^{\,2}(x_0,t_0') \ge 0 \quad ,$$

and so if $u_t(x_0,t_0') \le 0$, then we again arrive at a contradiction as in (2.5). If $u_t(x_0,t_0') > 0$, then there exists an $\varepsilon > 0$ such that

$$u_t(x_0,t) > 0 \quad \text{for} \quad t_0' - \varepsilon \le t \le t_0' \quad . \qquad (2.7)$$

Hence,

$$w_t(x_0,t) = \phi'(u(x_0,t))\cdot u_t(x_0,t) > 0 \quad \text{for} \quad t_0' - \varepsilon \le t < t_0'$$

by (2.6) and (2.7). This then implies that for some $\delta > 0$

$$\phi(u(x_0,t)) = w(x_0,t) < w(x_0,t_0') = \phi(\tfrac{-\beta}{2\alpha}) \quad \text{for} \quad t_0' - \delta \le t < t_0'$$

which contradicts $\phi(\frac{-\beta}{2\alpha})$ being a global minimum of $\phi(s)$.

Therefore, $w(x,t)$ cannot have a nonpositive minimum in Ω_T, and since $w(x,0) \geq 0$ and $w(\pm L,t) \geq 0$ by (2.2) and (2.3) we have $w > 0$ in Ω_T.

Finally, (2.2) and (2.3), along with $w > 0$ in Ω_T imply that $u(x,t) > 0$ in Ω_T. □

Lemma 2.2: Let $h(x,t)$ be bounded from above in Ω_T and let $u(x,t) \in C^0(\overline{\Omega}_T) \cap C^2(\Omega_T)$ satisfy

$$\phi(u)_{xx} - u_t + h(x,t)u \leq 0 \quad \text{in} \quad \Omega_T , \tag{2.8}$$

along with u_{xx} bounded from above in Ω_T and

$$u(x,0) \geq 0 , \quad -L \leq x \leq L , \quad \text{and}$$

$$u(\pm L,t) \geq 0 , \quad 0 \leq t \leq T .$$

Then $u(x,t) \geq 0$ in Ω_T.

Proof: Let $w = u + \varepsilon v$, where $\varepsilon > 0$ is an arbitrary constant and $v(t) = e^{\lambda t}$ where $\lambda > 0$ is a constant to be chosen. Then w satisfies

$$\begin{aligned}
&\phi(w)_{xx} - w_t + h(x,t)w \\
&\qquad = \phi(u)_{xx} - u_t + h(x,t)u + [\phi(u + \varepsilon v) - \phi(u)]_{xx} \\
&\qquad\qquad - \varepsilon v_t + \varepsilon h(x,t)v \\
&\qquad \leq [\phi(u + \varepsilon v) - \phi(u)]_{xx} - \varepsilon v_t + \varepsilon h(x,t)v ,
\end{aligned}$$

by (2.8).

We want to choose λ so that w satisfies a strict inequality of the type (2.1). Now, with $\phi(u) = (\alpha u + \beta)u$ we have

$$[\phi(u + \varepsilon v) - \phi(u)]_{xx} = [(\alpha u + \varepsilon\alpha v + \beta)(u + \varepsilon v) - (\alpha u + \beta)u]_{xx}$$

$$= [\varepsilon v(\beta + 2\alpha u + \varepsilon v)]_{xx}$$

$$= \varepsilon v 2\alpha u_{xx} .$$

Hence, w satisfies

$$\phi(w)_{xx} - w_t + h(x,t)w \le \varepsilon v[2\alpha u_{xx} - \lambda + h(x,t)]$$

$$< 0$$

if $\lambda > \sup_{(x,t) \in \Omega_T} \{2\alpha u_{xx} + h(x,t)\}$. Since $w(x,0) \ge 0$ and $w(\pm L,t) \ge 0$ we have that for any $\varepsilon > 0$,

$$u + \varepsilon v = w > 0 \text{ in } \Omega_T ,$$

by Lemma 2.1. Letting $\varepsilon \to 0$ gives $u \ge 0$ in Ω_T. □

We now prove the main result in this section.

<u>Theorem 2.3</u>: Let u and $v \in C^2(\Omega_T) \cap C^0(\overline{\Omega}_T)$ with u_{xx} and v_{xx} bounded from above in Ω_T and satisfying

$$\phi(u)_{xx} - u_t + f(u) \le \phi(v)_{xx} - v_t + f(v) \quad \text{in } \Omega_T ,$$

where $f \in C^1(\mathbb{R})$. Also assume that $u \ge 0$ and $v \ge 0$ in Ω_T along with

$$u(x,0) \ge v(x,0) \quad , \quad -L \le x \le L \quad , \text{ and}$$

$$u(\pm L,t) \ge v(\pm L,t) \quad , \quad 0 \le t \le T \quad .$$

Then $u \ge v$ in Ω_T with $u(x,t) > v(x,t)$ in Ω_T if there exists an $\overline{x} \in (-L,L)$ such that $u(\overline{x},0) > v(\overline{x},0)$.

Proof: Let $w = u - v$. Then w satisfies

$$d(x,t)w_{xx} + e(x,t)w_x - w_t + h(x,t)w \le 0 \quad \text{in} \quad \Omega_T ,$$

where

$$d(x,t) = \beta + \alpha(u + v) \quad ,$$

$$e(x,t) = 2\alpha(u_x + v_x) \quad , \quad \text{and}$$

$$h(x,t) = \begin{cases} \dfrac{f(u) - f(v)}{u - v} + 2\alpha(u_{xx} + v_{xx}) & , \quad \text{if} \quad u(x,t) \ne v(x,t) \\ f'(u) + 2\alpha(u_{xx} + v_{xx}) & , \quad \text{if} \quad u(x,t) = v(x,t). \end{cases}$$

Note that $h(x,t)$ is bounded from above in Ω_T, and since $u \ge 0$ and $v \ge 0$ in Ω_T, $d(x,t) > 0$ in Ω_T. Therefore, by the usual maximum principle (cf. [11]).

$$w(x,t) \ge 0 \quad \text{in} \quad \Omega_T \quad ,$$

with $w(x,t) > 0$ in Ω_T if $w(\bar{x},0) > 0$ for some $\bar{x} \in (-L,L)$. □

Using Lemma 2.2 and Theorem 2.3 we can prove a maximum principle for solutions of (1.1) - (1.3).

Theorem 2.4: Let $u(x,t) \in C^2(\Omega_T)$ with u_{xx} bounded from above in Ω_T, and satisfying

$$u_t = [(\alpha u + \beta)u]_{xx} + ug(u) \quad \text{in} \quad (-L,L) \times (0,\infty)$$

$$u(\pm L,t) = 0 \quad t \ge 0$$

$$u(x,0) \ge 0 \quad , \quad u(x,0) \not\equiv 0 \quad \text{in} \quad [-L,L] \quad ,$$

where $g \in C^1(\mathbb{R})$.

Then $u(x,t) > 0$ in $(-L,L) \times (0,\infty)$.

Proof: Lemma 2.2 implies $u(x,t) \geq 0$ in Ω_T (with $h(x,t) \equiv g(u(x,t))$). Then in Theorem 2.3, let $v(x,t) \equiv 0$. Since $u(\bar{x},0) > 0$ for some $\bar{x} \in (-L,L)$ we then have that $u(x,t) > 0$ in Ω_T. Since the argument works for any fixed $T > 0$, we have $u > 0$ in $(-L,L) \times (0,\infty)$. □

Remarks: Masuda and Mimura [7] have proved global existence and uniqueness for systems of equations involving nonlinear-diffusion terms of the type in (1.1). Their results show that solutions with nonnegative initial data always remain nonnegative. However, these results are only for one space dimension, while the results shown above work equally well for single equations of the form

$$u_t = \sum_{i=1}^{n} \phi(u)_{x_i x_i} + ug(u) \quad , \quad x \in D \quad , \quad t > 0$$

$$u(x,0) \geq 0 \quad x \in D \quad ,$$

$$u(x,t) = 0 \quad x \in \partial D \quad , \quad t \geq 0 \quad ,$$

where $D \subseteq \mathbb{R}^n$ is a bounded domain with smooth boundary.

The results in this section may prove useful in an existence theory for the above problem.

§3. STEADY-STATE SOLUTIONS

Here we investigate the number of positive steady-state solutions of (1.1) - (1.2); namely, we look for positive solutions of

$$[(\alpha v + \beta)v]'' + v(1 - v) = 0 \quad , \quad -L < x < L \quad , \tag{3.1}$$

$$v(\pm L) = 0 \quad . \tag{3.2}$$

We first make a transformation. Suppose $v(x)$ is a positive solution of (3.1) - (3.2). Let $w(x) \equiv (\alpha v(x) + \beta)v(x)$. Then $w(x) > 0$ and satisfies

$$w'' + \ell(w)(1 - \ell(w)) = 0 \quad , \quad -L < x < L \tag{3.3}$$

$$w(\pm L) = 0 \quad , \tag{3.4}$$

where

$$\ell(w) = \frac{2w}{\beta + \sqrt{\beta^2 + 4\alpha w}} \quad ,$$

and $v(x) = \ell(w(x))$. Conversely, if $w(x)$ is a positive solution of (3.3) - (3.4), then $v(x) \equiv \ell(w(x))$ is a positive solution of (3.1) - (3.2). Hence, if we can find all the positive solutions of (3.3) - (3.4), then we know all the positive solutions of (3.1) - (3.2).

Second, we note that in view of the comparison theorem 2.3, any solution of (1.1) - (1.3) with bounded nonnegative initial data can be bounded above by a solution of the ordinary differential equation

$$\frac{dU}{dt} = U(1 - U) \quad ,$$

$$U(0) = U_0 > 0 \quad ,$$

for some U_0, and $U(t) \to 1$ as $t \to \infty$. Therefore, for any solution $v(x)$ of (3.1) - (3.2) to be a solution of (1.1) - (1.3), it is necessary that $v(x) < 1$. Thus, we can restrict attention to positive solutions of (3.1) - (3.2) satisfying

$$0 < v(x) < 1 \quad , \quad -L < x < L \quad .$$

In terms of (3.3) - (3.4), this means

$$0 < w(x) < \alpha+\beta \quad , \quad -L < x < L \quad .$$

The boundary-value problem (3.3) - (3.4) has the general form

$$w'' + f(w) = 0 \quad , \quad -L < x < L \quad , \tag{3.5}$$

$$w(\pm L) = 0 \quad , \tag{3.6}$$

where the function $f(w)$ has the qualitative form

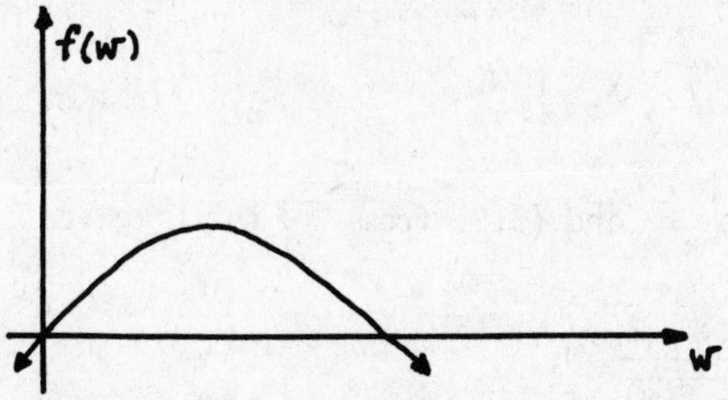

Several authors have studied (3.5) - (3.6) (cf. [4], [6], [12]). We describe below the so-called "time-map" associated with (3.5) - (3.6).

Let $w(x)$ be such that $0 < w(x) < \alpha+\beta$ and satisfy (3.5) - (3.6) with $f(w) = \ell(w)(1 - \ell(w))$. Then

$$0 = w'w'' + f(w)w'$$

or

$$c = \frac{1}{2}(w')^2 + F(w) \quad , \quad -L < x < L \quad , \tag{3.7}$$

where c is a constant and

$$F(w) = \int_0^w f(s)ds \quad .$$

Now, let $\tau \equiv \max_{-L \le x \le L} w(x)$. It follows easily that there exists a unique $\overline{x} \in (-L,L)$ with $\tau = w(\overline{x})$. Thus, $w'(\overline{x}) = 0$ and hence $c = F(\tau)$. Substituting into (3.7) yields

$$\frac{1}{2}(w')^2 = F(\tau) - F(w) \quad , \quad -L < x < L \quad .$$

Since $F'(w) = f(w)$ is positive for $0 < w < \tau < \alpha+\beta$, and since $w'(x) > 0$ on $-L < x < \overline{x}$ and $w'(x) < 0$ on $\overline{x} < x < L$ by the uniqueness of $\overline{x}$, we have

$$\sqrt{2} = \frac{w'}{\sqrt{F(\tau) - F(w)}} \quad , \quad -L < x < \overline{x} \quad , \tag{3.8}$$

and

$$-\sqrt{2} = \frac{w'}{\sqrt{F(\tau) - F(w)}} \quad , \quad \overline{x} < x < L \quad . \tag{3.9}$$

Integrating (3.8) from $-L$ to $\overline{x}$ and (3.9) from $\overline{x}$ to L gives

$$L + \overline{x} = \frac{1}{\sqrt{2}} \int_0^\tau \frac{dw}{\sqrt{F(\tau) - F(w)}} \quad , \tag{3.10}$$

and

$$L - \overline{x} = \frac{1}{\sqrt{2}} \int_0^\tau \frac{dw}{\sqrt{F(\tau) - F(w)}} \quad . \tag{3.11}$$

From (3.10) and (3.11) it is apparent that $\overline{x} = 0$ and thus we arrive at the relation

$$L = \frac{1}{\sqrt{2}} \int_0^\tau \frac{dw}{\sqrt{F(\tau) - F(w)}} \quad . \tag{3.12}$$

To justify the above procedure one must show that the integral in (3.12) really exists. First, note that

$$\frac{s}{\alpha + \beta} \le \ell(s) \le \frac{s}{\beta} \quad \text{for} \quad 0 < s < \alpha+\beta \quad , \tag{3.13}$$

and

$$\ell'(s) = \frac{2[\beta^2 + \beta\sqrt{\beta^2 + 4\alpha s} + 2\alpha s]}{(\beta + \sqrt{\beta^2 + 4\alpha s})^2 \sqrt{\beta^2 + 4\alpha s}} \tag{3.14}$$

$$> 0 \quad \text{for all} \quad s \ge 0 \quad .$$

Thus, for $0 < w < \tau < \alpha+\beta$,

$$
\begin{aligned}
F(\tau) - F(w) &= \int_w^\tau \ell(s)(1 - \ell(s))ds \\
&\geq \int_w^\tau \frac{s}{\alpha + \beta} (1 - \ell(s))ds \\
&\geq \quad (1 - \ell(\tau)) \int_w^\tau \frac{s}{\alpha + \beta} ds \\
&= \frac{(1 - \ell(\tau))}{2(\alpha + \beta)} (\tau^2 - w^2) \quad ,
\end{aligned}
$$

by (3.13) since $1 - \ell(s) \geq 1 - \ell(\tau)$ for $0 < w < s < \tau < \alpha+\beta$ by (3.14). Therefore,

$$
\begin{aligned}
\frac{1}{\sqrt{2}}\int_0^\tau \frac{dw}{\sqrt{F(\tau) - F(w)}} &\leq \sqrt{\frac{\alpha + \beta}{1 - \ell(\tau)}} \cdot \int_0^\tau \frac{dw}{\sqrt{\tau^2 - w^2}} \\
&= \sqrt{\frac{\alpha + \beta}{1 - \ell(\tau)}} \cdot \frac{\pi}{2} \ ,
\end{aligned}
$$

which shows that for any $\tau \in (0, \alpha + \beta)$ the integral in (3.12) actually exists.

We next show that the relationship (3.12) defines a 1-1 mapping between τ and L. Let

$$
\begin{aligned}
G(\tau) &= \frac{1}{\sqrt{2}} \int_0^\tau \frac{dw}{\sqrt{F(\tau) - F(w)}} \qquad (3.15) \\
&= \frac{1}{\sqrt{2}} \int_0^1 \frac{\tau du}{\sqrt{F(\tau) - F(\tau u)}} \quad ,
\end{aligned}
$$

letting $w = \tau u$.

<u>Lemma 3.1</u>: Let $G(\tau)$ be defined by (3.15). Then

$$
G(\tau) \geq \sqrt{\beta}\,\frac{\pi}{2} \quad \text{for} \quad 0 < \tau < \alpha + \beta \ , \qquad (3.16)
$$

and

$$
G(\tau) \quad \text{is strictly increasing for} \quad 0 < \tau < \alpha + \beta \ . \qquad (3.17)
$$

Proof: We have

$$F(\tau) - F(w) = \int_w^\tau \ell(s)(1 - \ell(s))ds$$

$$\le \int_w^\tau \frac{s}{\beta}\, ds = \frac{1}{2\beta}(\tau^2 - w^2) ,$$

by (3.13). Therefore,

$$G(\tau) \ge \frac{1}{\sqrt{2}} \int_0^\tau \frac{dw}{\sqrt{\frac{\tau^2 - w^2}{2\beta}}} = \sqrt{\beta}\,\frac{\pi}{2} ,$$

which proves (3.16).

To show (3.17), first let $0 < \tau_1 < \tau_2 < \alpha + \beta$. Then

$$G(\tau_2) - G(\tau_1) = \frac{1}{\sqrt{2}} \int_0^1 [g(\tau_2,u) - g(\tau_1,u)]du ,$$

where

$$g(\tau,u) = \frac{\tau}{\sqrt{F(\tau) - F(u)}} .$$

Thus, $G(\tau_2) > G(\tau_1)$ if for each fixed $u \in (0,1)$

$$\frac{\partial g}{\partial \tau}(\tau,u) > 0 \quad \text{for} \quad 0 < \tau < \alpha + \beta .$$

Now,

$$\frac{\partial g}{\partial \tau} = \frac{\sqrt{F(\tau) - F(\tau u)} - \frac{\tau}{2}\frac{f(\tau) - f(\tau u)u}{\sqrt{F(\tau) - F(\tau u)}}}{F(\tau) - F(\tau u)} ,$$

and since F is increasing for $0 < \tau < \alpha+\beta$, and $0 < u < 1$, the demoninator is positive. Therefore, $\frac{\partial g}{\partial \tau} > 0$ holds iff

$$F(\tau) - F(\tau u) > \frac{\tau}{2}[f(\tau) - f(\tau u)u] ,$$

or

$$F(\tau) - \frac{\tau}{2} f(\tau) > F(\tau u) - f(\tau u)\frac{\tau u}{2} \quad \text{for} \quad 0 < \tau < \alpha+\beta . \qquad (3.18)$$

Next, let $h(s) = F(s) - \frac{s}{2} f(s)$. Then (3.18) will be true if

$$h'(s) = \frac{1}{2}(f(s) - sf'(s)) > 0 \quad \text{for} \quad 0 < s < \alpha+\beta ,$$

or

$$\frac{f(s)}{s} > f'(s) \quad \text{for} \quad 0 < s < \alpha+\beta . \tag{3.19}$$

Thus, we have shown that (3.19) implies (3.17).

In our case, $f(s) = \ell(s)(1 - \ell(s))$, and so

$$f'(s) = \ell'(s)(1 - 2\ell(s)) .$$

Inequality (3.19) is then equivalent to

$$\frac{\ell(s)}{s}(1 - \ell(s)) > \ell'(s)(1 - 2\ell(s)) \quad \text{for} \quad 0 < s < \alpha+\beta . \tag{3.20}$$

Now, $0 < \ell(s) < 1$ for $0 < s < \alpha+\beta$, and so (3.20) is equivalent to

$$\frac{\ell(s)}{s} > \ell'(s) \qquad 0 < s < \alpha+\beta , \tag{3.21}$$

since $\frac{1 - 2\ell(s)}{1 - \ell(s)} < 1$ for $0 < s < \alpha+\beta$. Finally, (3.21) is easily seen to hold using the definition of $\ell(s)$ and (3.14). Therefore, (3.17) holds.□

From (3.17) it follows that (3.12) defines a 1 - 1 mapping between τ and L. Letting $w(x)$ be the unique solution of

$$w'' + f(w) = 0$$

$$w(0) = \tau , \quad 0 < \tau < \alpha+\beta \tag{3.22}$$

$$w'(0) = 0 ,$$

we see that $L = G(\tau)$ is the "time" it takes for the solution $w(x)$ to go from τ to 0, and that $w(x)$ is also a solution of (3.5) - (3.6) with $L = G(\tau)$. Secondly, given $L > \sqrt{\beta}\,\frac{\pi}{2}$, any solution of (3.5) - (3.6) is also the unique solution of (3.22) with $\tau = G^{-1}(L)$. This shows that for any $L > \sqrt{\beta}\frac{\pi}{2}$, (3.5) - (3.6) has a unique positive solution $w(x)$. Further, if $w(x)$ is any solution of (3.5) - (3.6) with $\tau = w(0) > 0$, then $L > \sqrt{\beta}\frac{\pi}{2}$. Therefore, the only nonnegative solution of (3.5) - (3.6) when $L < \sqrt{\beta}\frac{\pi}{2}$ is $w(x) \equiv 0$. Using the transformation given between (3.1) - (3.2) and (3.3) - (3.4) we have the following result:

Theorem 3.2: Let α and β be any positive constants. Then for $L < \sqrt{\beta}\frac{\pi}{2}$, there are no positive solutions of (3.1) - (3.2), and for each $L > \sqrt{\beta}\frac{\pi}{2}$, there is exactly one positive solution, denoted by $v(x;L)$.

§4. STABILITY RESULTS

In this section we first perform a stability analysis of the identically zero solution of (1.1) - (1.2), and then discuss the stability of the positive steady-state solutions of (1.1) - (1.2) found in §3.

Linearizing (1.1) about the solution $u(x,t) \equiv 0$ leads to the linear problem

$$\phi_t = \beta\phi_{xx} + \phi \quad , \quad -L < x < L \ , \quad t > 0 \ ,$$

$$\phi(\pm L,t) \equiv 0 \quad \text{for} \quad t \geq 0$$

$$\phi(x,0) = \phi^{\circ}(x).$$

This is easily seen to have the solution

$$\phi(x,t) = \sum_{n=1}^{\infty} e^{\lambda_n t} \phi_n \cdot \sin[\frac{n\pi}{2L}(x + L)]$$

where $\lambda_n = 1 - \beta \frac{n^2\pi^2}{4L^2}$ and

$$\phi_n = \frac{1}{L} \int_{-L}^{L} \phi^{\circ}(x)\sin[\frac{n\pi}{2L}(x + L)]dx \quad .$$

If $L < \sqrt{\beta}\,\frac{\pi}{2}$, then $\lambda_n < 0$ for all $n = 1, 2, \ldots$ and $\phi(x,t) \to 0$ as $t \to \infty$. Therefore, $u(x,t) \equiv 0$ is locally asymptotically stable in this case. When $L > \sqrt{\beta}\,\frac{\pi}{2}$, $\lambda_1 > 0$ and so $\phi(x,t)$ does not necessarily go to zero as $t \to \infty$. Thus, $u(x,t) \equiv 0$ is unstable for $L > \sqrt{\beta}\,\frac{\pi}{2}$.

We now prove the global stability of $u(x,t) \equiv 0$ when $L < \sqrt{\beta}\,\frac{\pi}{2}$. Let the functional $\Phi(t)$ be defined on functions $w(x,t)$ by

$$\Phi(t) = \int_{-L}^{L} \left(\frac{1}{2} w_x^2 - F(w)\right)dx \quad ,$$

where $F(w) = \int_0^w f(\ell(s))ds$. Next, suppose that $u(x,t)$ is a solution of (1.1) - (1.3) and let

$$w(x,t) = (\alpha u(x,t) + \beta)u(x,t) \; .$$

Then

$$\frac{d}{dt}\Phi(t) = \int_{-L}^{L} (w_x w_{xt} - f(\ell(w))w_t)dx$$

$$= \int_{-L}^{L} \frac{\partial}{\partial x}(w_x w_t)dx - \int_{-L}^{L} u_t w_t dx$$

$$= \cancelto{0}{w_x w_t \Big|_{-L}^{L}} - \int_{-L}^{L} u_t(2\alpha u + \beta)u_t dx$$

$$= -\int_{-L}^{L} u_t^2(2\alpha u + \beta)dx \le 0 \quad .$$

Therefore, $\Phi(t)$ is strictly decreasing on nonconstant solutions of (1.1) - (1.3), and hence (1.1) - (1.3) is gradient-like (cf. [3]).

<u>Theorem 4.1</u>: Let $L < \sqrt{\beta}\,\frac{\pi}{2}$ and let $u(x,t)$ be a solution of (1.1) - (1.3) with bounded nonnegative initial condition $u^{\circ}(x)$. Then $u(x,t) \to 0$ as $t \to \infty$.

Proof: Let $M > 0$ be chosen so that

$$0 \le u^\circ(x) \le M \quad , \quad -L \le x \le L \quad ,$$

and let $v(t)$ be the solution of

$$\frac{dv}{dt} = v(1 - v)$$

$$v(0) = M \; .$$

Then by the comparison result, Theorem 2.3,

$$0 \le u(x,t) \le v(t)$$

Since $v(t) \to 1$ as $t \to \infty$, $u(x,t)$ is bounded for all t.

System (1.1) - (1.3) gradient-like implies that all solutions must go to a rest state. (For a detailed discussion of the applicability of the Invariance Principle to the problem discussed here the reader is referred to the article by Aronson, Crandall, and Peletier [1A].) Therefore, since $u \equiv 0$ is the only rest state available when $L < \sqrt{\beta}\,\frac{\pi}{2}$, $u(x,t)$ must decay to zero as $t \to \infty$. □

Finally, we show the global stability of the rest state $v(x;L)$ of (1.1) - (1.2) when $L > \sqrt{\beta}\,\frac{\pi}{2}$.

Theorem 4.2: Let $L > \sqrt{\beta}\,\frac{\pi}{2}$ and let $v(x;L)$ be the unique positive solution of (3.1) - (3.2) found in §3. Let $u(x,t)$ be the solution of (1.1) - (1.3) with $u^\circ(x)$ bounded and not identically zero.
Then

$$u(x,t) \to v(x;L) \quad \text{as} \quad t \to \infty \; .$$

Proof: As in the proof of Theorem 4.1, $u(x,t)$ remains nonnegative and bounded for all $t > 0$. Again since the system is gradient-like the solution

$u(x,t)$ must go to a rest state. Since $L > \sqrt{\beta}\,\frac{\pi}{2}$ there are two choices, $u \equiv 0$ and $v(x)$. The proof will be complete if we can show that $u(x,t) \not\to 0$ as $t \to \infty$.

To this end, let $\ell \in (\sqrt{\beta}\,\frac{\pi}{2}, L)$ and for any $\varepsilon > 0$ define the function $z(x)$ by

$$z(x) = \varepsilon \sin(\frac{\pi}{2\ell}(x + \ell)) .$$

Then, after an easy calculation, one has

$$[(\alpha z + \beta)z]_{xx} + z(1 - z)$$
$$= \varepsilon [((1 - \beta \frac{\pi^2}{4\ell^2}) - \varepsilon (1 + \alpha \frac{\pi^2}{\ell^2})\sin(\frac{\pi}{2\ell}(x + \ell)))\cdot$$
$$\sin(\frac{\pi}{2\ell}(x + \ell)) + \frac{\alpha\pi^2}{2\ell^2}\varepsilon]$$

Since $\ell > \sqrt{\beta}\,\frac{\pi}{2}$, we can choose ε_0 small enough so that

$$[(\alpha z + \beta)z]_{xx} + z(1 - z) > 0 \quad \text{for} \quad -\ell < x < \ell ,$$

for all $0 < \varepsilon \le \varepsilon_0$.

By Theorem 2.4 we have that $u(x,t) > 0$ in $(-L,L)$ for any $t > 0$. Let $t_0 > 0$ be arbitrarily chosen. Then given any $\ell \in (\sqrt{\beta}\,\frac{\pi}{2}, L)$ there exists a $\delta > 0$ such that

$$u(x,t_0) \ge \delta > 0 \quad \text{in} \quad (-\ell,\ell) .$$

Let $z(x) = \varepsilon \sin(\frac{\pi}{2\ell}(x + \ell))$ with $\varepsilon = \min\{\delta, \varepsilon_0\}$. Then by what was shown above

$$[(\alpha u + \beta)u]_{xx} - u_t + u(1 - u) = 0 < [(\alpha z + \beta)z]_{xx} - z_t + z(1 - z)$$
$$\text{in} \quad (-\ell, \ell) \times (t_o, \infty)$$

with

$$u(x,t_0) \geq z(x) \quad -\ell < x < \ell$$

$$u(\pm\ell,t) \geq z(\pm\ell) = 0 \quad \text{for} \quad t > t_0 .$$

By uniqueness, $u(x, t+t_0)$ is the solution to (1.1) - (1.3) with $u(x,0+t_0) = u(x,t_0)$ as initial condition. Hence, by Theorem 2.3

$$u(x,t) \geq z(x) \quad -\ell < x < \ell, \quad t > t_0 ,$$

and therefore $u(x,t) \not\to 0$ as $t \to \infty$. □

REFERENCES.

1. D.G. Aronson, Density-dependent interaction-diffusion systems, Proceedings of the Advanced Seminar on Dynamics and Modelling of Reactive Systems, Academic Press, 1980.

1A. D. Aronson, M. Crandall, and L. Peletier, Stabilization of solutions of a degenerate nonlinear diffusion problem, Nonlinear Analysis, Theory, Methods and Appl., Vol. 6, No. 10, pp 1001-1022, 1982.

2. P.N. Brown, Reaction-diffusion equations in population biology, Ph.D. Dissertation, Tulane University, 1978.

3. C. Conley, Isolated invariant sets and the morse index, AMS Regional Conference Series in Mathematics, No. 38, 1978.

4. E. Conway, R. Gardner, and J. Smoller, Stability and bifurcation of steady-state solutions for predator-prey equations. Advances in Appl. Math. 3 (1982), 288-334.

5. M.E. Gurtin and R.C. MacCamy, On the diffusion of biological populations, Math. Biosciences 33, 35-49, 1977.

6. D. Ludwig, D.G. Aronson, and H.F. Weinberger, Spatial patterning of the spruce budworm, J. Math. Biology 8, 217-258 (1979).

7. K. Masuda and M. Mimura, On the existence of global solutions of density-dependent interaction-diffusion equations in ecology, (in preparation).

8. M. Mimura, Stationary pattern of some density-dependent diffusion system with competitive dynamics, Hiroshima Mathematical Journal, Vol. 11, No. 3, November 1981.

9. M. Mimura and K. Kawasaki, Spatial segregation in competitive interaction-diffusion equations, J. Math. Biol. 9, 49-64 (1980).

10. A. Okubo, Diffusion and ecological problems: mathematical models, Biomathematics, Vol. 10, Springer-Verlag 1980.

11. M.H. Protter and H.F. Weinberger, Maximum principles in differential equations, Prentice-Hall, 1967.

12. J. Smoller and A. Wasserman, Global bifurcation of steady-state solutions, J. Diff. Eqns. 39 (1981).

Peter N. Brown

University of Houston

E D CONWAY

Diffusion and predator–prey interaction: pattern in closed systems

1. INTRODUCTION

In this essay we summarize the results to date in the search for and understanding of patterned solutions of the system

$$\begin{aligned} u_t &= d_1 \Delta u + uf(u) - uv\phi(u) \\ & \qquad\qquad\qquad\qquad x \in \Omega , \quad t > 0 \qquad\qquad (1.1) \\ v_t &= d_2 \Delta v - vg(v) + uv\phi(u) \end{aligned}$$

subject to homogeneous Neumann boundary condition

$$\frac{\partial u}{\partial n} = 0 , \ \frac{\partial v}{\partial n} = 0 \qquad \text{for} \ \ x \in \partial\Omega , \ \ t > 0 . \qquad (1.2)$$

This system can be considered a model of two populations which are continuously distributed throughout the bounded spatial region Ω and which simultaneously undergo simple (Fickian) diffusion and interact as predator (v) and prey (u). In Section 2 below we give a rather thorough description of the system from this point of view.

The system (1.1) is an example of a reaction-diffusion (RD) system,

$$U_t = D\Delta U + F(U)$$

where $U(x,t) \in \mathbb{R}^N$, $F : \mathbb{R}^N \to \mathbb{R}^N$ and $D \geq 0$ is diagonal. Modern interest in such systems dates from the paper of Turing (TURING 1952) in which the central issue was the question of "pattern" which in our context will be interpreted to be a spatially inhomogeneous solution which is either constant in time ($\equiv$ steady state) or periodic in time. It is also desirable that such solutions be stable in some appropriate sense. In PRIGOGINE-NICOLIS (1967) the term dissipative structure was coined for such solutions and their significance was stressed.

It makes sense to distinguish three classes of such "structured" or "patterned" solutions only one of which will be considered here. First, the system (1.1) can be considered for $\Omega = \mathbb{R}^n$, $n = 1,2$ or 3, so that there are no boundary effects whatsoever. In such a situation, structures can take the form of travelling waves, wave trains, etc. FIFE (1979) and SMOLLER (1983) should be consulted for orientation in this context. Secondly, the system (1.1) can be considered with Dirichlet or Robin (membrane-type) boundary conditions which allow flux at the boundary of Ω. Typically this gives rise to a much larger class of patterned solutions which we can refer to as flux-driven structures. For example, if $F(U_0) = 0 = F(U_1)$ and U_1 is asymptotically stable, then if $U = U_0$ on $\partial\Omega$ we can imagine a steady state $U(x)$ such that $U(x) = U_0$ for $x \in \partial\Omega$ while (for large Ω), $U(x) \approx U_1$ for x well away from the boundary. For an example of such a situation, we refer to CONWAY-GARDNER-SMOLLER (1982). Finally, we shall restrict our attention to the case of diffusion and reaction in a "closed" system which allows no flux through the boundary. This is modeled by the boundary condition (1.2). That stable structures can exist in such a situation is not at all obvious and, in fact, they cannot do so in the scalar case $(N = 1)$ nor in many systems (cf. Section 3 below). Rather, one of the primary tasks is to fully understand just what features of RD systems are necessary and sufficient for pattern.

In this paper we survey what is known to date for system (1.1). Preoccupation with such special systems is a hallmark of most recent work in RD systems and with good reason. Although small amplitude pattern (cf Section 4) is determined only by the local behavior of F in a neighborhood of a critical point P, $F(P) = 0$, it is generally true that pattern involves the global behavior of $F(U)$ for U in some extended portion of phase space. Thus, the currently common taxonomy of functions based upon differentiability or integrability properties is irrelevant to our task. Until we have a classification based upon global qualitative or structural properties we must confine ourselves to studies of particular systems.

The predator-prey system (1.1) is particularly rich and has received much attention. Its solutions exhibit all major types of phenomena uncovered in other RD systems of two equations. Most works that are applicable to this system have concentrated on particular aspects and have imposed conditions on the system with those objects in mind. It thus seemed worthwhile to summarize just what has been achieved while keeping in view the system as it arises naturally as a model of two interacting populations. This should help indicate what new work is needed for a fuller understanding of the system. I hope that the expository benefits of surveying a reasonably substantial body of work from a unified perspective justifies what is largely a commentary on the work of others.

In Section 2 we discuss the system (1.1)-(1.2) as a model ecological system of predator-prey type. We introduce three classes of interactions which together include most examples in the ecological literature. We have tried also to indicate the significance of the various conditions imposed. In Section 3 we survey results showing that under certain conditions pattern cannot occur. In the next two sections, we outline the two methods currently available for analytically demonstrating the existence of pattern. In Section 4 we give a detailed discussion of diffusive or Turing instability of spatially homogeneous steady states and the concommitant bifurcation of small amplitude pattern. In Section 5 we introduce Fife's construction of solutions with transition layers. It should be clear from these two sections that although Turing instability has received the greater attention, it is Fife's ideas that come closest to providing a general understanding of pattern. Finally, in Section 6 we briefly indicate a few questions that seem worthwhile pursuing.

2. DESCRIPTION OF THE MODEL REACTION-DIFFUSION SYSTEM

We shall be concerned with models of the following general form:

$$\begin{aligned} u_t &= d_1 \Delta u + uM(u,v) \\ & \qquad\qquad\qquad\qquad t > 0 , \quad x \in \Omega \\ v_t &= d_2 \Delta v + vN(u,v) \end{aligned}$$

$$\frac{\partial u}{\partial n} = 0 = \frac{\partial v}{\partial n} \quad \text{for} \quad t > 0 , \quad x \in \partial\Omega . \tag{2.1}$$

Here Ω is a bounded domain in $\mathbb{R}^n$ and $\partial\Omega$ is its boundary which we assume to be smooth. The functions u and v are interpreted as population densities of two interacting populations continuously distributed throughout the spatial region Ω. We refer to OKUBO (1980) for a general discussion of continuum models in population dynamics and to MAYNARD-SMITH (1974) and May (1976) for more general discussions of population models. The terms Δu and Δv model dispersal by means of simple diffusion interpreted in a "Fickian" sense (a more general model of diffusive-like dispersal is put forth in COHEN AND MURRAY (1981)). The boundary conditions (2.1) are interpreted as an assumption that both populations are confined to Ω, i.e. there is no migratory "flux" across the boundary of Ω.

We have written the interaction terms i.e. the vector field $(uM(u,c), vN(u,v))$ in Kolmogorov form (cf. MAY (1973)). As in the case when u and v are independent of x, this has the consequence that if M and N are smooth functions, then the positive quadrant of phase space $[u \geq 0, v \geq 0]$ is invariant. In other words, if u and v are non-negative for $t = 0$, then they remain so for t 0. The book SMOLLER (1983) has a good discussion of this set of ideas.

The nature of the classical interaction of the two populations is determined by how the functions M and N, considered as per capita growth rates for u and v respectively, depend upon v and u respectively. More specifically,

$$\begin{aligned} &M_v < 0 , \quad N_u > 0 \quad : \quad \text{Predator-Prey} \quad (u \text{ is prey}) \\ &M_v > 0 , \quad N_u < 0 \quad : \quad \text{Predator-Prey} \quad (v \text{ is prey}) \\ &M_v < 0 , \quad N_u < 0 \quad : \quad \text{Competition} \\ &M_v > 0 , \quad N_u > 0 \quad : \quad \text{Symbiosis} \end{aligned} \tag{2.2}$$

where $M_v \equiv \partial M/\partial v$, etc. Of course, at each point (u,v) of phase space (if we rule out zero derivatives) one of these conditions must hold. The real restriction here is that one and the same condition is to hold for all positive values of u and v.

In this paper we shall restrict our attention to the predator-prey interaction and shall let u denote the concentration of prey, i.e. we assume the first of the conditions (2.2). However, this still leaves us with too general a model. To explain the further restrictions we wish to impose, it is helpful to change notation:

$$\begin{aligned} f(u) &\equiv M(u,0) , \quad -v\phi(u,v) \equiv M(u,v) - M(u,0) \\ -g(v) &\equiv N(0,v) , \quad u\phi(u,v) \equiv N(u,v) - N(0,v) . \end{aligned} \tag{2.3}$$

Notice that in defining ϕ and ψ we are taking advantage of our assumption that M and N are smooth. Our equations now take the form

$$\begin{aligned} u_t &= d_1 \, \Delta u + uf(u) - uv\phi(u,v) \\ v_t &= d_2 \, \Delta u - vg(v) + uv\psi(u,v) \, . \end{aligned} \tag{2.4}$$

In discussing the interaction further, we shall need only consider the special solutions of (2.1)-(2.4) which are independent of x, i.e., $u(x,t) = U(t)$, $v(x,t)$, $v(x,t) = v(t)$ where

$$\begin{aligned} \frac{dU}{dt} &= Uf(U) - UV\phi(U,V) \\ \frac{dV}{dt} &= -Vg(V) + UV\psi(U,V) \, . \end{aligned} \tag{2.5}$$

Notice that every solution of (2.5) is a solution of (2.1)-(2.4).

Assumptions concerning f. Notice that if $V = 0$ for one value of t, then it vanishes for all values of t and U is then a solution of

$$\frac{dU}{dt} = Uf(U) \, .$$

Thus, $f(U)$ may be interpreted as the "per capita" or "normalized" growth rate of the prey U in the absence of the predator V. We assume a limitation to growth expressed by $F(U) < 0$ for $U > K$ where K is a constant referred to as the "carrying capacity." We choose units for U so that $K = 1$. Thus our minimal assumptions concerning f are as follows:

$$f(1) = 0 \, , \quad f(U) < 0 \quad \text{for } U > 1 \, . \tag{2.6}$$

Beyond this we must distinguish between three classes of functions f which, taken together, account for the majority of models appearing in the

ecological literature

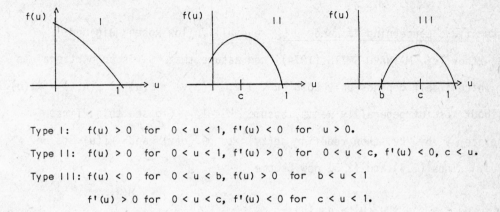

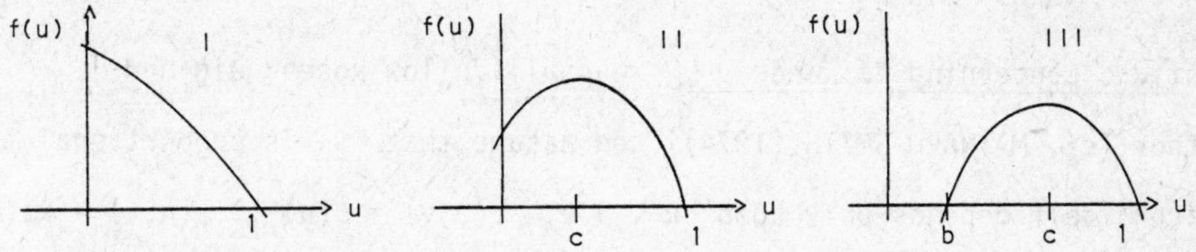

Type I: $f(u) > 0$ for $0 < u < 1$, $f'(u) < 0$ for $u > 0$.

Type II: $f(u) > 0$ for $0 < u < 1$, $f'(u) > 0$ for $0 < u < c$, $f'(u) < 0$, $c < u$.

Type III: $f(u) < 0$ for $0 < u < b$, $f(u) > 0$ for $b < u < 1$

$f'(u) > 0$ for $0 < u < c$, $f'(u) < 0$ for $c < u < 1$.

Figure 2.1. The three classes of functions f

An example of type I is $f(u) = 1 - u$ which gives rise to the classic logistic S-shaped curve of saturated growth. Both types II and III exhibit the "Allee effect" of a fall-off in per capita growth rate at low population densities. The very strong form of fall-off in type III has been termed "asocial growth" and is discussed in PHILIP (1957). Its most obvious feature is a threshold effect; unless U exceeds b then it will decay to zero.

Assumption Concerning g. In an analogous manner, we see that $-g$ can be interpreted as the normalized growth rate for the predator in the absence of the prey. Our basic assumption is

$$g(0) = \mu, \quad g'(v) \geq 0 \quad \text{for} \quad v > 0 . \tag{2.7}$$

Thus, $V(t) \to 0$ in the absence of the prey, U; this rules out the case of a predator with alternative prey. The most common assumption is that g is constant which results in exponential decay of V. We wish to allow for an even faster fall-off.

Assumptions Concerning ϕ and ψ. We shall follow Rosenzweig and MacArthur (cf. MAYNARD-SMITH (1974)) and assume that ψ is proportional to ϕ which itself depends only upon u, i.e. $\phi(u,v) = \phi(u)$, $\psi(u,v) = k\phi(u)$. Without loss of generality we may assume $k = 1$. (To see this, first replace v by kv then redefine $g(kv)$ to be $g(v)$ and $k\phi(u)$ to be $\phi(u)$.) Thus (2.4) and (2.5) now become

$$\begin{aligned} u_t &= d_1 \Delta u + uf(u) - uv\phi(u) \\ v_t &= d_2 \Delta v - vg(v) + uv\phi(u) \end{aligned} \qquad (2.8) \equiv (1.1)$$

$$\begin{aligned} \dot{U} &= Uf(U) - UV\phi(U) \\ \dot{V} &= -Vg(V) + UV\phi(U) \end{aligned} \qquad (2.9)$$

The most common examples of ϕ are $\phi(u) = c$ and $\phi(u) = c/(1+du)$ where c and d are constant. Notice that if these equations are to satisfy the predator-prey conditions (1.2), then we must have

$$\phi(u) > 0 \ , \ [u\phi(u)]' \equiv \phi(u) + u\phi'(u) > 0 \ \text{ for } \ u > 0 \ . \qquad (2.10a)$$

The above two examples also satisfy

$$\phi'(u) \leq 0 \ \text{ for } \ u > 0 \qquad (2.10b)$$

as do all "Holling type 2" functional responses (cf. MAYNARD-SMITH (1974)). Finally, we shall always assume

$$\phi(1) \geq g(0) = \mu \ . \qquad (2.10c)$$

If this were false, then even when the prey is at the carring capacity of the ecosystem $(U = 1)$ we would have $\dot{V}/V$ negative so that the predator would not grow.

Spatially homogeneous solutions. It will be helpful to have in mind the qualitative features of the phase plane describing solutions of (2.9). Basic to this is the location of the equilibrium points:

$$\dot{U} = 0 \longleftrightarrow U = 0 \quad \text{or} \quad V = h(U) \quad f(U)/\phi(U) \tag{2.11a}$$

$$\dot{V} = 0 \longleftrightarrow V = 0 \quad \text{or} \quad U\phi(U) = g(V) \tag{2.11b}$$

The condition (2.10a) insures that the last condition defines U as a function of V. Since $\phi > 0$, we see that $h(1) = 0$ and if, as in most examples, we have $\phi'' \geq 0$ then it, like f, is either monotone decreasing or has a single maximum in $U \geq 0$. Figure 2.2 thus depicts the three typical cases

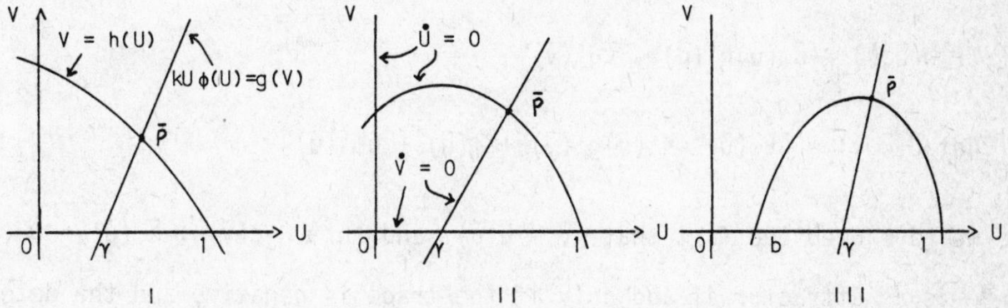

Figure 2.2. Null-clines for System 2.9

In all cases $(0,0)$, (1.0) and $(\bar{u},\bar{v}) \equiv \bar{P}$ are equilibrium points; in case III there is a fourth, $(b,0)$. The number γ is defined by $\gamma\phi(\gamma) = \mu$ and it follows from (2.10) that $0 < \gamma < 1$. In case III, we haven't bothered to consider the situation where $0 < \gamma < b$ for it seems rather unlikely from an ecological point of view.

The stability character of the various rest points is determined by the linearization of the vector field $(Uf(U) - UV\phi(U)\ ,\ -Vg(V) + UV\phi(U))$ about the point in question:

$$J(\hat{u},\hat{v}) = \begin{bmatrix} f(\hat{u}) + \hat{u}f'(\hat{u}) - \hat{v}[\hat{u}\phi(\hat{u})]' & -\hat{u}\phi(\hat{u}) \\ \hat{v}[\hat{u}\phi(\hat{u})]' & -g(\hat{v}) - \hat{v}g'(\hat{v}) + \hat{u}\phi(\hat{u}) \end{bmatrix} .$$

We see that $J(0,0)$ is diagonal with entries $f(0)$ and $-g(0) < 0$. Thus it is a saddle when f (or h) is of type I or II but is an attractor in case f (or h) is of type III.

Similarly, we see that $\det J(1,0) = f'(1)[\phi(1) - \mu] < 0$ so that $(1,0)$ is a saddle in all three cases. Also, $\det J(b,0) = bf'(b)[b\phi(b) - g(0)]$ is negative because $f'(b) > 0$ and $b\phi(b) - g(0) < \gamma\phi(\gamma) - g(0)$ so that $(b,0)$ is also a saddle.

In the case of $\overline{P}$ we have

$$\begin{aligned} \operatorname{Tr} J(\bar{u},\bar{v}) &= \bar{u}\phi(\bar{u})h'(\bar{u}) - \bar{v}g'(\bar{v}) \\ \det J(\bar{u},\bar{v}) &= \bar{u}\bar{v}\phi(\bar{u})[-h'(\bar{u})g'(\bar{v}) + \phi(\bar{u}) + \bar{u}\phi'(\bar{u})] \end{aligned} \tag{2.12}$$

where we have used the fact that $\bar{v} = h(\bar{u})$ and that $-g(\bar{v}) + \bar{u}\phi(\bar{u}) = 0$. Now $\overline{P}$ is an attractor if and only if the trace is negative and the determinant is positive so that

$$h'(\bar{u}) < 0 \longrightarrow (\bar{u},\bar{v}) \text{ is an attractor} \tag{2.13a}$$

$$\text{If } g'(\bar{v}) = 0 \text{, then } h'(\bar{v}) < 0 \longleftrightarrow (\bar{u},\bar{v}) \text{ is an attractor.} \tag{2.13b}$$

We also see that

$$\dot{U} < Uf(U) < 0 \quad \text{for} \quad U > 1 , \quad V > 0$$

so that every orbit satisfies $U(t) < 3/2$ for all sufficiently large values of t. Also,

$$\frac{d}{dt}(V + U) = Uf(U) - Vg(V) < \text{constant} - \mu V$$

for $U \leq 3/2$, $V > 0$. Thus every orbit in the positive quadrant enters the set

$$R_K = \{(U,V) : 0 \leq U \leq 3/2 \ ; \ 0 \leq V + U \leq K\}$$

in finite time. Here, K is some constant dependent upon f and μ and we notice that for all sufficiently large K, the set R_K is invariant for the vector field. It then follows from (2.13) and the Poincare-Bendixon theorem that in case I the equilibrium point $\overline{P}$ is a global attractor for the entire positive quadrant, i.e. if $U(0) > 0$, $V(0) > 0$, then $(U(t), V(t)) \to \overline{P}$ as $t \to +\infty$. In case II, $\overline{P}$ is a global attractor whenever it is stable as it is whenever it is to the right of the "hump" on the $V = h(U)$ curve (cf. Figure 2.2). Again from the Poincare-Bendixon theorem, it follows that in case II when $\overline{P}$ is unstable there must be an attracting limit cycle containing $\overline{P}$ in its interior. In general, this limit cycle is unique although this is not immediately obvious.

It is interesting to examine the transition from stable $\overline{P}$ to unstable $\overline{P}$. For example, we might fix f and ϕ but vary g perhaps by translation of its graph from right to left. Near the transition the eigenvalues of $J(\bar{u},\bar{v})$ are complex and at transition the eigenvalues cross the imaginary axis. It is easy to check that they do so transversely so that the limit cycle appears through a Hopf bifurcation (cf. HASSARD et al, 1981).

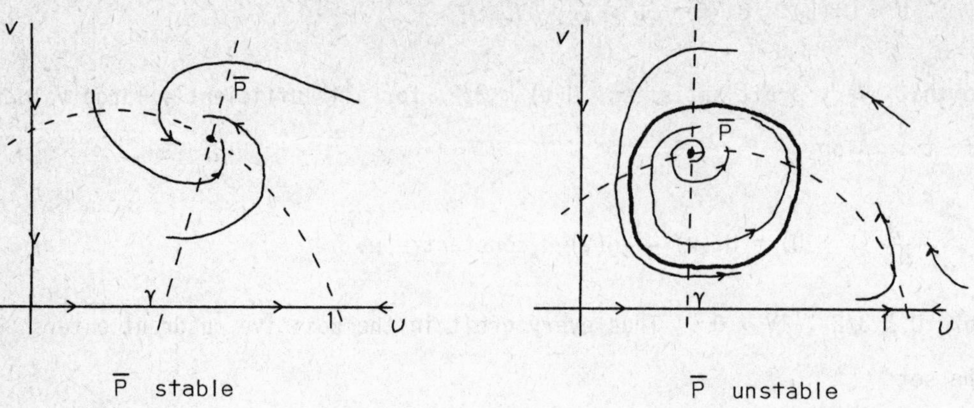

Figure 2.3. Phase plane in Case II

In case III the situation is somewhat more complex. The origin $(0,0)$ is now an attractor so that when $\overline{P}$ is also stable they divide the positive quadrant into two regions whose orbits tend to one or the other equilibrium point. It is not too hard to show that the separatrix is the stable manifold for the saddle point $(b,0)$. At the transition from $\overline{P}$ stable to $\overline{P}$ unstable there is again a Hopf Bifurcation to an attracting limit cycle. However, for all γ sufficiently near to b , there is no limit cycle and $(0,0)$ is now a global attractor for the positive quadrant. Unfortunately, we do not have the space to give a detailed demonstration of these features.

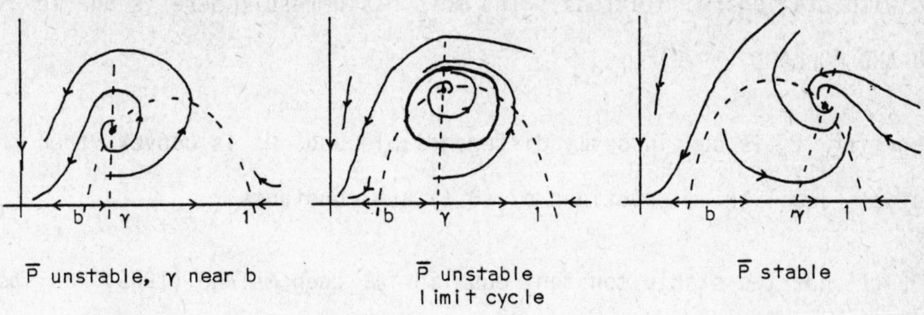

Figure 2.4. Phase plane in Case III

3. NEGATIVE RESULTS

In this section we survey a number of results which identify conditions under which pattern cannot obtain.

A. Scalar Case. The reader should be aware that generally pattern cannot occur in the case of a single equation, i.e. problems of the form

$$\begin{aligned} u_t &= \Delta u + f(u) \quad , \quad x \in \Omega \, , \quad t > 0 \\ \frac{\partial u}{\partial n} &= 0 \quad , \quad t > 0 \, , \quad x \in \partial\Omega \, . \end{aligned} \tag{3.1}$$

where f is some smooth real-valued function. We first note that if

$$\Phi(t) = \frac{1}{2}\int_\Omega |\nabla u|^2 \, dx - \int_\Omega F(U) \, dx$$

where $F' = f$, then

$$\frac{d\Phi}{dt} = -\int_\Omega u_t^2 \, dx \le 0 \, .$$

This rules out the possibility of time-periodic solutions so we concern ourselves with stationary solutions. The most basic result here is due to CASTEN AND HOLLAND (1978):

<u>Theorem</u>. If f is continuously differentiable and Ω is convex, then the only stable equilibrium solutions of (3.1) are constants.

If (3.1) has two stable constant equilibria, then MATANO (1979) has shown that for some regions Ω it is possible to have stable non-constant equilibria. However, the regions approximate two disconnected sets. Also, as COHEN AND MURRAY (1981) argue, if a non-Fickian model for diffusion is considered, then patterned distributions of even a single reacting and diffusing substance are possible.

<u>B. Type I Phase Planes</u>. When the function h is monotone decreasing, there is a global (in the positive quadrant) Lyapunov function for the system of ODE's (2.9) given by

$$\ell(u,v) = \frac{1}{2} u^2 - \bar{u}\phi(\bar{u}) \int \frac{1}{\phi(u)} du + v - \bar{v} \log(v)$$

This in turn defines a Lyapunov functional for the PDE's (1.1) - (1.2):

$$\ell(u,v) = \int_\Omega \ell(u,v)dx$$

i.e. the system (2.8) - (2.1) is "gradient-like." Using this observation, DE MOTTONI AND ROTHE (1979) show that every non-negative solution of the PDE with homogeneous Neumann boundary conditions converges uniformly to the constant state $(\bar{u},\bar{v})$. MIMURA (1979) had also discovered this but his discussion was not quite as complete. Special cases of this result were earlier proved by WILLIAMS AND CHOW (1978) and LEUNG (1978).

One minor point: De Mottoni and Rothe do not allow $g(v)$ to be constant.

This is undoubtedly due to their reliance upon "invariant rectangles" so that the restriction can probably be removed by using the interpolation argument in ALIKAKOS (1979) to get uniform bounds on the predator density, $v(x,t)$.

C. Large diffusion coefficients. If both diffusion coefficients are large enough or, equivalently, if the spatial domain is small enough, then there can be no pattern because every solution will converge to a spatially homogeneous solution of the ODE's. This is discussed by CONWAY, HOFF and SMOLLER (1978). A special case was discovered earlier by OTHMER (1977). To be more precise, convergence to spatial homogeneity will occur if both d_1 and d_2 are greater than λ/M . Here λ is the smallest positive eigenvalue of $-\Delta$ with homogeneous Neuman Conditions on Ω while M is the maximum of the Euclidean norm of $J(u,v)$ over $0 \leq u \leq 1$, $0 \leq v \leq m$ where m is such that $v(x,t) \leq m$ for x in Ω and t sufficiently large.

4. SMALL AMPLITUDE PATTERN FROM TURING INSTABILITY

The earliest method of analytically demonstrating the presence of pattern was to show that, as a certain parameter associated with the problem varied, a spatially inhomogeneous equilibrium solution bifurcated from an equilibrium which was constant in space, i.e. from a critical point of the interaction considered as a vector field in u-v phase space. The bifurcation theorems used (e.g. Chapter 13 of SMOLLER (1983)) are local, depending only upon the linearization of the equations about the constant solutions, and thus the solutions obtained are only small perturbations of the constant solutions. On the other hand, this approach can yield information concerning local uniqueness and stability of the bifurcating solutions.

Bifurcations of this type occur only at values of the parameter for which zero is in the spectrum of the linearized problem. We are interested in stable patterns, thus we look for situations where, as the parameter varies, the constant solution changes from stable to unstable as an eigenvalue crosses the origin from left to right. To first order, the new solution bifurcates from the old in the direction of the eigenfunction corresponding to the eigenvalue that becomes positive. In order to get pattern then, this eigenfunction must itself be spatially inhomogeneous and thus, as we shall see below, the critical eigenvalue must be associated with the higher modes. Since the lowest order eigenvalue remains negative and since its eigenfunctions are constant, this means that the constant solution, the critical point, remains stable with respect to spatially homogeneous perturbations. In other words, we must look for a critical point that <u>is stable as an equilibrium solution of the ODE (2.9) but is unstable as a solution of the PDE (2.8)</u>.

We first rule out the origin from consideration. In type III interactions it is a stable rest point of the ODE (2.9) but as shown in CONWAY-SMOLLER (1977a), it must then also be stable as a solution of the PDE (2.8). Since (1,0) and (b,0) are saddle points for the ODE, this leaves $(\bar{u},\bar{v})$ as the only possible source of bifurcating pattern. The basic question is, "Under what conditions is $(\bar{u},\bar{v})$ stable as a solution of the ODE but unstable as a solution of the PDE?" It turns out that this has a very simple answer.

As shown by CASTEN-HOLLAND (1977), whose discussion we briefly summarize, the stability of $(u(t,x), v(t,x)) \equiv (\bar{u},\bar{v})$ can be determined by a linear stability analysis. The linearization of (2.1) - (2.8) is

$$\begin{aligned} z_t &= D\Delta z + Az \ , \quad t > 0 \ , \quad x \in \Omega \\ \frac{\partial z}{\partial n} &= 0 \ , \quad t > 0 \ , \quad x \in \partial\Omega \end{aligned} \qquad (4.1)$$

where $A = J(\bar{u},\bar{v})$, $D = \text{diag}\{d_1,d_2\}$ and $z(x,t)$ can be thought of as an approximation to $(u(x,t) - \bar{u}, v(x,t) - \bar{v})$. The constant solution $(\bar{u},\bar{v})$ is an asymptotically stable solution of (2.1) - (2.8) if and only if every solution of (4.1) converges to zero as $t \to +\infty$. To obtain a useable criterion for this stability we make an eigenfunction expansion,

$$z(x,t) = \sum_{j=0}^{\infty} y_j(t)\psi_j(x)$$

where ψ_j is the <u>jth</u> (scalar) eigenfunction of $-\Delta$,

$$\Delta\psi_j + \lambda_j\psi_j = 0 \ , \quad \frac{\partial\psi_j}{\partial n} = 0 \quad \text{on} \quad \partial\Omega \ .$$

Here $0 = \lambda_0 \leq \lambda_1 \leq \ldots \to +\infty$ and we normalize the ψ_j's to be an orthonormal basis in $L_2(\Omega)$. From (4.1) we see that each y_j must satisfy the ODE

$$\frac{dy_j}{dt} = (A - \lambda D)y_j$$

so that we have the following criterion for stability.

If each matrix, $A - \lambda_j D$, $j = 0,1,\ldots,$ has both eigenvalues with negative real parts, then $(\bar{u},\bar{v})$ is (an asymptotically) stable equilibrium solution of the PDE (1.1) - (1.2). (4.2a)

If any one of the matrices $A - \lambda_j D$ has an eigenvalue with positive real part, then $(\bar{u},\bar{v})$ is unstable. (4.2b)

Letting $A(\lambda) = A - \lambda D$, we see from our earlier expression for $A = J(\bar{u},\bar{v})$ that

$$\text{Tr } A(\lambda) = \text{Tr } A - \lambda(d_1 + d_2)$$

$$\det A(\lambda) = d_1 d_2\lambda^2 + [d_1\bar{v}g'(\bar{v}) - d_2\bar{u}\phi(\bar{u})h'(\bar{u})]\lambda + \det A \ . \qquad (4.3b)$$

We are looking for a "Turing Instability" so that $(\bar{u},\bar{v})$ must be stable as a solution of the ODE but unstable as a solution of the PDE. The ODE stability forces $\mathrm{Tr}\, A < 0$ and $\det A > 0$. Since $d_1 + d_2 > 0$, this means that $\mathrm{Tr}\, A(\lambda) < 0$ for all $\lambda \geq 0$. Thus, in order to have PDE instability, we must have $D(\lambda) \equiv \det A(\lambda)$ be negative for λ equal to one of the λ_j's. Since $D(0) = \det A > 0$, we see that it can only be a higher mode, $j \geq 1$, that is unstable. (N.b. Notice that a Hopf Bifurcation is ruled out by the fact that the trace of $A(\lambda_j)$ is negative for all j .) We have thus reduced our problem to determining conditions under which $D(\lambda)$ takes on negative values at one of the positive λ_j's .

Since the coefficient of λ^2 in $D(\lambda)$ is positive, we see that if

$$D'(0) = d_1\bar{v}g'(\bar{v}) - d_2\bar{u}\phi(\bar{u})h'(\bar{u})$$

is positive or zero, then $D(\lambda) > 0$ for all $\lambda > 0$ so that $(\bar{u},\bar{v})$ will be stable for the PDE as well. Since $g'(v) \geq 0$ we see then that $h'(u) > 0$ is necessary for $D(\lambda)$ to be negative. Likewise, if $g'(\bar{v}) = 0$, then $(\bar{u},\bar{v})$ can be asymptotically stable only if $h'(\bar{u}) < 0$ (cf (2.13b)) so that $g'(\bar{v}) > 0$ is also necessary. Thus,

Turing instability cannot occur unless
both $h'(\bar{u})$ and $g(\bar{v})$ are positive.

Now also from (2.12) we see that, since the trace of A is negative, we must have $D'(0) > (d_1 - d_2)\bar{u}\phi(\bar{u})h'(\bar{u})$ so that

Turing instability cannot occur unless $d_1 < d_2$,

for otherwise $D'(0) \geq 0$.

To obtain sufficient conditions we must consider the domain Ω through its eigenvalues λ_j . First, let $\alpha = \bar{u}\phi(\bar{u})h'(\bar{u})$ and $\beta = \bar{v}g'(\bar{v})$ which are now assumed to be positive. The minimum value of D is assumed at $\lambda = \bar{\lambda}$ where

$$2\bar{\lambda} = \alpha/d_1 - \beta/d_2$$

and this minimum value can be expressed by

$$D(\bar{\lambda}) = -\frac{d_2}{d_1}\frac{\alpha^2}{4} - \frac{d_1}{d_2}\frac{\beta^2}{4} + \frac{1}{2}\alpha\beta + \det A \ .$$

Thus, there is a $\bar{\delta} > 1$ such that for d_2/d_1 greater than $\bar{\delta}$ we have $\bar{\lambda} > 0$ and $D(\bar{\lambda}) < 0$ so that the graph of $D(\lambda)$ is as in figure 4.1.

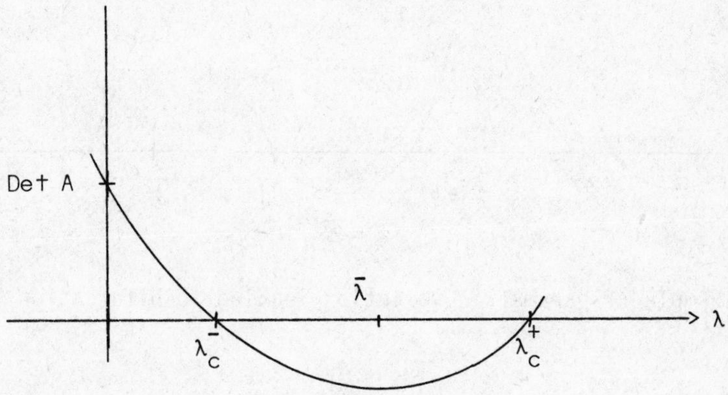

Figure 4.1. Graph of $D(\lambda)$ for $d_2 > \bar{\delta}d_1$

We have a Turing instability precisely when one of the λ_j's is in the interval $(\lambda_c^-, \lambda_c^+)$. By varying one or several parameters (e.g. d_1, d_2, γ, ϕ, etc.) we can place $(\lambda_c^-, \lambda_c^+)$ wherever we wish on the positive λ axis so it is clear that there is Turing Instability on any domain Ω for appropriate values of the parameters in the equations. Also, no matter what the values of $\bar{\lambda} > 0$ and $D(\bar{\lambda}) < 0$ there are domains for which one or more

λ_j's lie within $(\lambda_c^-, \lambda_c^+)$. In fact, we need only take advantage of the dependence of the eigenvalues upon the size of the domain Ω (COURANT-HILBERT, 1953) to see this.

Bifurcation. We now imagine a one parameter family of problems such that the polynomial $D(\lambda)$ and/or the values of $\lambda_1, \lambda_2, \ldots$ vary as the parameter varies. We can anticipate bifurcation of new solutions from $(\bar{u}, \bar{v})$ at those values of the parameter for which one of the λ_j's, $\lambda_{\bar{j}}$ say, lies on the graph of $D(\lambda)$ (and therefore is a root of $D(\lambda)$) while all other λ_j's lie below the graph. (Notice that at most two "modes" i.e. two λ_j's,

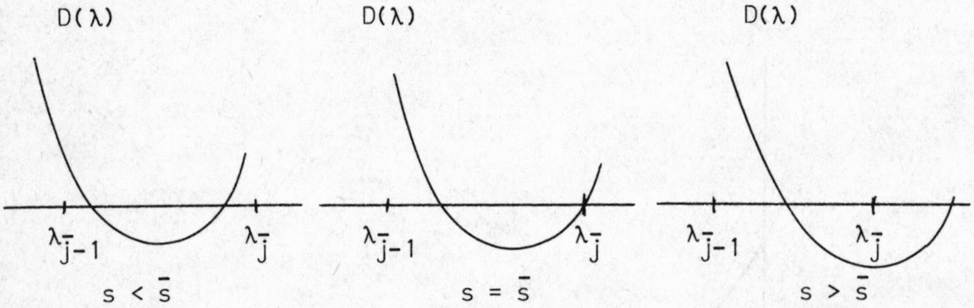

Figure 4.2. Example of parametric variation leading to bifurcation

can simultaneously satisfy this condition and in such a case the two modes must be consecutive.) Even at this level of generality it is quite simple to verify that bifurcation does take place at least when $\lambda_{\bar{j}}$ is a simple eigenvalue of $-\Delta$, when it crosses the graph of $D(\lambda)$ in a "transverse" manner and when all other λ_j's are strictly below the graph of $D(\lambda)$. If we write the right-hand-side of equation (1.1) in vector form,

$$F(P) = D\Delta P + G(P)$$

where $P = (u,v)$ and G is the vector field $(uf(u)-uv\phi(u),-vg(v)+uv\phi(u))$. Then $F(\overline{P}) = 0$ and the right-hand-side of the linearization (4.1) gives us

$$F_P(\overline{P}) = D\Delta + A .$$

where again $A = G'(\overline{P}) = J(\bar{u},\bar{v})$. The spectrum of $F_P(\overline{P})$ consists only of eigenvalues,

$$\mu_{ij} , \; i = 1,2 ; \; j = 0,1,... \tag{4.5a}$$

with associated eigenfunctions

$$\psi_j(x)\zeta_{ij} , \; i = 1,2 ; \; j = 0,1,2,... .$$

The scalar functions ψ_j are the above mentioned eigenfunctions of $-\Delta$ while for each $j \geq 0$, μ_{1j} and μ_{2j} are the two eigenvalues of the 2x2 matrix $A(\lambda_j) = A - \lambda_j D$ and ζ_{1j} and ζ_{2j} are the corresponding eigenvectors. Since Trace $A(\lambda_{\bar{j}}) < 0$ while Det $A(\lambda_{\bar{j}}) = 0$, it follows that either $\mu_{1\bar{j}}$ or $\mu_{2\bar{j}}$ is zero while the other is negative. Without loss of generality we assume $\mu_{1\bar{j}} = 0$.

Since, by assumption, $\lambda_{\bar{j}}$ is simple as an eigenvalue of $-\Delta$ it follows that $\mu_{1\bar{j}} = 0$ is a simple eigenvalue of $F_P(\overline{P})$. i.e. the null space of $F_P(\overline{P})$ is one-dimensional. Using the completeness of the eigenfunctions (4.5b) it is also easy to see that the range of $F_P(\overline{P})$ is one-dimensional. In fact, the range consists of those functions whose expansion has no component along $\psi_{\bar{j}}\,\zeta_{1\bar{j}}$. Thus, as is shown in CRANDALL-RABINOWITZ (1971), there will be bifurcation if

$$F_{Ps}(\overline{P})\psi_{\bar{j}}\zeta_{1\bar{j}} \notin \text{Range } F_P(\overline{P}) \tag{4.6}$$

Here the subscripts denote the second order mixed Frechet derivative with respect to P and the bifurcation parameter s and it is understood that F_{Ps} is evaluated at the critical value, $s = \bar{s}$.

In verifying this last requirement for bifurcation we shall first treat the case where the matrices D and A depend on the parameter s but the λ_j's do not. (Of course, by a simple length scale change this includes the case of Ω growing or shrinking with s by the same amount in all dimensions.) Letting primes denote differentiation with respect to s and setting $\bar{\psi} = \psi_{\bar{j}}$, $\bar{\mu} = \mu_{1\bar{j}}$, we have

$$F_{Ps}\,\bar{\psi}\bar{\zeta} = (D'\Delta + A)\bar{\psi}\bar{\zeta} = (A' - \bar{\lambda}D')\bar{\psi}\bar{\zeta} .$$

But, since $\bar{\mu} = 0$ we have

$$[A' - \bar{\lambda}D']\bar{\zeta} + [A - \bar{\lambda}D]\bar{\zeta}' = \bar{\mu}\bar{\zeta}' + \bar{\mu}'\bar{\zeta} = \bar{\mu}'\bar{\zeta}$$

and

$$[A - \lambda D]\bar{\zeta}' = \alpha\zeta_{2\bar{j}} .$$

Thus,

$$F_{P_s}\psi_{\bar{j}}\zeta_{1\bar{j}} = \mu'_{1\bar{j}}\,\psi_{\bar{j}}\zeta_{1\bar{j}} - \alpha\psi_{\bar{j}}\zeta_{2\bar{j}} .$$

Since our transversality assumption requires that $\mu'_{1\bar{j}} \neq 0$, we see that condition (4.6) is satisfied.

Alternatively, we can consider a situation where D and A are fixed so that the D has two positive roots, λ_c^- and λ_c^+ , as in figure 4.1. We then consider a one parameter family of regions, $\Omega(s)$, such that as s varies the eigenvalues $\lambda_j(s)$ vary smoothly. As s increases past $\bar{s}$ exactly one eigenvalue $\lambda_{\bar{j}}$ moves into the interval $(\lambda_c^-, \lambda_c^+)$ while all other λ_j's remain outside. To verify (4.6) in this case, we assume that in a neighborhood of $s = \bar{s}$ there is a one parameter family of diffeomorphisms of $\Omega(s)$ onto $\Omega(\bar{s}) \equiv \Omega$. In this neighborhood of $\bar{s}$ our problem can then be considered on a fixed domain but now the Laplace operator Δ becomes a variable coefficient, uniformly elliptic, second order operator whose coefficients depend upon the parameter s. We then have

$$F_{Ps}\psi_{\bar{j}}\zeta_{1\bar{j}} = D\Delta'\psi_{\bar{j}}\zeta_{1\bar{j}} .$$

Now

$$\begin{aligned}\Delta'\psi_{\bar{j}} &= -\Delta\psi'_{\bar{j}} - \lambda_{\bar{j}}\psi'_{\bar{j}} - \lambda'_{\bar{j}}\psi_{\bar{j}} \\ &= -(\Delta + \lambda_{\bar{j}})\,(\Sigma\, c_k\psi_k) - \lambda'_{\bar{j}}\psi_{\bar{j}} \\ &= \sum_{k\neq j} (\lambda_k - \lambda_{\bar{j}})c_k\psi_k - \lambda'_{\bar{j}}\,\psi_{\bar{j}}\end{aligned}$$

where we have made an eigenfunction expansion of $\psi'_{\bar{j}}$. Thus,

$$F_{Ps}\psi_{\bar{j}}\zeta_{1j} = -\lambda'_{\bar{j}}\psi_{\bar{j}}D\zeta_{1\bar{j}} + \sum_{k\neq j} (\lambda_k - \lambda_{\bar{j}})c_k\psi_k(D\zeta_{1\bar{j}}) .$$

Since transversality now demands $\lambda'_{\bar{j}} \neq 0$ we see that (4.6) is satisfied as long as $D\zeta_{1\bar{j}}$ is not scalar multiple of $\zeta_{2\bar{j}}$. We can regard this last as an additional (generic) condition which we impose on D and A.

Thus, in each of these situations we see that bifurcation to a new solution occurs as $\lambda_{\bar{j}}$ crosses the graph of $D(\lambda)$. Since to first order in $(s-\bar{s})$ we have both $u(x) - \bar{u}$ and $v(x) - \bar{v}$ proportional to $\psi_{\bar{j}}(x)$ we see that this new equilibrium is indeed inhomogeneous.

<u>Summary</u>. <u>Nonconstant equilibria bifurcate from</u> $(\bar{u},\bar{v})$ <u>when the first</u> λ_j <u>to cross the graph of</u> $D(\lambda)$ <u>does so transversely, is simple and</u> $j \geq 1$. <u>This can happen only when</u> $\mathrm{Tr}\,A < 0$, $h'(\bar{u}) > 0$ <u>and</u> $g'(\bar{u}) > 0$, i.e.

when

$$\bar{v}g'(\bar{v}) > \bar{u}\phi(\bar{u})h'(\bar{u}) > 0 \ . \tag{4.7}$$

This necessary condition is also sufficient in the sense that if (4.7) is satisfied, then bifurcation for any simple λ_j can be effected by varying the diffusion coefficients d_1 and d_2.

It should be clear that (4.7) is a rather severe restriction upon the interaction terms. It is not satisfied by the most common models which have $g'(v) \equiv 0$. If we take $g'(v) \equiv \varepsilon > 0$, then for reasonable choices of f and ϕ the condition (4.7) will be satisfied only if $\bar{u}$ lies in a small interval to the left of the "hump", c_1, of h (see Figure 4.3).

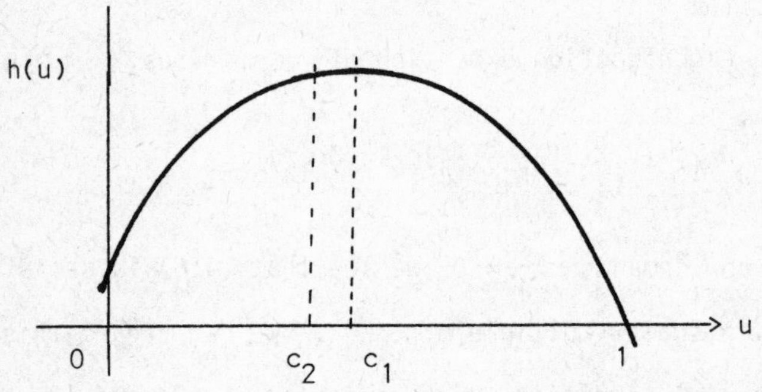

Figure 4.3. Zone of Turing instability

If $\bar{u} > c_1$, then $(\bar{u},\bar{v})$ is stable as a solution of the PDE. If $0 < \bar{u} < c_2$, then $(\bar{u},\bar{v})$ is unstable for the ODE, i.e. trace $A > 0$.

Stability. The rarity of bifurcation at Turing Instability is compounded by the fact that the non-constant bifurcating solutions need not be stable. The calculations required to demonstrate stability or instability are quite complicated and we shall not attempt them. In the case of one space dimension MIMURA-NISHIURA-YAMAGUTI (1979) contains results which allow us to conclude that both stability and instability can occur.

Literature. The idea of diffusion acting as a destabilizing factor goes back (at least) to TURING (1952) and was developed by various people in various contexts e.g. OTHMER-SCRIVEN (1969) or PRIGOGINE-NICOLIS (1967). The first discussions in the context of population dynamics are in SEGEL-JACKSON (1972), LEVIN-SEGEL (1976) and SEGEL-LEVIN (1976). An attempt to determine the stability of the bifurcating solutions was made in BROWN (1978). The most complete discussion of pattern in predator-prey systems is MIMURA-NISHIURA-YAMAGUTI (1979). NISHIURA (1982) studies the global behavior of the bifurcating branch of solutions and, under some conditions, shows that it can eventually make contact with the large amplitude pattern discussed in the next section. Finally, a local analysis when two eigenvalues, λ_j and λ_{j+1}, simultaneously cross $D(\lambda)$ is given in FUJII-MIMURA-NISHIURA (1982).

5. LARGE AMPLITUDE PATTERN: TRANSITION LAYERS

In this section we will exhibit stationary patterns in a way completely different from that in the previous section. The basic ideas of this approach are due to Paul Fife and appeared in a number of papers in 1976 and 1977, most notably in FIFE (1976 A) where a general program was mapped out. Only a few steps in Fife's program have been carried out in a rigorous fashion: (1) a basic existence theorem for the Dirichlet Problem in one dimension was proved in FIFE (1976 B); (2) a similar theorem for the Neumann Problem was proved in MIMURA-TABATA-HOSONO (1980) by reducing the problem to that treated by Fife; (3) an analogous problem on the infiinte line was solved in GARDNER-SMOLLER (1983) resulting in periodic travelling waves. The argumentation in the last work is based upon Conley's generalization of the Morse Index and is quite different in spirit from the first two.

Fife's approach differs from the Turing bifurcation method of the preceding section in several significant respects. The Turing approach yields solutions of small oscillation because it depends upon the local structure of the interaction terms in a neighborhood of a point in phase space. In fact, it depends only upon the linearization of the interaction terms and is applicable to linear equations. In contrast, Fife's method is inherently nonlinear, depends upon certain global features of the interaction terms and yields solutions of macroscopic oscillation. We shall also see that Fife's method is more general in that it will exhibit patterned solutions for systems for which Turing instability cannot occur. On the other hand, Fife's mechanism has been rigorously verified only in one space dimension and, even in that case, the stability of the solution is unproved although numerical evidence leaves a little doubt that the pattern is quite robust.

In the preceding section we obtained pattern only when $d_2 < d_1$. In this section we shall consider the case of d_1 arbitrarily small and we begin with the limit case $d_1 = 0$. Thus we seek non-negative solutions of the steady-state problem,

$$0 = uf(u) - uv\phi(u) = u\phi(u)[h(u) - v] \tag{5.1a}$$

$$0 = v'' - vg(v) + uv\phi(u) \quad \text{for} \quad t > 0, \quad |x| < L \tag{5.1b}$$

and

$$v'(\pm L,t) = 0, \qquad t > 0.$$

We are restricting ourselves to the case of one space dimension because, as mentioned earlier, there are no rigorous results in higher dimensions and because this allows a much more concrete exposition. We have also changed length scales so as to set $d_2 = 1$ and will let the variability of L express the variability of d_2 as well. In Section 3 we saw that functions h of type I allow only spatially homogeneous steady states so we shall assume h to be of type II or type III (cf. Figure 2.2).

If (u,v) is a solution of (5.1)-(5.2), then at any point x, $|x| < L$, the values $u(x)$ and $v(x)$ must satisfy the algebraic relation (5.1a). i.e. either $u(x) = 0$, $u(x) = k(v(x))$ or $u(x) = \ell(v(x))$ where ℓ and k are the two branches of the inverse of the function $v = h(u)$. In Figure 5.1 we have illustrated a situation where h is the type II and denoted

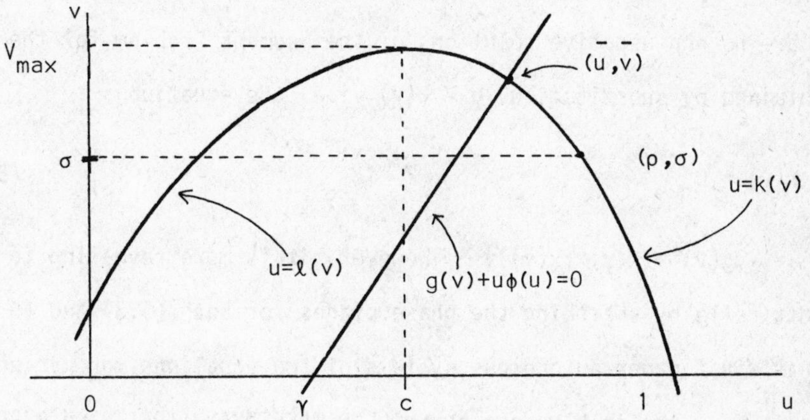

Figure 5.1

The maximum value of h by V_{max} so that k and ℓ are defined on the interval $v \leq V_{max}$. As a first step then, we eliminate u from (5.1b) by setting u equal to zero or one of the inverse functions k and ℓ. The branch $u = \ell(v)$ can give rise only to solutions which are unstable. We'll say a bit more about this at the end of this section but let us now concentrate our attention on $u \equiv 0$ and $u = k(v)$. First we look for non-negative solutions of

$$v'' - vg(v) = 0 \tag{5.3}$$

subject to the Neumann boundary condition (5.2). Such a function v together with $u \equiv 0$ would be a solution of the system (5.1)-(5.2). However, there are no such solutions, for if v is a solution of (5.3) such that $v'(-L) = 0$ and $v(-L) \geq 0$, then

$$v'(L) = \int_{-L}^{+L} v(s)g(v(s))ds > 0$$

since $g(v)$ is strictly positive on $v \geq 0$. An analogous argument shows that there are no non-negative solutions of the Neumann Problem for the equation obtained by substituting $u = k(v)$ i.e. the equation,

$$v'' + K(v) = 0 \tag{5.4}$$

where $K(v) = -vg(v) + k(v)\phi(k(v))v$. However, it is more revealing to argue geometrically by sketching the phase planes for both (5.3) and (5.4) considered as first order autonomous systems of two equations for v and v' . (This is very profitably done by SMOLLER-WASSERMAN (1981) and GARDNER-SMOLLER (1983)). In Figure 5.2(b) we have pictured the case of $c < \bar{u} < 1$ so that K has a zero at $\bar{v}$ in addition to that at $v = 0$. In the case where $\bar{u} < c$ then $(v,v') = (0,0)$ remains a center so that our conclusions are not altered.

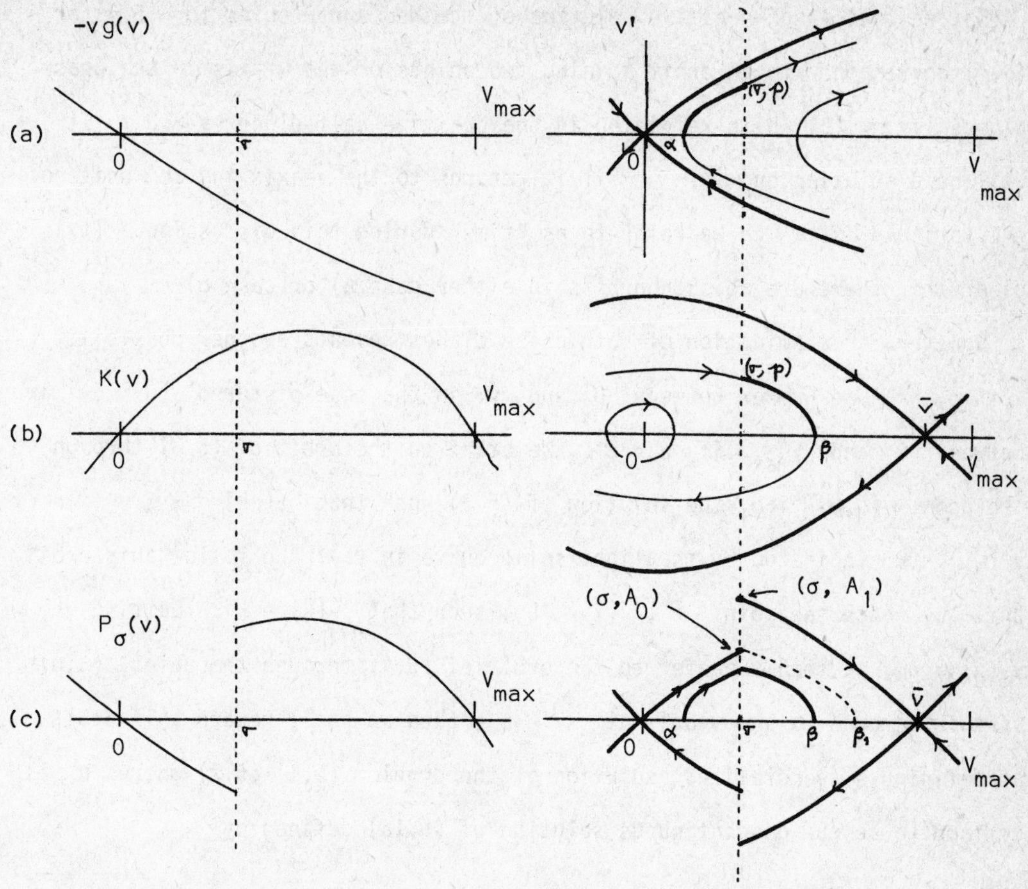

Figure 5.2. Phase planes for (5.3), (5.4) and (5.7)

This is Fife's notion of a singular solution with an internal "transition layer" at $x = \bar{x}$. Now remember that (5.1)-(5.2) is

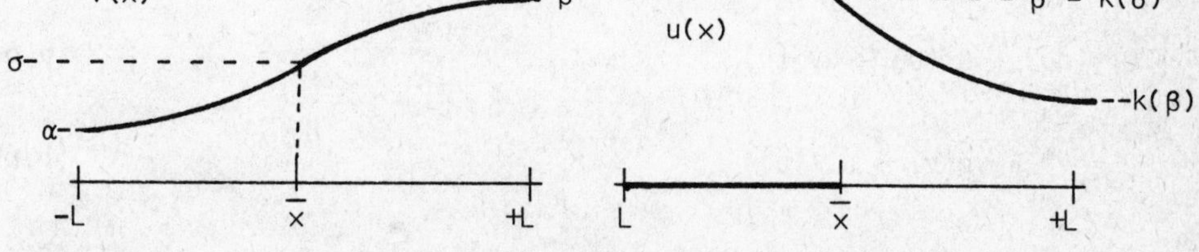

Figure 5.3. Graphs of generalized solution of (5.1)-(5.2)

Notice that a non-negative solution of the Neumann Problem for (5.3) or (5.4) corresponds to an orbit joining two points on the v-axis in the phase planes (a) or (b) while remaining in the positive half-plane $v > 0$. It will be a solution on $|x| < L$ if it returns to the v-axis in $2L$ units of the variable x which we refer to as "time" during this discussion. It is clear that there are no such orbits in either case a) or case b).

However, an examination of both phase planes reveals another possibility. Let σ be some number between 0 and $\bar{v}$ in the case pictured $(\bar{u} > c)$ or between 0 and V_{max} if $\bar{u} < c$. We start on the orbit of (5.3) through the point $(\alpha,0)$ i.e. the solution of (5.3) such that $v(-L) = \alpha$, $v'(-L) = 0$ indicated by the light solid curve in (a). We follow this orbit until we reach the point $\bar{x}$, $|\bar{x}| < L$, such that $v(\bar{x}) = \sigma$. Letting $v'(\bar{x}) = p$, we then transfer to the orbit of (5.4) through the point (σ,p) . If this returns to the v-axis at $x = +L$, then we shall regard this orbit as defining a (generalized) solution of the problem (5.1)-(5.2) where u is defined to be the discontinuous solution of (5.1a) defined by

$$u(x) = \begin{cases} 0 & \text{for} \quad -L < x < \bar{x} \\ k(v(x)) & \text{for} \quad \bar{x} < x < +L \end{cases} \tag{5.5}$$

the singular case $d_1 = 0$; the significant aspect is that the generalized solution pictured in Figure (5.3) is to be an approximation to the (smooth) solution of the full equation (5.6),

$$0 = du'' + uf(u) - uv\phi(u) \tag{5.6a}$$

$$|x| < L , \quad t > 0$$

$$0 = v'' - vg(v) + uv\phi(u) \tag{5.6b}$$

$$0 = u' = v' \quad \text{at} \quad x = \pm L , \quad t > 0 \tag{5.6c}$$

for sufficiently small values of $d_1 = d > 0$. This is pictured in Figure (5.4).

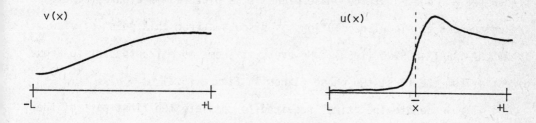

Figure 5.4. Graphs of solutions of (5.6) for small $d > 0$

But now let us summarize the important questions that must be faced.

A. Under what conditions can we actually carry out the construction of such singular solutions?

B. What can we say concerning the properties of these solutions?

C. Under what conditions will these singular solutions be the limits of solutions of (5.6) as $d \downarrow 0$?

D. Are such solutions stable as stationary solutions of the original system (1.1)-(1.2)?

A. Construction of Singular Solutions. We can procedd in one of several ways. One is to consider the equation

$$v'' + P_\sigma(v) = 0 \tag{5.7}$$

where

$$P_\sigma(v) = \begin{cases} -vg(v) & \text{if } v < \sigma \\ K(v) & \text{if } v > \sigma . \end{cases}$$

P_σ is a discontinuous function of v but it is easy to see (e.g. by "mollifying" P_σ) that (5.7) has generalized solutions which are smooth but piecewise C^2 and that the phase plane is as pictured in Figure (5.3c). The uniqueness of singular solutions is discussed from this point of view in MIMURA-TABATA-HOSONO (1980). However, it seems more transparent to argue directly from the first two phase planes in Figure (5.3).

Let $T_0(p)$ denote the "time" required to complete the first part of the journey from $(\alpha,0)$ to (σ,p) and $T_1(p)$ denote the time required to go from (σ,p) to $(\beta,0)$. Thus, if $v(-L) = \alpha$, $v'(-L) = 0$, then $v(-L + T_0(p)) = \sigma$, $v'(-L + T_0(p)) = p$, etc. We let $T = T_0 + T_1$ denote the time to complete the entire journey. It follows from SMOLLER-TROMBA-WASSERMAN (1980) that T_0 and T_1 and hence T are monotonically increasing functions of p. $T_0(p)$ is defined for $0 < p < A_0$ where (σ,A_0) is on the unstable manifold through the saddle $(0,0)$. We see then that $T_0(p) \to +\infty$ as $p \to A_0$. Similarly, $T_1(p)$ is defined for $0 < p < A_1$ where (σ,A_1) is on the stable manifold for $(\bar{v},0)$ if $c \leq \bar{u}$ or is on the orbit through $(V_{max},0)$ if $c < \bar{u}$. In the first case $T_1(p) \to +\infty$ as $p \to A_1$ while in the second case the limit is finite. The function $T(p) = T_0(p) + T_1(p)$ is defined for $0 < p < A_0 \wedge A_1$ and thus also monotonically increases to $+\infty$ unless $\bar{u} < c$ and $A_1 < A_0$. This information is summarized in Figure 5.5. It is now clear that, in case (A), for each positive L there is precisely one p such that $T(p) = L$ and thus on the interval $|x| < L$, there is precisely one solution of the type pictured in Figure 5.3. In case (B) we have one such solution for each L, $0 < L < \bar{L}$. Following MIMURA-TABATA-HOSONO (1980) we shall refer to these solutions as (generalized) solutions of mode 1. The solutions constructed above have $v' > 0$ but because the phase plane is symmetric about the v-axis, we see

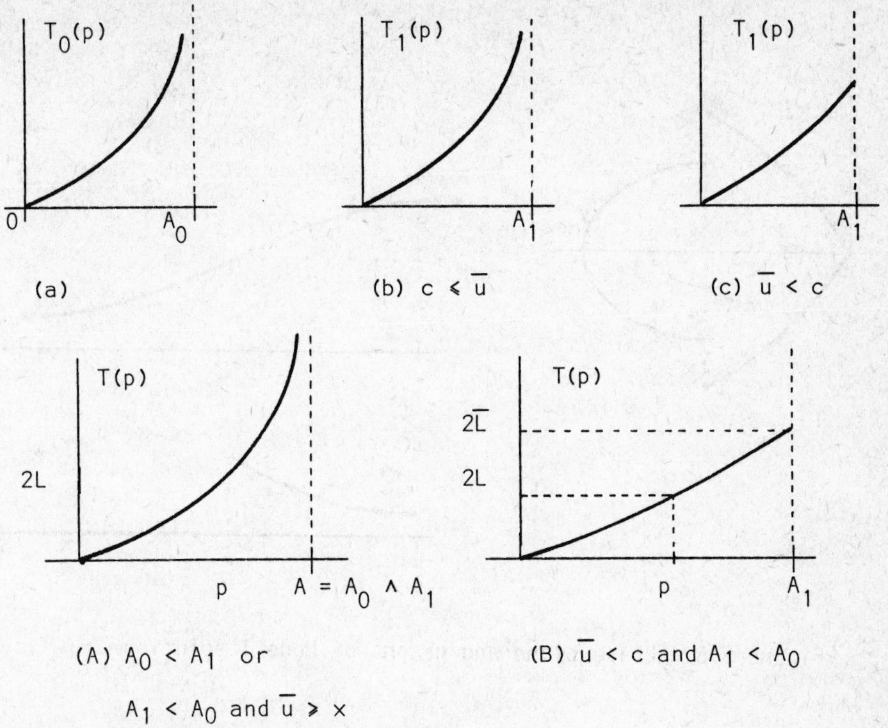

Figure 5.5. Graphs of time-maps

that there is a second family of solutions of mode 1 for which $v' < 0$. Such a solution is pictured in Figure 5.6 in a case (B) phase plane.

We refer to the solutions below as having mode 1 because v assumes the value σ at one point in the interval and therefore there is one transition layer for u . It should be clear that there are solutions with arbitrarily many transitions that correspond to "orbits" in the patched phase plane which circle the point $(\sigma,0)$.

For example (see Figure 5.7) we can consider a solution of mode 3 starting from the point $(\alpha_3,0)$ through (σ,p_3), $(\beta_3,0)$, $(\sigma,-p_3)$, $(\alpha_3,0)$, (σ,p_3) and terminating at $(\beta_3,0)$. The "time" taken to complete this journey is $T_{(3)} = 3T_0 + 3T_1$ and for this to define a solution we must have $T_{(3)} = 2L$.

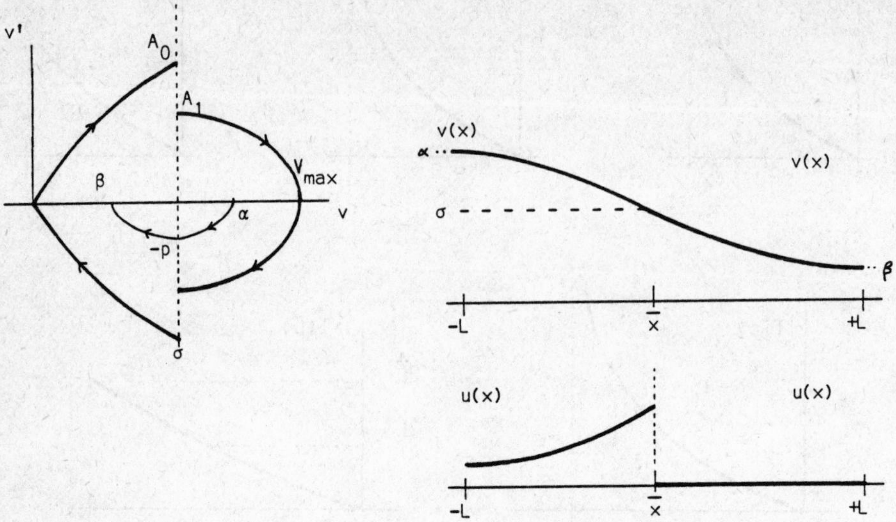

Figure 5.6. Phase plane and graphs of Mode 1 solution

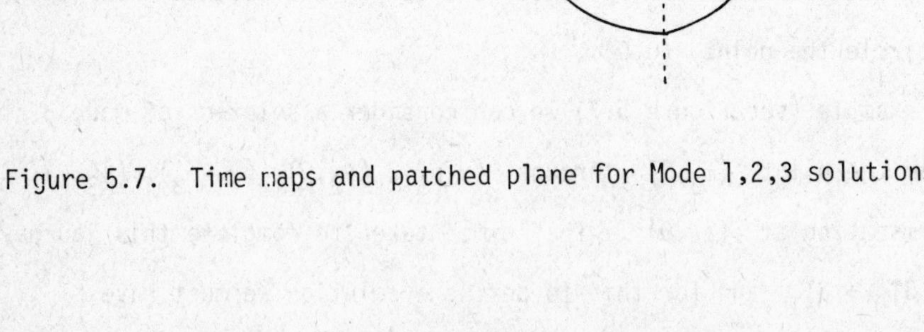

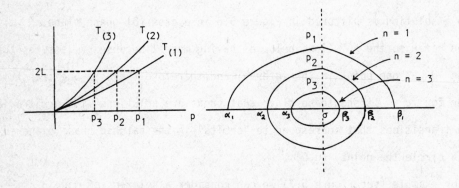

Figure 5.7. Time maps and patched plane for Mode 1,2,3 solutions

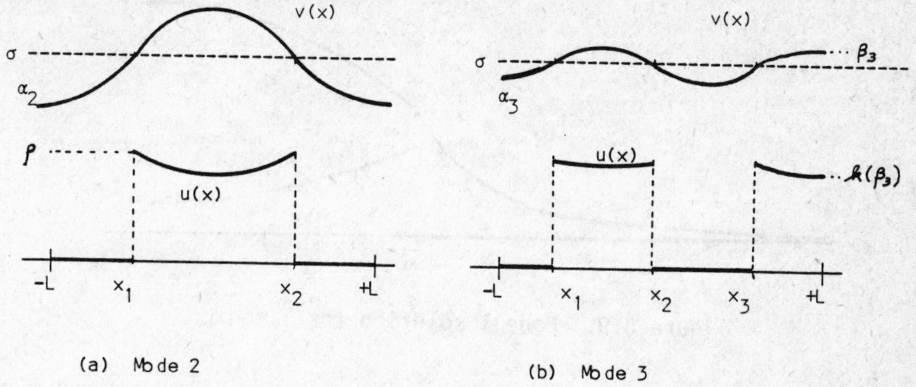

Figure 5.8. Solutions of Modes 2 and 3

Again, $T_{(n)}$ is a monotone increasing function of p which allows an unambiguous choice of orbits for a given L once the mode and parity (as in mode 1 we can have $v(-L) > \sigma$ or $v(-L) < \sigma$) are chosen.

B. Properties. Certain additional properties can be easily determined from an examination of the "patched" phase plane and the time maps. For example, it is clear that for a fixed interval the solutions v of higher mode have smaller oscillation and, in fact, $v_n \to \sigma$ as $n \to \infty$. As a second example we look at the problems which have a solution of fixed mode on arbitrarily large intervals (i.e. case (A) in Figure 5.5). In such a case we can easily determine the limiting orbit in the patched plane and from this learn a good deal concerning the behavior of solutions of large L . As an example consider a mode 1 solution in the situation depicted in Figure 5.2, i.e. $A_0 < A_1$. For very large L the solution is constructed from an orbit that approximates the segment of the unstable manifold from $(0,0)$ to (σ,A_0) joined to the uniquely determined orbit joining (σ,A_0) to $(\beta_1,0)$.

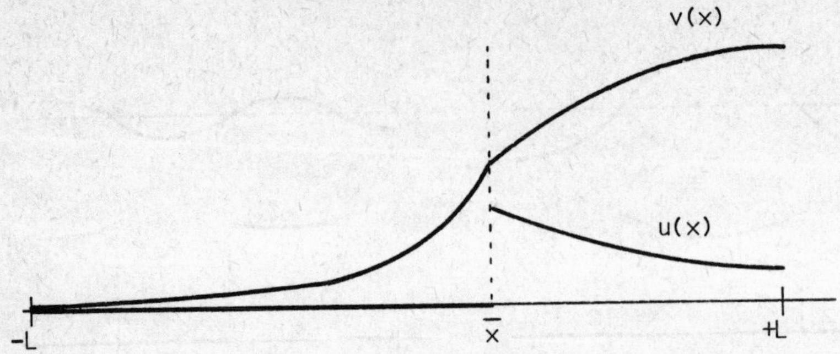

Figure 5.9. Mode 1 solution for large L

Such a solution is depicted in Figure 5.9. As L increases so does $\bar{x}$ in such a way that $L - \bar{x} \to T_1(A_0)$ as $L \to \infty$. We also see that if we fix a point x_1, $|x_1| < L$, then $u(x_1) = 0$ for all sufficiently large L and $v(x_1) \to 0$ as $L \to \infty$.

C. Singular Solutions as Limits of Solutions for $d > 0$. The above construction proceeded for any σ in the interval $(0, \bar{v})$ or $(0, V_{max})$. However, it is a striking feature of this process that these singular solutions can be limits of solutions of (5.6) as $d \to 0$ only if σ is a certain value which is completely determined by f and ϕ. To see this, we examine the solutions of (5.6a) in the small interval, $|x-\bar{x}| < \eta$, about a transition point, $\bar{x}$, for the singular solution.

In the case depicted in Figure 5.10, we wish $u(\bar{x}-\eta;d) \to 0$ and $u(\bar{x}+\eta;d) \to \bar{u}(x+\eta)$ as $d \downarrow 0$ and if η is small, $u(x+\eta) \approx \rho = k(\sigma)$. Under the change of scale $y = (x-\bar{x})/\sqrt{d}$, (5.6a) becomes

$$0 = U'' + Uf(U) - UV\phi(U)$$

where $U(y) = u(x;d)$. If η is small and we confine our attention to $|x-x| < \eta$, then, since v is continuous at $\bar{x}$, we can approximate V by

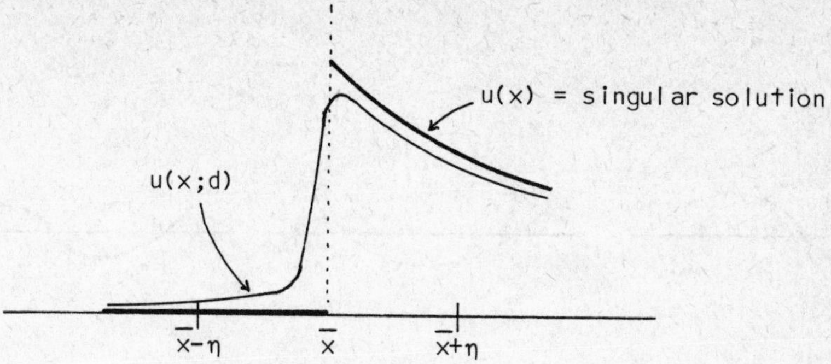

Figure 5.10. Transition layer

σ . Thus, to first approximation U should satisfy

$$0 = U'' + Uf(U) - \sigma U\phi(U) \equiv U'' + m(U;\sigma) \tag{5.9a}$$

on $|y| < \eta\sqrt{d}$. Moreover, as $d \downarrow 0$, $U(-\eta/\sqrt{d}) \to 0$ and $U(\eta/\sqrt{d}) \to u(\bar{x}+\eta) \approx \rho$ for small η . Thus, U should be near a solution of (5.9) which is defined on all of $\mathbb{R}$ and which satisfies

$$U(-\infty) = 0 \ , \quad U(+\infty) = \rho \tag{5.9b}$$

It turns out that such solutions <u>do not exist</u> unless

$$0 = \int_0^\rho m(s;\sigma)ds = \int_0^\rho [sf(s) - \sigma s\phi(s)]ds \tag{5.10}$$

which, together with $\rho f(\rho) - \sigma\rho\phi(\rho) = 0$, uniquely specifies ρ and σ . To see this most simply, we sketch the phase plane for (5.9a) making the usual observation that $\frac{1}{2}(U')^2 + M(U;\sigma)$ is constant along orbits where

$$M(u;\sigma) = \int_0^u m(s;\sigma)ds \ .$$

The function $m(\cdot;\sigma)$ vanishes at $u = 0$ and, for $h(0) < \sigma < V_{max}$, at two positive values the larger of which is ρ (cf. Figure 5.1).

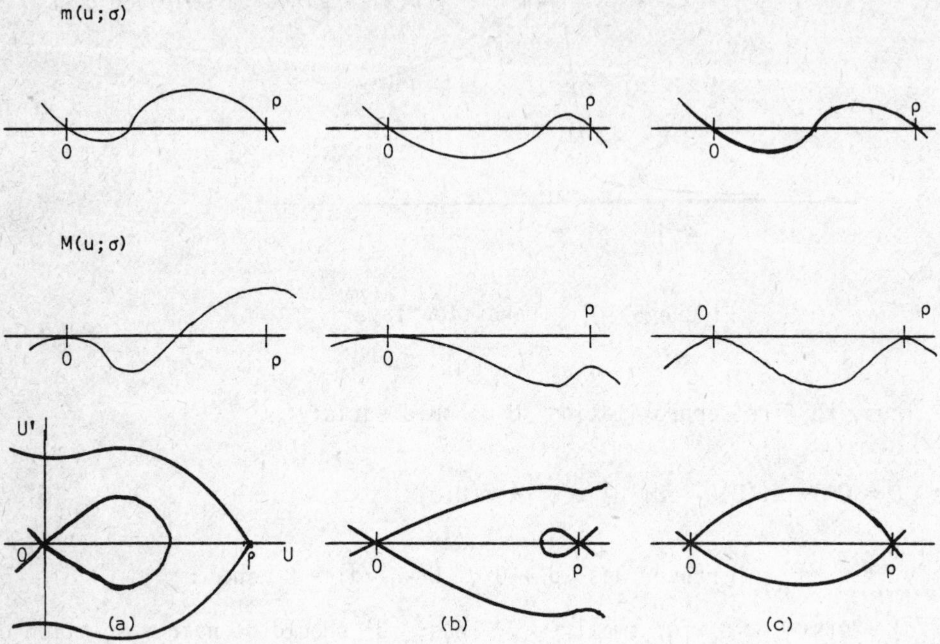

Figure 5.11. Phase plane for (5.9a): (a) σ near $h(0)$
(b) σ near V_{max} (c) σ 'just right'

Thus, the phase planes are as depicted in the bottom row of Figure 5.11 and we see indeed that connecting orbits between $(0,0)$ and $(\rho,0)$ can exist only if (5.10) is satisfied.

But of course this is only a necessary condition; it must be shown that if (5.10) is satisfied that there are solutions of (5.6) which converge, in a sense to be specified, to the singular solutions. This has been done by Fife in FIFE (1976B) in the case of Dirichlet Problems and later the case of Neumann Problems was reduced to that of Dirichlet in MIMURA-TABATA-HOSONO

(1980). The latter paper treats a special case of our problem but can most probably be extended without major modification. In any case we shall not go into the details of their argument.

D. Stability. The question of the stability of these solutions considered as equilibrium solutions of (2.1)-(2.8) is as yet unresolved. Fife offers an heuristic argument for stability in FIFE (1967A) and extensive computer simulation certainly bears him out. In a related problem, H. Weinberger has proved asymptotic stability in the limiting case $d = 0$ (WEINBERGER (1983)) but it is not clear if his method can be extended to the case of $d \quad 0$.

6. OPEN PROBLEMS

A. We have already mentioned that the stability of the large amplitude pattern discussed in the previous section is as yet unresolved. However, going beyond mere stability, FIFE (1967A and 1967C) has offered heuristic arguments for a very precise description of how the steady states are approached by time dependent solutions. His picture, which is consistent with extensive computer simulation, involves the very rapid formation of "fronts" with a slower migration of the fronts to the final equilibrium configuration. Fife's arguments are based upon formal asymptotic analysis and a rigorous validation of his ideas is difficult to imagine. But it might be worthwhile to examine the relationship to travelling waves on the infinite line to the steady state on very long intervals.

B. There is no really effective procedure for determining the destabilization of limit cycles in the unambiguous way in which we discussed Turing instability of critical points. A first step has been taken in MAGINU (1979) but his criteria cannot be verified in our problem.

C. Computer simulation indicates that as the diffusion coefficients vary, the large amplitude spatially inhomogeneous steady states discussed in Section 5 can undergo a Hopf-like bifurcation to a spatially inhomogeneous time-periodic solution in which the oscillations at different points in the interval are not in phase. In KEENER (1983), an analogous bifurcation was constructed by a formal perturbation calculation in a neighborhood of a critical point undergoing Turing instability of two adjacent modes. Not much is known about such solutions.

D. Do the above two sections capture all steady states or are there others that are the result of entirely different processes? This question is somewhat nebulous but we can reinterpret it by asking if all steady states are "connected" in some sense. In FUJII-MIMURA-NISHIURA (1982) it is shown that, under certain conditions, the spatially inhomogeneous steady state bifurcating from the constant steady state as in Section 4 can be continued, as d_2 approaches zero, to a large amplitude steady state discussed in Section 5. In that study the interaction remained fixed and only the diffusion coefficients varied. Now we know that some interactions do not allow Turing instability even though they do allow the transition-layer type of steady states, but might it be true that every stable steady state is the continuation, under variation of the diffusion coefficients, of one of the singular, transition-layer type solutions constructed when $d_2 = 0$? What about continuation under homotopies of the interaction terms?

E. Fife's construction of solutions with internal transition layers is valid only in one (space) dimension. It is important that this be extended to two and three dimensions. The scalar case in two dimensions is discussed in FIFE-GREENLEE (1974).

REFERENCES

1. N. Alikakos, L_p Bounds of Solutions of Reaction-Diffusion Equations, Comm. Partial Diff. Equations, 4 (8), (1979) 827-828.

2. S.C. Bhargava and R.P. Saxena, Stable Periodic Solutions of the Reactive-Diffusive Volterra Systems of Equations. J. Theor. Biol. 67 (1977) 399-406.

3. E. Bradford and J.P. Phillip, Stability of Steady Distributions of Asocial Populations Dispersing in One Dimension, J. Theor. Biol. 29 (1970) 13-26.

4. E. Bradford and J.P. Phillip, Note on Asocial Populations Dispersing In Two Dimensions, J. Theor. Biol. 29 (1970), 27-33.

5. P.N. Brown, Reaction Diffusion-Equations in Population Biology, Ph.D. Dissertation, Tulane University, 1978.

6. P.N. Brown, Decay to Uniform States in Ecological Interactions, SIAM J. Appl. Math. Vol. 38 (1980).

7. R.G. Casten and C.J. Holland, Stability Properties of Solutions to Systems of Reaction-Diffusion Equations, SIAM J. Appl. Math. Vol. 33 (1977).

8. R. Casten and C. Holland, Instability Results for Reaction-Diffusion Equations with Neumann Boundary Conditions, J. Diff. Equations 27 (1978) 266-273.

9. P.L. Chow and W.C. Tam, Periodic and Travelling Wave Solutions to Volterra-Lotka Equations with Diffusion. Bull. Math. Biol. 38 (1976) 643-658.

10. W.C. Clark, D.D. Jones and C.S. Holling, Patches, Movements, and Population Dynamics in Ecological Systems: A Terrestrial Perspective. In: Spatial Pattern in Plankton Communities, J.H. Steele (Ed.), New York, Plenum Press (1978) 385-432.

11. D.S. Cohen and J.D. Murray, A Generalized Diffusion Model for Growth and Dispersal in a Population, J. Math. Biol. 12 (1981) 237-249.

12. Conway, E., Diffusion and the Predator-Prey Interaction: Steady States with Flux at the Boundaries, in Nonlinear Partial Differential Equations, J.A. Smoller (Ed.), Contemporary Mathematics Vol. 17. American Math. Soc., Providence, R.I. (1983).

13. E. Conway, R. Gardner and J. Smoller, Stability and Bifurcation of Steady-State Solutions for Predator-Prey Equations, Advances in Appl. Math. 3 (1982) 288-334.

14. E. Conway, D. Hoff and J. Smoller, Large Time Behavior of Solutions of Systems of Nonlinear Reaction--Diffusion Equations, SIAM J. Appl. Math. 35 (1978) 1-16.

15. E. Conway and J. Smoller, Diffusion and Predator-Prey Interaction, SIAM J. Appl. Math. 33 (1977) 673-686.

16. E. Conway and J. Smoller, Diffusion and the Classical Ecological Interactions: Asymptoties, in Nonlinear Diffusion, W. Fitzgibbon and H. Walker (eds.), Pitman, London--San Francisco--Melbourne (1977).

17. R. Courant and D. Hilbert, <u>Methods of Mathematical Physics</u>, Vol. 1, Wiley-Interscience, New York (1953).

18. M.G. Crandall and P.H. Rabinowitz, Bifurcation from simple eigenvalues, J. Funct. Analysis 8 (1971) 321-340.

19. M. Crandall and P. Rabinowitz, Bifurcation, Perturbation of Simple Eigenvalues, and Linearized Stability, Arch. Rat. Mech. Anal. 52 (1973) 161-180.

20. P. de Mottoni and F. Rothe, Convergence to Homogeneous Equilibrium State for Generalized Volterra-Lotka Systems with Diffusion, SIAM J. Appl. Math. Vol 37 (1979).

21. P.C. Fife, <u>Mathematical Aspects of Reacting and Diffusing Systems</u>, Lecture Notes in Biomathematics, No. 28, Springer-Verlag, New York (1979).

22. P.C. Fife, Boundary and Interior Transition Layer Phenomena for Pairs of Second Order Differential Equations, J. Math. Anal. Appl., 54 (1976B) 497-521.

23. P.C. Fife, Pattern Formation in Reacting and Diffusing Systems, J. Chem. Phys. Vol. 64 (1976A) 554-564.

24. P.C. Fife, Singular Perturbation and Wave-front Techniques in Reaction-Diffusion Problems, in <u>Asymptotic Methods and Singular Perturbations</u> Symposium in Applied Mathematics, SIAM-AMS Proceddings, Vol. 10 (1976C).

25. P.C. Fife, Stationary Patterns for Reaction-Diffusion Equations, in <u>Nonlinear Diffusion</u>, Research Notes in Mathematics 14, Pitman, London 1977.

26. P. Fife and W.M. Greenlee, Interior Transition Layers for Elliptic Boundary Value Problems with a Small Parameter, Russian Math Surveys 14 (1974) 103- .

27. H. Fujii, M. Mimura and Y. Nishiura, A Picture of the Global Bifurcation Diagram in Ecological Interacting and Diffusing Systems. Phy. D 5 (1982), no. 1, 1-42.

28. R.A. Gardner and J.A. Smoller, The Existence of Periodic Travelling Waves for Singularly Pertruded Predator-Prey Equations Via the Conley-Index, J. Diff. Equations 47, (1983) 133-161.

29. N.S. Goel, S.C. Maitra and E.W. Montroll, On the Volterra and Other Nonlinear Models of Interacting Populations. Rev. Modern Phys. 43, (1971) 231-276.

30. M.E. Gurtin and MacCamy, On the Diffusion of Biological Populations. Math. Biosciences 33 (1977) 35-49.

31. K.P. Hadeler, U. an der Heiden, and F. Rothe, Nonhomogenous Spatial Distributions of Populations. J. Math. Biol. 1 (1974), 165-176.

32. B. Hassard, N. Kazarinoff and Y. Wan, Theory and Application of Hopf Bifurcation, Cambridge University Press, Cambridge (1981).

33. J.P. Keener, Diffusion Induced Chaos, in Nonlinear Partial Differential Equations. J. A. Smoller, ed., Contemporary Math. Vol. 17. American Math. Society, Providence, R.I. (1983).

34. K. Kishimota, The Diffusive Lotka-Volterra System with Three Species Can Have a Stable Non-Constant Equilibrium Solution, J. Math. Biology 16 (1982) 103-112.

35. A. Leung, Limiting Behavior for a Prey-Predator Model with Diffusion and Crowding Effects, J. Math. Biol. 6 (1978) 87-93.

36. A. Leung, Monotone Schemes for Semilinear Elliptic Systems Related to Ecology, Math. Meth. in the Appl. Sci. Vol. 4 (1982) 272-285.

37. A. Leung and D. Clark, Bifurcations and Large-Time Asymptotic Behavior for Prey-Predator Reaction-Diffusion Equations with Dirichlet Boundary Data, J. Differential Equations 35 (1980) 113-127.

38. S. Levin, Spatial Patterning and the Structure of Ecological Communities, in Some Mathematical Questions in Biology, VII (S. Levin, Ed.) Lectures on Mathematics in the Life Sciences, Vol. 8, Amer. Math. Soc., Providence, R.I., (1976).

39. S.A. Levin, Pattern Formation in Ecological Communities. In Spatial Pattern in Plankton Communities, J.H. Steele (Ed.), New York, Plenum Press (1978) 433-465.

40. S.A. Levin and L.A. Segel, Hypothesis for Origin of Planktonic Patchiness. Nature 259 (1976), 659.

41. D. Ludwig, D.G. Aronson and H.F. Weinberger, Spatial Patterning of the Spruce Budworm, J. Math Biology 8 (1979) 217-258.

42. K. Maginu, Stability of Spatially Homogeneous Periodic Solutions of Reaction-Diffusion Equations, J. Diff. Equations 31 (1979) 130-138.

43. H. Matano, Asymptombolic Behavior and Stability of Solutions of Semilinear Diffusion Equations, Publ. RIMS, Kyoto Univ. 15 (1979) 401-454.

44. R.M. May, Stability and Complexity in Model Econsystems. Princeton, N.J., Princeton Univ. Press (also 2nd Edition, 1974).

45. R.M. May, (Ed.) Theoretical Ecology: Principles and Applications. Philadelphia, Toronto: W.B. Saunders Co. (1976).

46. J. Maynard-Smith, Models in Ecology, Cambridge Univ. Press, Cambridge (1974).

47. R. McMurtrie, Persistence and Stability of Single-Species and Prey-Predator Systems in Spatially Heterogeneous Environments. Math. Biosciences 39 (1978) 11-51.

48. M. Mimura, Asymptotic Behavior of a Parabolic System Related to a Planktonic Prey and Predatory System. SIAM J. Appl. Math., Part C, Vol. 37 (1979) 449-512.

49. M. Mimura, Spatial Structures in Nonlinear Interaction-Diffusion Systems, Kokyuroku 317, RIM, Kyoto Univ., Kyoto, Japan, (1978) 17-29.

50. M. Mimura and J.D. Murray, On a Diffusive Prey-Predator Model Which Exhibits Patchiness, J. Theoret. Biol., 75 (1979) 249-262.

51. M. Mimura and T. Nishida, On a Certain Semilinear Parabolic System Related to Lotka-Volterra's Ecological Model, Publ. Research Inst. Math. Sci., Kyoto Univ. 14 (1978) 269-282.

52. M. Mimura and Y. Nishiura, Spacial Patterns for Interaction-Diffusion Equations in Biology, Proc. International Symposium on Mathematical Topics in Biology, 1978, 136-146.

53. M. Mimura, Y. Nishiura and M. Yamaguti, Some Diffusion Prey and Predator Systems and Their Bifurcation Problems, Ann. Y.Y Acad. Sci. 316 (1979) 490-510.

54. M. Mimura, M. Tabata and Y. Hosono, Multiple Solutions of Two-Point Boundary Value Problems of Neumann Type with a Small Parameter, SIAM J. Math. Anal. Vol. II (1980) 613-631.

55. W.W. Murdoch and A. Oaten, Predation and Population Stability, in Advances in Ecological Research. A. Macfayden, ed. Academic Press, London 1975.

56. J.D. Murray, Non-Existence of Wave Solutions for the Class of Reaction-Diffusion Equations Given by the Volterra Interacting-Population Equations with Diffusion. J. Theor. Biol. 52 (1975) 459-469.

57. G. Nicolis and I. Prigogine, Self-Organization in Nonequilibrium Systems. New York, London, Sydney, Toronto: J. Wiley & Sons. 1977.

58. Y. Nishiura, Global Structure of Bifurcating Solutions of Some Reaction-Diffusion System, SIAM J. Math. Anal. 13 (1982) 555-593.

59. A. Okubo, Diffusion and Ecological Problems: Mathematical Models. Springer-Verlag, Berlin--Heidelberg--New York (1980).

60. H.G. Othmer, Current Problems in Pattern Formation, In: Lectures on Mathematics in the LIfe Sciences. Levin, S.A. (ed.), Vol. 9, 57-85. Providence: Amer. Math. Soc.

61. H.G. Othmre and L.E. Scriven, Instability and Dynamic Pattern in Cellular Networks. J. Theor. Biol. 32 (1971) 507-537.

62. H.G. Othmer and L.E. Scriven, Interactions of Reaction and Diffusion in Open Systems, Industrial and Engineering Chemistry Fund. 8 (1969) 302-313.

63. J.R. Philip, Sociality and Sparce Populations. Ecology 38, (1957) 107-111.

64. I. Prigogine and G. Nicolis, On Symmetry-Breading Instabilities in Dissipative Systems, J. Chem. Physics 46 (1967) 3542-3550.

65. F. Rothe, Convergence to the Equilibrium State in the Volterra-Lotka Diffusion Equations, J. Math. Biology Vol. 3 (1976) 319-324.

66. L.A. Segel and J.L. Jackson, Dissipative Structure: An Explanation and an Ecological Example, J. Theor. Biol. 37 (1972) 545-559.

67. L.A. Segel and S.A. Levin, Application of Nonlinear Stability Theory to the Study of the Effects of Diffusion on Predator-Prey Interactions. In: Topics in Statistical Mechanics and Biophysics: A Memorial to Julius L Jackson, AIP Conf. Proc. No. 27, 123-152. Piccirelli, R.A., (ed.). New York: Amer. Inst. Phys., 1976.

68. J.A. Smoller, Shock Waves and Reaction Diffusion Equations. Springer Verlag, Berlin--Heidelberg--New York (1983).

69. J. Smoller, A. Tromba and A. Wasserman, Non-Degenerate Solutions of Boundary Value Problems, J. Nonlinear Anal. 4 (1980) 207-216.

70. J. Smoller and H. Wasserman, Global Bifurcation of Steady-State Solution, J. Differential Equations 39 (1981) 269-290.

71. A.M. Turing, The Chemical Basis of Morphogenesis, Phil. Trans. Royal Soc. B 237 (37) (1952) 37-72.

72. H.F. Weinberger, A Simple System with a Continuum of Stable Inhomogeneous Steady States, Preprint. 1983.

73. S. Williams and P.L. Chow, Nonlinear Reaction-Diffusion Models for Interacting Populations, J. Math. Anal. Appl. 62 (1978) 157-169.

E. D. Conway

Tulane University

C M DAFERMOS

Stabilizing effects of dissipation

1. INTRODUCTION

The conservation laws of isothermal, isentropic or adiabatic thermoelasticity in all their standard variants lead to systems of quasilinear hyperbolic equations. Even when the initial data is very small, the formation of shock waves prevents the system from producing smooth global solutions to the Cauchy problem. However, global solutions exist in the class of functions of bounded variation in the sense of Tonelli-Cesari. When the material is viscous and/or heat may diffuse dissipative mechanisms emerge in the system of conservation laws which manifest themselves through the appearance of "parabolic" or "memory" terms. The same phenomenon also arises in the context of the theory of chemically reacting media with dissipation induced by diffusion.

A dissipative mechanism may, in general, affect the asymptotic behavior as well as the smoothness of solutions. Ranked according to their effectiveness, dissipative mechanism may be classified into the following categories:

I. Those which are so powerful to smoothen out even rough initial data and always yield smooth solutions.

II. Those that preserve the smoothness of the initial data but are incapable of smoothing out rough initial data.

III. Those capable of preserving the smoothness of smooth and "small" initial data but cannot prevent the breaking of smooth waves of large amplitude - i.e. there is a threshold phenomenon.

IV. Those that are not capable of preventing the breaking of waves of small amplitude.

In this note we outline results which demonstrate the effect of dissipative mechanisms on systems of conservation. In Section 2 we exhibit results of this type within the context of a simple model. Section 3 and 4 survey results for more complicated systems of physical interest. The final sections overview recent work [7], [10] concerning the existence of globally defined smooth thermomechanical processes in nonlinear thermoviscoelasticity.

2. A SIMPLE MODEL

We consider wave phenomena governed by the Hopf equation

$$u_t + uu_x = 0 \qquad -\infty < x < \infty \; ; \; t \geq 0 \tag{2.1}$$

$$u(x,0) = u_0(x) \qquad -\infty < x < \infty \, .$$

The initial datum is assumed to be smooth. The characteristics satisfy

$$dx/dt = u \quad \text{and} \quad du/dt = 0$$

and consequently are straight lines. If we set $du/dx = v$ then v satisfies

$$v_t + uv_x + v^2 = 0$$

Thus along a characteristic

$$dv/dt + v^2 = 0$$

Therefore, if $v(0) < 0$ then $v(t) \to -\infty$ in finite time.

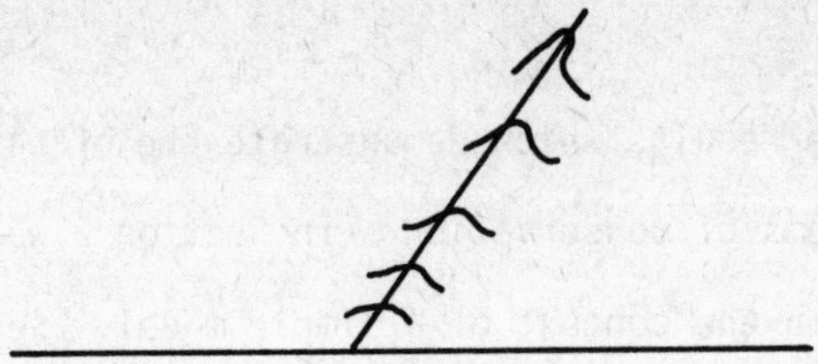

Figure 2.2

From Figure 2.2 we see that the wave gets more and more steep and finally breaks. Thus the equation does not admit smooth global solutions and provides an example of case IV. If we add diffusion to 2.1 we obtain Burgers' equation

$$u_t + uu_x = \epsilon u_{xx} \qquad \epsilon > 0, -\infty < x < \infty, \quad t > 0 \tag{2.3}$$

$$u(x,0) = u_0(x), \qquad -\infty < x < \infty$$

In sharp contrast to the preceeding example the initial value problem for this problem with rough, bounded, measurable initial data admits a smooth solution. This is frequently called parabolic damping and equations of this type provide examples of Case I.

If we add frictional damping to (2.1) rather than diffusion we obtain

$$u_t + uu_x + \mu u = 0 \qquad \mu > 0 \text{ (a constant)}, -\infty < x < \infty, \ t > 0 \tag{2.4}$$

$$u(x,0) = u_0(x) \qquad -\infty < x < \infty$$

The characteristics for this equation satisfy,

$$dx/dt = u \qquad du/dt = -\mu u$$

Setting $u_x = v$ we observe that v satisfies

$$v_t + uv_x + v^2 + \mu v = 0$$

Thus along characteristics

$$dv/dt + v^2 + \mu v = 0$$

If $v(0) \geq -\mu$ the solution remains bounded. However, if $v(0) < -\mu$ then the solution blows up. Thus if $|u_x(x,0)| \leq \mu$ one obtains smooth global solutions. This is an example of the threshold phenomena of Case III.

The method of characteristics has the dual virtues of simplicity and elegance; however, we need to develop more versatile techniques. We apply the energy method to the previous example (2.4). We compute our first estimate by differentiating our equation with respect to x to obtain

$$u_{xt} + (uu_x)_x + \mu u_x = 0$$

We multiply the above equation by $2u_x$ and integrate over $(-\infty, \infty) \times [0,s]$ thus obtaining

$$\int_{-\infty}^{\infty} u_x^2 (x,s)\, dx + \int_0^s \int_{-\infty}^{\infty} (2\mu + u_x)\, u_x^2\, dx\, dt$$

$$= \int_{-\infty}^{\infty} u_x^2 (x,0)\, dx$$

Clearly if $u_x(x,t) \geq -2\mu$ then $||u_x(\cdot,s)||_{L^2} \leq ||u_x(\cdot,0)||_{L^2}$.

To obtain a pointwise bound in terms of an L^2 bound we observe

$$u_x^2 (x,s) = \int_{-\infty}^{x} \frac{\partial}{\partial \xi} (u_x(\xi,s))^2\, d \;= 2\int_{-\infty}^{x} u_{xx}(\xi,s)u_x(\xi,s)\, d\xi$$

$$\leq 2||u_x||_{L^2}||u_{xx}||_{L^2}$$

Bounds for $||u_{xx}||_L 2$ are produced by differentiating our original equation twice with respect to x

$$u_{xxt} + (uu_x)_{xx} + \mu u_{xx} = 0$$

Multiplying the above by $2u_{xx}$ and integrating over $(-\infty,\infty) \times [0,s]$ we obtain

$$\int_{-\infty}^{\infty} u_{xx}^2 (x,s)\, dx + \int_0^s \int_{-\infty}^{\infty} (2\mu + 5u_x) u_{xx}^2 \, dxdt = \int_{-\infty}^{\infty} u_{xx}^2 (x,0) dx$$

and thereby observe that $||u_{xx}(\cdot,t)||_L 2$ can be bounded in terms of initial data if $u_x(x,s) \geq (-2/5)\mu$. Utilizing the foregoing methods of estimation we can show that (2.4) has bounded solution provided that the initial data is such that $||u_x(x,0)||_L 2 \cdot ||u_{xx}(x,0)||_L 2$ is bounded by $4\mu^2/25$.

3. NONLINEAR ELASTICITY

We begin by considering a nonlinear wave equation of the form

$$u_{tt} - (\sigma(u_x))_x = 0$$

Here σ is an increasing function whose graph looks like

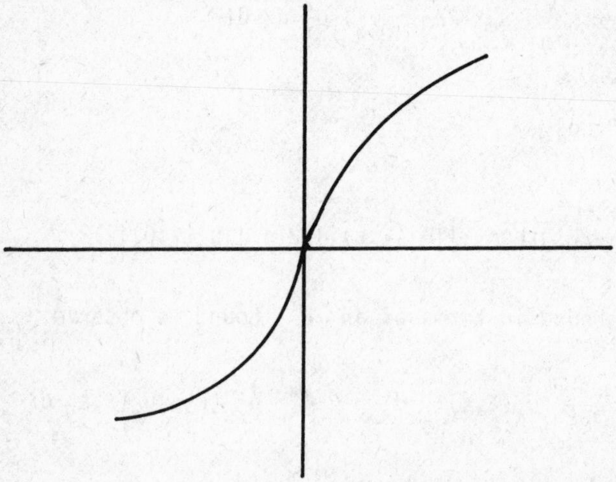

Figure 3.2

If we set $u_t = v$ and $u_x = w$ we can observe that our equation is equivalent to the following first order system

$$w_t - v_x = 0$$

$$v_t - \sigma(w)_x = 0$$

One can show that the solution blows up by monitoring the derivatives of the Riemann invariants along the characteristics. If we introduce a frictional damping into the equation we have

$$u_{tt} - \sigma(u_x)_x + \mu u_t = 0 \qquad (\mu > 0)$$

This produces a system of the form

$$w_t - v_x = 0$$

$$v_t - \sigma(w)_x + \mu v = 0$$

Nishida shows the existence of smooth solutions under smooth and "small" initial data by using the method of characteristics. Energy estimates can be used to obtain a similar result.

The following equation arises in the study of n-dimensional nonlinear elasticity

$$u_{tt} - \mathrm{Div}\ T(\nabla u) = 0$$

The function u is an n-dimensional vector. T takes values in the class of $n \times n$ matrices and is obtained from the stored energy function W by

$$T(F) = \frac{\partial W(F)}{\partial F} \ .$$

We require that $\mathrm{Div}\, T(\nabla u)$ be strongly elliptic, i.e. that

$$\sum_{i,j,\alpha,\beta} \frac{\partial^2 W}{\partial F_{i\alpha} F_{j\beta}} \xi_i \xi_j \xi_\alpha \xi_\beta > 0 \quad \text{for all} \quad \xi,\zeta \in R^n, \ |\xi| = |\zeta| = 1$$

Hence we are concerned with a n-dimensional hyperbolic system. If n is small, shock waves always develop. However, if n is large (too large for physical application), the characteristics do not crowd and shock waves are avoided in the initial value problem on R^n, provided the initial data are smooth and small. Nevertheless, shock waves will again develop when the body is bounded and/or the initial data are large. If we add dissipation to the system we obtain

$$u_{tt} - \mathrm{Div}\, T(\nabla u) + \mu\, u_t = 0$$

Once again we can observe the threshold phenomena of case III.

4. MATERIALS WITH FADING MEMORY

Here dissipation is induced by the presence of memory terms. The dissipative mechanism is viscosity. Such problems arise in the mechanics of materials with memory. We consider integrodifferential equations of the form

$$u_{tt} - (\sigma(u_x))_x - \int_{-\infty}^{t} k'(t-\tau)(g(u_x))_x \, d\tau = 0 \qquad (4.1)$$

The natural physical assumptions are to require that $\sigma' > 0$ and $g' > 0$. The kernel $k(\ )$ is of relaxation type, i.e. it is a nonnegative convex function which goes to zero rapidly enough to be integrable on the positive semiaxis. A typical kernel would look like

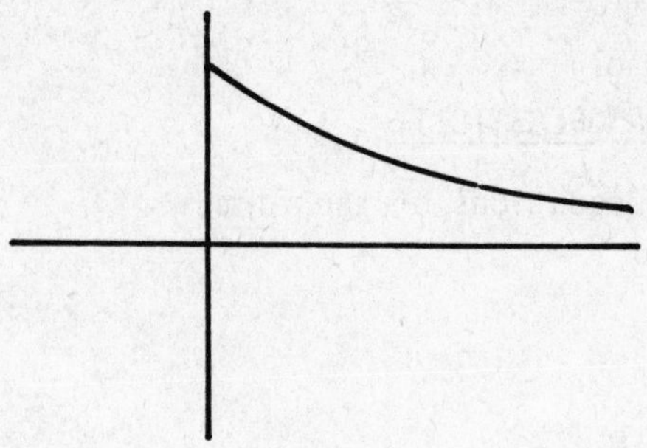

Figure 4.2

Examination of the equilibrium states leads to the extra condition

$$[\ \sigma - k(0)g\ ]' > 0$$

MacCamy [24] has considered the special case where $\sigma = g$. It is easy to analyze the case where

$$\sigma = g \quad k(s) = e^{-s}$$

and the equation is of the form

$$u_{tt} - \sigma(u_x)_x + \int_{-\infty}^{t} e^{-(t-\tau)}\sigma(u_x)_x \, d\tau = 0$$

The two rightmost terms are viewed as a Volterra operator acting on $\sigma(u_x)_x$. This operator may be inverted to produce an equivalent problem of the form

$$u_{tt} - (\sigma(u_x))_x + u_t = 0$$

As previously discussed this equation yields smooth global solutions when the initial conditions are sufficiently smooth and small.

The techniques used can be extended to multi-dimensional problems. Equations where the viscosity is induced by memory have been considered by a number of mathematicians. The reader is referred to [16], [24], [25], [31], [32], [8], [9].

5. ONE DIMENSIONAL THERMOELASTICITY

We consider a system of equations of the form:

(i) $u_t - v_x = 0$

(ii) $v_t - \sigma(u,\theta)_x = 0$

(iii) $[e(u,\theta) + \frac{1}{2} v^2]_t - \sigma(u,\theta)v_{\ x} + q(\theta_x)_x = 0$

The symbols represent the following physical quantities

v = velocity

θ = temperature

u = deformation gradient

σ = stress

e = internal energy

q = heat flux

The heat flux obeys Fourier's law and hence $q(\theta_x) = -k\theta_x$. Clearly the system is part hyperbolic and part parabolic; equations (i) and (ii) are "hyperbolic" and equation (iii) is "parabolic". The third equation is a mathematical formulation of the principle of conservation of energy and it can be shown to be equivalent to

(iii) $\theta\eta(u,\theta)_t + q(\theta_x)_x = 0$

where η denotes entropy and η and θ satisfy

$$\sigma_\theta + \eta_u = 0\ , \quad \sigma_u > 0\ , \quad \eta_\theta > 0$$

In a long recent paper Slemrod [31] uses energy methods to consider the system. It is shown that the dissipation stabilizes the hyperbolic part of the problem and global smooth solutions are obtained provided the initial data are smooth and small. It remains an open problem to show whether thermal dissipation is sufficient to stabilize the system when $n = 2$ or 3. Our conjecture is that it is not.

We can introduce viscosity into our system by replacing equation (ii) with

$$v_t - \sigma(u,\theta)_x = [\mu(u)v_x]_x$$

are replacing equation (iii) by

$$[e(u,\theta) + \tfrac{1}{2} v^2]_t - [(\sigma + \mu(u)v_x)v]_x - [k(u,\theta)\theta_x]_x = 0$$

This system will be discussed in the second part of these notes. It turns out that our new system behaves similarly to the system

$$u_t - v_x = 0$$

$$v_t - \sigma(u)_x = v_{xx}$$

which has been studied by many authors. (cf [2], [14], [15]).

6. ONE DIMENSIONAL NONLINEAR THERMOVISCOELASTICITY

The referential (Lagrangean) form of the conservation laws of mass, momentum and energy for one dimensional materials with reference density $\rho_0 = 1$ is

(i) $u_t - v_x = 0$

(ii) $v_t - \sigma_x = 0$

(iii) $[e + \frac{1}{2} v^2]_t - [\sigma v]_x + q_x = 0$

where u is the deformation gradient (specific volume, the inverse of density), v denotes velocity, σ stands for stress, e is internal energy and q denotes heat flux. The above list of balance laws is supplemented with the second law of thermodynamics which is expressed through the Clausius-Duhem inequality

$$\eta_t + (q/\theta)_x \geq 0$$

where η denotes specific entropy and θ is absolute temperature. It should be noted that η, θ and e may take on only positive values.

We consider linearly viscous materials which have constitutive relations of the form

$$e = \hat{e}(u,\theta)$$

$$\sigma = -\hat{p}(u,\theta) + \hat{\mu}(u)v_x$$

$$q = -\hat{k}(u,\theta)\theta_x$$

where the viscosity $u\hat{\mu}(u)$ is uniformly positive, i.e.

$$u\hat{\mu}(u) \geq \mu_o > 0 \qquad 0 < u < \infty$$

In the special case of an ideal, linearly viscous gas with constant specific heat these relations become

$$e = c\theta$$

$$\sigma = -R\,\theta/u + \mu_o v_x/u$$

$$q = -k_o \theta_x/u$$

We assume that $\hat{e}(u,\theta)$, $\hat{p}(u,\theta)$, $\hat{\mu}(u)$ and $\hat{k}(u,\theta)$ are twice continuously differentiable on $0 < u < \infty$, $0 \le \theta < \infty$. The relationship

$$\hat{e}_u(u,\theta) = -\hat{p}(u,\theta) + \theta\hat{p}_\theta(u,\theta)$$

assures compatibility with the second law of thermodynamics. We require that the elastic part of the stress be compressive at high density and tensile at low density at any temperature, i.e. that there are $0 < \tilde{u} \le \tilde{U} < \infty$ such that

$$\hat{p}(u,\theta) \ge 0 \quad 0 < u < \tilde{u} \quad 0 \le \theta < \infty$$

$$\hat{p}(u,\theta) \le 0 \quad \tilde{U} < u < \infty \quad 0 \le \theta < \infty$$

We assume that for any $0 < \bar{u} < \overline{U} < \infty$ there are positive constants ν, k, and N possibly depending on $\bar{u}$ and/or $\overline{U}$ such that for any $\bar{u} < u < \overline{U}$, $0 \le \theta < \infty$,

$$\hat{e}(u,\theta) \ge 0 \quad \nu \le \hat{e}_\theta(u,\theta) < N(1+\theta^{1/3})$$

$$|\hat{p}_u(u,\theta)| \le N(1+\theta^{4/3}), \quad |\hat{p}_\theta(u,\theta)| < N(1+\theta^{1/3})$$

$$N \ge \hat{k}(u,\theta) \ge k_o > 0, \quad |\hat{k}_u(u,\theta)| \le N, \quad |\hat{k}_{uu}(u,\theta)| \le N$$

A typical problem is to consider a body with reference configuration the interval [0,1] whose boundary is stress free and thermally insulated

$$\sigma(0,t) = \sigma(1,t) = 0 \quad 0 \le t < \infty$$

$$q(0,t) = q(1,t) = 0$$

and determine the thermomechanical process described by the functions $\{u(x,t),\ v(x,t),\ \theta(x,t)\}$, $0 \le x \le 1$, $0 \le t < \infty$ under prescribed initial conditions

$$u(x,0) = u_o(x),\ v(x,0) = v_o(x),\ \theta(x,0) = \theta_o(x) \quad 0 \le x \le 1$$

We shall seek a solution to our system in the space of Hölder continuous functions. $C^\alpha[0,1]$ denotes the Banach space of functions on $[0,1]$ which are uniformly Hölder continuous with exponent α while $C^{\alpha,\alpha/2}(Q_T)$ represents the Banach space of functions on $Q_T = [0,1] \times [0,T]$ which are uniformly continuous with exponents α in x and $\alpha/2$ in t. We have the following existence theorem.

<u>Theorem</u> Consider the initial-boundary value problem for the system under the above assumptions on $\hat{e}$, $\hat{p}$, $\hat{k}$. Assume that $u_o(x)$, $u_o'(x)$, $v_o(x)$, $v_o'(x)$, $v_o''(x)$, $\theta_o(x)$, $\theta_o'(x)$, $\theta_o''(x)$ belong to $C^\alpha[0,1]$. Then there exists a unique admissible solution $\{u(x,t),\ v(x,t),\ \theta(x,t)\}$ on $[0,1] \times [0,\infty)$ such that $u(x,t)$, $u_x(x,t)$, $u_{xt}(x,t)$, $v(x,t)$, $v_x(x,t)$, $v_t(x,t)$, $v_{xx}(x,t)$, $\theta(x,t)$, $\theta_x(x,t)$, $\theta_t(x,t)$, $\theta_{xx}(x,t)$ belong to $C^{\alpha,\alpha/2}(Q_T)$ for any $T > 0$.

Using the constitutive relations to substitute in our original system we obtain,

$$u_t - v_x = 0$$

$$v_t + \hat{p}_x = [\hat{\mu}(u)v_x]_x$$

$$[\hat{e}(u,\theta) + \tfrac{1}{2} v^2]_t + [(\hat{p}(u,\theta) - \hat{\mu}(u)v_x)v]_x - [\hat{k}(u,\theta)\theta_x]_x = 0$$

Our existence theorem is established via the aid of the classical Leray-Schauder fixed point theorem which we record here for easy reference.

Theorem Let B be a Banach space and $P: [0,1] \times B \to B$ be a map with the following properties

(i) For any fixed $\lambda \in [0,1]$, $P(\lambda,\cdot): B \to B$ is completely continuous

(ii) For every bounded subset C of B, the family of maps $P(\cdot,x): [0,1] \to B$, $x \in C$, is uniformly equicontinuous

(iii) There is a bounded subset C of B such that any fixed point in B of $P(\lambda,\cdot)$, $0 \leq \lambda \leq 1$, is contained in C

(iv) $P(0,\cdot)$ has precisely one fixed point in B

Then $P(1,\cdot)$ has at least one fixed point in B.

For our purposes, B will be the Banach space of functions $\{u(x,t), v(x,t), \theta(x,t)\}$ on Q_T with $u, v, v_x, \theta, \theta_x$ in $C^{1/3,1/6}(Q_T)$ with norm $|||(u,v,\theta)|||_B \overset{\text{def}}{=} |||u|||_{C^{1/3,1/6}} + |||v|||_{C^{1/3,1/6}} + |||\theta|||_{C^{1/3,1/6}} + |||v_x|||_{C^{1/3,1/6}} + |||\theta_x|||_{C^{1/3,1/6}}$.

Our existence theorem is proved by the following procedure: Solutions are visualized as fixed points of a map P on the Banach space $C^{1/3,1/6}(Q_T)$ and existence follows from application of the Leray-Schouder theorem. The map P carries $\{U(x,t), V(x,t), \Theta(x,t)\}$ into the solution of a complicated linear parabolic system obtained by linearizing the system about $\{U(x,t), V(x,t), \Theta(x,t)\}$. By virtue of the smoothing action of linear parabolic systems, P is completely continuous and its range is contained in the set of functions $\{u(x,t), v(x,t), \theta(x,t)\}$ with u, u_x, u_t, u_{xt}, v, v_x, v_t, v_{xx}, θ, θ_x, θ_{xx} in $C^{\alpha,\alpha/2}(Q_T)$. What remains to be done in order to complete the list of requirements for the Leray-Schauder fixed point theorem is to show that any possible fixed point of P, i.e. any solution $\{u(x,t), v(x,t), \theta(x,t)\}$ of the system satisfies the admissibility conditions and is contained in an priori bounded set of B. In what follows we shall establish these bounds.

We begin by remarking that if we know that there exists a generic constant Λ so that

$$\max_{[0,T]} \int_0^1 v_{xx}^2(x,t)\, dx \le \Lambda$$

$$\max_{[0,T]} \int_0^1 v_t^2(x,t)\, dx \le \Lambda$$

$$\int_0^T \int_0^1 v_{xt}^2(x,t)\, dxdt \le \Lambda$$

Then v, v_x are bounded in $C^{1/3,1/6}(Q_T)$. This is obtained by standard application of interpolation theory techniques.

We observe that the total energy of the system remains constant, i.e.

$$\int_0^1 [e(x,t) + \tfrac{1}{2} v^2(x,t)]\, dx = E_0$$

Rewriting the third equation, the energy equation, of our system as

$$\hat{e}_\theta \theta_t + \theta \hat{p}_\theta v_x - \mu v_x^2 = [\hat{k}\theta_x]_x$$

we invoke the maximum principle to insure that $\theta(x,t) > 0$ for $t > 0$. The next lemma shows that density remains positive.

Lemma There exist constants $0 < \bar{u} < \overline{U} < \infty$ such that $u < \bar{u}(x,t) < \overline{U}$ for $0 \le x \le 1$ and $0 \le t < \infty$.

Idea of proof The second equation of our system can be rewritten as

$$v_t + \hat{p}_x = \hat{M}(u)_{xt}$$

where

$$\hat{M}(u) \overset{\text{def}}{=} \int_1^u \hat{\mu}(w)\, dw$$

By virtue of the positivity assumption on the viscosity, $\hat{M}(u)$ is a strictly increasing function mapping $(0,\infty)$ onto $(-\infty,\infty)$. Integration of the equation on $[0,x] \times [0,t]$ yields

$$\hat{M}(u(x,t)) = \hat{M}(u(x,0)) + \int_0^t \hat{p}(u(x,\tau),\, \theta(x,\tau))d\tau + \int_0^x v(y,t)dy$$
$$- \int_0^x v(y)dy$$

The result will follow from the above equality and the assumption that $\hat{p} \geq 0$ for u small and $\hat{p} \leq 0$ for u large.

We now observe that $\max\limits_{[0,T]} \int_0^1 \theta(x,t)\, dx \leq \Lambda$ and we proceed to obtain additional estimates.

Lemma

$$\int_0^T \int_0^1 \theta^{-4/3} \theta_x^{\,2}\, dxdt < \Lambda$$

$$\int_0^T \int_0^1 \theta^{8/3}\, dxdt < \Lambda$$

$$\int_0^T \max_{[0,1]} \theta^{5/3}(\cdot,t)\, dt < \Lambda$$

Proof We define

$$\hat{H}(u,\theta) = \int_0^\theta \xi^{-1/3}\, \hat{e}_\theta(u,\xi)\, d\xi$$

and observe that

$$|\hat{H}(u,\theta)| \leq 2N(1 + \theta)$$

After a short calculation we deduce

$$\hat{H}_\theta(u,\theta) = \theta^{-1/3}\,\hat{e}_\theta(u,\theta)$$

$$\hat{H}_u(u,\theta) = \theta^{2/3}\,\hat{p}_\theta(u,\theta) - \hat{G}(u,\theta)$$

where

$$\hat{G}(u,\theta) \overset{\text{def}}{=} 2/3 \int_0^\theta \xi^{-1/3}\,\hat{p}_\theta(u,\xi)\,d\xi$$

Thus setting $H(x,t) = \hat{H}(u(x,t), \theta(x,t))$ and multiplying the energy equation by $\theta^{-1/3}$ we get

$$H_t + \hat{G}(u,\theta)v_x - \theta^{-1/3}\hat{\mu}(u)v_x^{\,2} - \theta^{-1/3}[\hat{k}(u,\theta)\theta_x]_x = 0$$

Integrating over $[0,1] \times [0,T]$ and after an integration by parts with respect to x we obtain

$$\int_0^T \int_0^1 \hat{\mu}(u)\theta^{-1/3}\, v_x^{\,2}\, dxdt + 1/3 \int_0^T \int_0^1 \hat{k}(u,\theta)\theta^{-4/3}\, dxdt$$

$$= \int_0^1 H(x,T)dx - \int_0^1 H(x,0)dx + \int_0^T \int_0^1 \hat{G}(u,\theta)v_x\, dxdt$$

Because $|\hat{G}(u,\theta)| \le 2N(1+\theta)$ we can use the Cauchy-Schwarz inequality together with growth rates to reduce

$$\frac{\mu_o}{2}\,\overline{U}^{\,-1} \int_0^T \int_0^1 \theta^{-1/3}\, v_x^{\,2}\, dxdt + \frac{k_o}{3} \int_0^T \int_0^1 \theta^{-4/3}\theta_x^{\,2}\, dxdt$$

$$\le \Lambda + \epsilon \int_0^T \int_0^1 \theta^{8/3}\, dxdt$$

On the other hand

$$\int_0^T \int_0^1 \theta^{8/3} \, dxdt \le \lambda \int_0^T \max_{[0,1]} \theta^{5/3}(\cdot,t)\, dt$$

$$\le \lambda + \lambda \int_0^T \int_0^1 \theta^{2/3} |\theta_x| \, dxdt$$

$$\le \lambda + \tfrac{1}{2} \int_0^T \int_0^1 \theta^{8/3} \, dxdt + \lambda \int_0^T \int_0^1 \theta^{-4/3} \theta_x^{\,2} \, dxdt$$

The desired result is obtained by combining the last two estimates.

We now employ the bounds obtained thus far to estimate by interpolation the square of various derivatives of the solution in terms of low powers of

$$Y \overset{\text{def}}{=} \max_{[0,T]} \int_0^1 \theta_x^{\,2}(x,t)\, dx$$

$$Z \overset{\text{def}}{=} \max_{[0,T]} \int_0^1 v_{xx}^{\,2}(x,t)\, dx$$

By multiplying the momentum equation by $M(u)_x - v$ and integrating over $[0,1] \times [0,T]$ we obtain

$$\max_{[0,T]} \int_0^1 u_x^{\,2}(x,t)\, dx \le \Lambda + \Lambda\, Y^{1/9}$$

We next show that

$$Y \le \Lambda + \Lambda\, Z^{3/4}$$

$$\int_0^T \int_0^1 \theta_t^{\,2}(x,t) dxdt \le \Lambda + \Lambda\, Z^{3/4}$$

This is done by setting $\hat{Q}(u,\theta) = \int_0^\theta k(u,\xi)\, d\xi$; multiplying the energy equation by Q_t and integrating. We now differentiate the momentum equation with respect to t and multiply by v_t to deduce, after a long estimation,

$$\max_{[0,T]} \int_0^1 v_t^2(x,t)\,dx \le \Lambda + \Lambda Z^{11/12}$$

$$\int_0^T \int_0^1 v_{xt}^2(x,t)\,dxdt \le \Lambda + \Lambda Z^{11/12}$$

Utilizing $v_t + \hat{p}_x = [\hat{\mu} v_x]_x$ and solving for v_{xx} we obtain

$$Z \le \Lambda + \Lambda Z^{11/12}$$

and consequently

$$Z \le \Lambda$$

We differentiate the energy equation formally with respect to t, multiply by $\hat{e}_\theta \theta_t$, integrate over $[0,1] \times [0,t]$, $0 < t \le T$, and integrate by parts. A long sequence of computations produces

$$\max_{[0,T]} \int_0^1 \theta_t^2(x,t)\,dx \le \Lambda$$

$$\int_0^T \int_0^1 \theta_{xt}^2(x,t)\,dxdt \le \Lambda$$

$$\max_{[0,T]} \int_0^1 \theta_{xx}^2(x,t)\,dx \le \Lambda$$

The desired conclusions now follow via interpolation theory arguments.

We conclude with an open problem. The following asymptotic result is conjectured:

$$\theta(x,t) \to \theta \quad \text{as} \quad t \to \infty$$

$$v(x,t) \xrightarrow[wk]{L^2} 0$$

$$u(x,t) \xrightarrow[wk]{L^\infty} \hat{u}(x)$$

Acknowledgment. I wish to thank my colleagues at the University of Houston for their invitation and their warm hospitality. I would also like to thank the editor for assistance in preparation of these notes.

REFERENCES.

1. S. Agmon, Lectures on Elliptic Boundary Value Problems. Van Nostrand, Princeton, NJ, 1965.

2. G. Andrews, On the existence of solutions to the equation $u_{tt} = u_{xxt} + \sigma(u_x)_x$. J. Diff. Eqs. 35 (1980), 200-231.

3. G. Andrews and J.M. Ball, Asymptotic behavior and changes of phase in one-dimensional nonlinear viscoelasticity. J. Diff. Eqs. 44 (1982), 306-341.

4. A.J. Chorin, A numerical model for flame propagation (to appear).

5. C.M. Dafermos, The mixed initial-boundary value problem for the equations of nonlinear one-dimensional viscoelasticity. J. Diff Eqs. 6 (1969), 71-86.

6. C.M. Dafermos and L. Hsiao, Quasilinear integrodifferential conservation laws (to appear).

7. C.M. Dafermos and L. Hsiao, Global smooth thermomechanical processes in one-dimensional nonlinear thermoviscoelasticity. Nonlinear Analysis 6 (1982), 435-454.

8. C.M. Dafermos and J.A. Nohel, Energy methods for nonlinear hyperbolic Volterra integrodifferential equations. Comm. P.D.E. 4 (1979), 219-278.

9. C.M. Dafermos and J.A. Nohel, A nonlinear hyperbolic Volterra equation in viscoelasticity. Am. J. Math. Special issue dedicated to P. Hartman (1981), 87-116.

10. C.M. Dafermos, Global smooth solutions to the initial-boundary value problem for the equations of one-dimensional thermoviscoelasticity, SIAM J. Math. Anal. 13 (1982), 397-408.

11. R.J. Diperna, Singularities of solutions of nonlinear hyperbolic systems of conservation laws. Arch. Rat. Mech. Anal. 60 (1975), 75-100.

12. A. Friedman, Partial Differential Equations of Parabolic Type, Prentice Hall, Englewood Cliffs, NJ, 1964.

13. J. Glimm, Solutions in the large for nonlinear hyperbolic systems of equations. Comm. Pure Appl. Math. 18 (1965), 697-715.

14. J.M. Greenberg, R.C. MacCamy, and V.J. Mizel, On the existence, uniqueness and stability of solutions of the equation $\sigma'(u_x)u_{xx} + \lambda u_{xtx} = \rho_0 u_{tt}$. J. Math. Mech. 17 (1968), 707-728.

15. J.M. Greenberg, On the existence, uniqueness and stability of solutions of the equation $\rho_0 X_{tt} = E(X_x)_x + \lambda X_{xxt}$. J. Math. Anal. Appl. 25 (1969), 575-591.

16. M.E. Gurtin, and A.C. Pipkin, A general theory of heat conduction with finite wave speeds. Arch. Rat. Mech. Anal. 31 (1968), 113-126.

17. A.V. Kazhykhov, Aur la solubilite globale des problemes mono-dimensionnels aux valeurs initiales-limitees pour les equations d'un gaz visqueux et calorifere. C. R. Acad. Sc. Paris 284 (1977), Ser. A, 317-320.

18. A.V. Kazhykhov, and V.V. Shelukhin, Unique global solution with respect to time of initial-boundary value problems for one-dimensional equations of a viscous gas. PMM 41 (1977), 282-291. English translation: Appl. Math. Mech. 41 (1977), 273-282.

19. S. Klainerman, Global existence for nonlinear wave equations, Comm. Pure Appl. Math. 33 (1980), 43-101.

20. W. Kosinski, Gradient catastrophe in the solution of non-conservative hyperbolic systems. J. Math. Anal. Appl. 61 (1977), 672-688.

21. O.A. Ladyzenskaja, V.A. Solonnikov, and N.N. Ural'ceva, Linear and Quasilinear Equations of Parabolic type, (Translated from the Russian by S. Smith), American Math. Society, Providence, RI, 1968.

22. P.D. Lax, Development of singularities of solutions of nonlinear hyperbolic partial differential equations. J. Math. Phys. 5 (1964), 611-613.

23. T.P. Liu, Solutions in the large for equations of non-isentropic gas dynamics. Indiana U. J. 26 (1977), 137-168.

24. R.C. MacCamy, An integrodifferential equation with applications in heat flow, Q. Appl. Math. 35 (1977), 1-19.

25. R.C. MacCamy, A model for one-dimensional, nonlinear viscoelasticity. Q. Appl. Math. 35 (1977), 21-33.

26. A. Matsumura, Global existence and asymptotics of the solutions of the second order quasilinear hyperbolic equations with first order dissipation, Publ. Res. Inst. Math. Sci. Kyoto U., Ser. A, 13 (1977), 349-379.

28. T. Nishida, Global smooth solutions for the second order quasilinear wave equations with the first order dissipation (Unpublished).

29. M.H. Protter, and H.F. Weinberger, Maximum Principles in Differential Equations. Prentice Hall, Englewood Cliffs, NJ, 1967.

30. M. Slemrod, Instability of steady shearing flows in a nonlinear viscoelastic fluid. Arch. Rat. Mech. Anal. 68 (1978), 211-225.

31. M. Slemrod, Global existence, uniqueness and asymptotic stability of classical smooth solutions in one-dimensional non-linear thermoelasticity. Arch. Rat. Mech. Anal. 76 (1981), 97-133.

32. O. Staffans, On a nonlinear hyperbolic Voltarra equation, SIAM J. Math. Anal. 11 (1980), 793-812.

33. C.A. Truesdell, and W. Noll, The Nonlinear Field Theories of Mechanics Handbuch der Physik, Vol. III/3. Springer-Verlag, Berlin 1965.

Constantine M. Dafermos
Brown University

J DOUGLAS, JR

Approximation of convection-dominated diffusion problems

In many diffusion processes arising in physical problems convection essentially dominates diffusion and it is natural to seek numerical methods which reflect this almost hyperbolic nature. We shall consider combining the method of characteristics with finite difference techniques to treat the model problem given by

$$\text{a)}\quad \varphi(x)\partial c/\partial t + u(x)\partial c/\partial x - \partial/\partial x(d(x)\partial c/\partial x) = 0,\ x \in R,\ t > 0 \tag{1.1}$$

$$\text{b)}\qquad c(x,0) = 0$$

We begin by giving a brief description of the method of characteristics. We set

$$\psi(x) = [\varphi(x)^2 + u(x)^2]^{\frac{1}{2}} \tag{1.2}$$

and let the characteristic direction associated with the operator be denoted by $\tau = \tau(x)$, where

$$\frac{\partial}{\partial\tau(x)} = \frac{\varphi(x)}{\psi(x)}\frac{\partial}{\partial t} + \frac{u(x)}{\psi(x)}\frac{\partial}{\partial x} \tag{1.3}$$

Equation (1.1a) can now be put in the form

$$\psi(x)\frac{\partial c}{\partial\tau(x)} - \frac{\partial}{\partial x}(d(x)\ \partial c/\partial x) = 0 \tag{1.4}$$

The characteristic derivative will be approximated in the following manner. Let

$$\overline{x} = x - u(x)\, \Delta t/\varphi(x) \tag{1.5}$$

and observe that

$$\left(\psi \frac{\partial c}{\partial \tau}\right)(x,t) \approx \psi(x) \frac{c(x,t) - c(\overline{x}, t - \Delta t)}{[(x - \overline{x})^2 + (\Delta t)^2]^{\frac{1}{2}}} \tag{1.6}$$

$$= \varphi(x) \frac{c(x,t) - c(\overline{x}, t - \Delta t)}{\Delta t}$$

In what follows we consider a time step $\Delta t > 0$ and approximate the solution at times $t^n = n\Delta t$. We begin by discussing a finite difference method for a uniform grid. We set

$$\text{a)}\quad x_i = ih \tag{1.7}$$

$$\text{b)}\quad z(x_i, t^n) = z_i^n$$

$$\text{c)}\quad \overline{x}_i = x_i - u_i \Delta t/\varphi_i$$

The centered, weighted second difference in x of a grid function z is denoted by

$$\delta_{\overline{x}}\,(a\delta_x z_i) = h^{-2}[a_{i+\frac{1}{2}}(z_{i+1} - z_i) - (a_{i-\frac{1}{2}}(z_i - z_{i-1})] \tag{1.8}$$

where $a_{i+\frac{1}{2}} = a(x_{i+\frac{1}{2}}) = a((i + \frac{1}{2})h)$. Let w_i^n denote the grid values of the appropriate solution. Let $\overline{w}_i^{n-1} = w^{n-1}(\overline{x}_i)$ where $w^{n-1}(x)$ is the extension of the grid values to the function obtained via piecewise linear interpolation. The basic finite difference approximation scheme for (1.1a-b) is given by the equations

$$\text{a)}\quad \varphi_i \frac{w_i^n - \overline{w}_i^{n-1}}{\Delta t} - \delta_{\overline{x}}\,(a\delta_x w^n)_i = 0\ , \quad -\infty < i < \infty,\ n \ge 1\ . \tag{1.8}$$

b) $w_i^0 = c_0(x_i)$, $-\infty < i < \infty$

It is clear that there exists a unique solution to (1.8).

We let $\zeta(x,t) = c(x,t) - w(x,t)$ denote the difference between solutions to (1.1) and approximate solutions provided by (1.8). Thus ζ_i^n denotes the error at (x_i, t^n). The following theorem which is obtained via application of a maximum principle argument insures convergence and provides a error bound.

<u>Theorem 1.</u> Let the solution of (1.1) belong to $W^{2,\infty}(R \times [0,T]) \cap L^\infty(0,T; W^{3,\infty}(R))$ and let the approximate solution w_i^n be defined by the finite difference (1.8) where $\overline{w}_i^{n-1} = w^{n-1}(\overline{x}_i) = w^{n-1}(x_i - u_i\Delta t/\varphi_i) = (I_1\{w_j^{n-1}\})(\overline{x}_i)$, the piecewise - linear interpolant of $\{w_j^{n-1}\}$. Then, the error $\zeta = c - w$ satisfies

$$\max|\zeta_i^n| \le \text{const.} * \{ \sup_{R\times[0,T]} \left|\frac{\partial^2 c}{\partial\tau(x)}\right| \Delta t + \tag{1.9}$$

$$||c||_{L^\infty(W^{2,\infty})} \min(h, h^2/\Delta t) +$$

$$Q(d)\ ||c||_{L^\infty(W^{\infty,3})} h\}$$

where $Q(d) \le \text{const.} * ||d||_{W^{2,\infty}}$.

It is not difficult to generalize this method to time dependent, variable - spacing grids. We shall indicate this flexibility. As before let $t^n = n\Delta t$ and let

$$\delta^n = \{\ldots, x_{-2}^n, x_{-1}^n, x_0^n, x_1^n, x_2^n, \ldots\},\ x_i^n - x_{i-1}^n = h_i^n > 0\ , \tag{1.10}$$

$$x_j^n \to \pm\infty \text{ as } j \to \pm\infty$$

be the grid employed at time t^n.

Let the approximation to $(dc_x)_x$ be given by $(c_i^n = c(x_i^n, t)$, $d_{i+\frac{1}{2}}^n = d((x_i^n + x_{i+1}^n)/2)$

$$\delta_{\bar{x}}(d\ \delta_x c)_i^n = \frac{2}{h_i^n + h_{i+1}^n} [d_{1+\frac{1}{2}}^n \frac{c_{i+1}^n - c_i^n}{h_{i+1}^n} - d_{i-\frac{1}{2}}^n \frac{c_i^n - c_{i-1}^n}{h_i^n} \tag{1.10}$$

$$= (dc_x)_x(x_i^n, t^n) + O(||u^n||_{W^{3,\infty}([x_{i-1}^n, x_{i+1}^n]} (h_i^n + h_{i+1}^n))$$

with the constant depending on $d(x)$ but not any properties of δ^n. Again we interpret w^{n-1} as $I_1\{w_j^{n-1}\}$, the piecewise - linear interpolant of $\{w_j^{n-1}\}$, now over the grid δ^{n-1}. Let

$$\bar{x}_1^n = x_i^n - u(x_i^n)\Delta t/\varphi(x_i^n) \tag{1.11}$$

the natural generalization of equation (1.8) is seen to be

$$\varphi(x_i^n) \frac{w_i^n - w^{n-1}(\bar{x}_i^n)}{\Delta t} - \delta_{\bar{x}} (d\ \delta_x\ w)_i^n = 0 \quad n \geq 1 , \quad \forall_i \tag{1.12}$$

$$w_i^0 = c_0(x_i^0) \quad \forall_i$$

If $\bar{x}_i^n \in [x_{j-1}^{n-1}, x_j^{n-1}] \doteq J_i^n$ then it can be shown that

$$\max_i |\zeta_i^n| \leq \max_1 |\zeta_i^{n-1}| + K[\left|\left|\frac{\partial^2 u}{\partial \tau^2}\right|\right|_{L^\infty(R\times[t^{n-1}, t^n]} \Delta t \tag{1.13}$$

$$+ \sup_i ||u^{n-1}||_{W^{2,\infty}(J_i^n)} \min(|J_i^n|, |J_i^n|^2/\Delta t)$$

$$+ \sup_i ||u^n||_{W^{3,\infty}([x_{i-1}^n, x_{i+1}^n]} (h_i^n + h_{i+1}^n))\Delta t$$

and consequently that

$$\max_{0\le t^n\le T}\ \max_i|\zeta_i^n| \le K\Big[\Big\|\frac{\partial^2 u}{\partial\tau^2}\Big\|_{L^\infty(R\times[0,T]}\Delta t \tag{1.14}$$

$$+ \sum_{n=0}^{[T/\Delta t]-1} \sup_i\|u^{n-1}\|_{W^{2,\infty}(J_i^n)} \min(|J_i^n|,\ |J_i^n|^2/\Delta t)\cdot\Delta t$$

$$+ \sum_{n=1}^{[T/\Delta t]} \sup_i\|u^n\|_{W^{3,\infty}([x_{i-1}^n,\ x_{i+1}^n]}(h_i^n + h_{i+1}^n)\Delta t\Big]$$

No relation was assumed between the grids of successive time steps in the derivation of the error bound (1.14). Clearly a reasonably optimal error bound is achieved by selecting δ^n to have close spacing where the solution has large higher derivatives in space and sparse spacing where it is nearly linear in x. Practically, a selection of δ^n, given δ^{n-1} and $\{w_i^{n-1}\}$ can be based on the following concepts:

i) Let the initial selection of δ^n be given by

$$x_i^n = x_i^{n-1} + u(x_i^{n-1})\Delta t/\varphi(x_i^{n-1})\ . \tag{1.15}$$

Reorder the grid points where necessary and maintain some small minimum spacing.

ii) Compute $\{w_i^n\}$ using the current selection of δ^n and then use finite differences to approximate the second and third derivations of the solution. Use these approximations to check that the second and third error terms in (1.13) are sufficiently small. If not, refine the grid where required. Coarsen the grid where it can be done. If refinement was required, repeat this step.

At points where rule (1.15) is retained the error term involving $|J_i^n|$ and $|J_i^n|^2/\Delta t$ can be modified. We observe that

$$\bar{x}_i^{n-1} - x_i^n = O((\Delta t)^2) \tag{1.16}$$

and thus the interpolation error at $\bar{x}_i^n$ can be virtually eliminated. This has the consequence that the second term on the right hand side of (1.14) can be replaced by the expression

$$\sup\{||u^{n-1}||_{W^{2,\infty}(J_i^n)} |J_i^n|\Delta t : \ i \ \text{ such that} \tag{1.17}$$

$$x_i^n = x_{i-1}^n + u(x_i^{n-1})\ \Delta t/\varphi(x_i^{n-1})\}$$

$$+ \sup\{||u^{n-1}||_{W^{2,\infty}(J_i^n)} \ \min(|J_i^n|, \ |J_i^n|^2/\Delta t) :$$

$$x_i^n \neq x_i^{n-1} + u(x_i^{n-1})\Delta t/\varphi(x_i^{n-1})\}$$

This observation is important because (1.17) measures the error in the interpolation arising from the transport term while the third term in (1.14) measures the error in the diffusion term and will have a smaller multiplicative constant when transport seriously dominates diffusion.

Jim Douglas, Jr.
University of Chicago

J A GOLDSTEIN*

Remarks on the Feynman–Kac formula

1. INTRODUCTION

The Feynman-Kac formula is an integral representation formula for the solution of the problem

$$\alpha \partial u/ t = \frac{-\Delta}{2m} u + V(x)u \quad (x \in \mathbb{R}^{\ell}, t > 0), \tag{1}$$

$$u(0,x) = f(x) \quad (x \in \mathbb{R}^{\ell}). \tag{2}$$

When $\alpha = i$ [resp. $\alpha = -1$] this is the Schrödinger [resp. heat] equation with a potential. The formula was discovered formally in the Schrödinger case by Feynman in 1948 in a deservedly famous paper [5]. For its standard interpretation by physicists, see Feynman-Hibbs [6]. In 1951 Kac [13] found and interpreted the formula in the context of the heat equation; in this case the integral formula involves a Wiener integral.

In the above discussion the mass parameter m was assumed to be positive. Taking $\alpha = -1$ and $m = m_0 + im_1$ with $m_0 > 0$, $m_1 < 0$, (1) is still parabolic; but letting $m_0 \to 0$ we get the Schrödinger equation in the limit. Thus one can think of the Schrödinger equation as the limiting boundary value in the analytic mass parameter of the heat equation.

The first rigorous interpretation of the Feynman integral formula was given by Nelson in a beautiful paper [15] in 1964. He used the Trotter product formula [19]; he also discussed the analytic continuation in the mass parameter. Later, Chernoff [2] found a new product formula which contained the Trotter formula as a simple special case.

*Partially supported by an NSF grant

We shall survey this whole connected group of ideas in this paper. We shall also give a generalization of the Chernoff formula and show how it gives an especially simple approximation of the Wiener integral formula for the heat equation. This result does not apply directly to the Schrödinger equation, but it has an indirect connection because of the analytic continuation in the mass parameter circle of ideas.

This paper is partly an expository account of our paper [11], but we have included some background in an attempt to make it a self-contained survey. Additionally, we present some new results.

2. THE CHERNOFF FORMULA

Let X be a Banach space. By a <u>contraction</u> on X we mean a linear operator on X of norm at most one.

<u>Lemma [2]. Let L be a contraction on X. Then for all $f \in X$ and $n = 1,2,3,\ldots$,</u>

$$||L^n f - e^{n(L-I)} f|| \leq \sqrt{n}\,||Lf - f|| .$$

<u>Proof.</u> $$||e^{n(L-I)}f - L^n f|| = e^{-n} || \sum_{m=0}^{\infty} (m!)^{-1} n^m (L^m f - L^n f)||$$

$$\leq e^{-n} \sum_{m=0}^{\infty} (n^m/m!)\,||L^{|m-n|} f - f||$$

$$\leq e^{-n} \sum_{m=0}^{\infty} (n^m/m!)^{1/2} [(n^m/m!)(m-n)^2]^{1/2} ||Lf - f|| \qquad (3)$$

$$\leq e^{-n} (\sum_{m=0}^{\infty} n^m/m!)^{1/2} (\sum_{m=0}^{\infty} (n^m/m!)(m^2 - 2mn + n^2))^{1/2} ||Lf - f|| \quad (4)$$

$$= e^{-n}(e^n)^{1/2}(ne)^{1/2} ||Lf - f|| = \sqrt{n}\,||Lf - f|| .$$

(3) holds because $||L^j f - f|| = ||\sum_{k=0}^{j-1} (L^{k+1} f - L^k f)|| \leq j\,||Lf - f||$ (telescoping series). (4) holds by the Schwarz inequality. The equality

following (4) holds by an elementary calculation involving infinite series; alternatively, it follows from the formulas for the first two moments of the Poisson distribution. □

Chernoff's formula [2]. *Let* $\{F(t) : t \geq 0\}$ *be a family of contractions on* X *such that* $F(0) = I$ *and for* f *in some dense set* $D \subset X$,

$$F'(0)f = \lim_{h \to 0} h^{-1}(F(h)f - f) = Af$$

where A, *the closure of its restriction to* D, *generates a strongly continuous (or* (C_0)*) semigroup* $T = \{T(t) : t \geq 0\}$ *of contractions on* X. *Then for each* $f \in X$,

$$\lim_{n \to \infty} F(t/n)^n f = T(t)f$$

holds uniformly for t *in bounded intervals of* $[0,\infty)$.

Proof. We will not bother proving the uniformity assertion since we do not use it in the sequel. Let $f \in D$, $t > 0$, $L = F(t/n)$. Then

$$F(t/n)^n f - T(t)f = [L^n f - e^{n(L-I)}f] + [e^{n(L-I)}f - T(t)f].$$

The first bracketed term is bounded in norm by

$$\sqrt{n}||Lf - f|| = \sqrt{n}\,\frac{t}{n}\left|\left|\frac{F(t/n)f - f}{t/n}\right|\right| \to 0 \cdot ||Af|| = 0$$

as $n \to \infty$ by the Lemma. The second bracketed term is

$$\exp\{t[\frac{F(t/n) - I}{t/n}]\}f - T(t)f$$

which tends to zero as $n \to \infty$ by the semigroup approximation theorem (cf. e.g. Kato [14]) because the generators converge, by hypothesis. The theorem follows. □

Trotter product formula [19]. *Let* A_1, A_2, *and* $A_3 = \overline{A_1 + A_2}$ *generate* (C_0) *contraction semigroups* $T_j = \{T_j(t) : t \geq 0\}$ *on* X, $j = 1,2,3$. *Then for each* $f \in X$,

$$T_3(t)f = \lim_{n\to\infty}(T_1(t/n)T_2(t/n))^n f ,$$

uniformly for t *in bounded intervals of* $[0,\infty)$.

Proof. Apply the Chernoff formula with

$$F(t) = T_1(t)T_2(t). \quad \square$$

Remarks. (i) The Trotter formula (in the finite dimensional case) goes back to the work of Lie in the late nineteenth century.

(ii) In the Chernoff formula the hypothesis that $||F(t)|| \leq 1$ can be weakened to $||F(t)|| \leq e^{\omega t}$ for small t by working with $e^{-\omega t}F(t) = \tilde{F}(t)$ instead of $F(t)$.

3. THE FEYNMAN-KAC FORMULA

Following Nelson [15] (now and throughout this section) we apply the Trotter product formula to (1), (2). Rewrite (1) as

$$du/dt = A_1 u + A_2 u$$

where

$$A_1 = \frac{\beta\Delta}{2m}, \; A_2 = -\beta V(x)$$

and $\beta = -\alpha^{-1}$. Thus $\beta = 1$ for the heat equation and $\beta = i$ for the Schrödinger equation. When $\beta = 1$ assume $\mathrm{Re}\, V(x) \geq 0$ for all x and $X = L^p(\mathbb{R}^\ell)$ for $1 \leq p < \infty$ or $BUC(\mathbb{R}^\ell) = \{g \in L^\infty(\mathbb{R}^\ell) : g$ is uniformly continuous$\}$. When $\beta = i$ assume $\mathrm{Im}\, V(x) \leq 0$ and $X = L^2(\mathbb{R}^\ell)$. A_2 is interpreted as a multiplication operator on X. Then A_1, A_2, and $A_3 = A_1 + A_2$ all generate (C_0) contraction semigroups on X if $V = V_1 + V_2$ where $V_1 \in L^\infty(\mathbb{R}^\ell)$, $V_2 \in L^q(\mathbb{R}^\ell)$ where $q \geq 2$, $q > \ell/2$, and V is uniformly continuous if $X = BUC(\mathbb{R}^\ell)$. (For more general conditions see [17].) The semigroups corresponding to A_1, A_2 are given by

$$T_1(t)f(x) = (2\pi t\beta/m)^{-\ell/2} \int_{R^\ell} \exp[-\beta m|x-y|^2/2t] f(y)dy,$$

$$T_2(t)f(x) = \exp(-\beta tV(x))f(x)$$

for $f \in X$ and $x \in \mathbb{R}^\ell$. Induction yields

$$[T_1(t/n)T_2(t/n)]^n f(x_0) = c_n \underbrace{\int_{\mathbb{R}^\ell} \cdots \int_{\mathbb{R}^\ell}}_{n} \exp\{-\beta S(x_0,\ldots,x_n;t\} f(x_n) \cdot \cdot dx_1 \ldots dx_n \tag{5}$$

where

$$S(x_0,\ldots,x_n;t) = \sum_{j=1}^{n} \left\{ \frac{m}{2} \frac{|x_j - x_{j-1}|^2}{(t/n)^2} - V(x_j) \right\} \left(\frac{t}{n}\right) \tag{6}$$

is a Riemann sum for the classical action integral

$$S(\omega;t) = \int_0^t [\tfrac{m}{2}|\omega'(s)|^2 - V(\omega(s))]ds$$

for a particle of mass m travelling along a path ω in $\mathbb{R}^\ell$ under the influence of the potential V, where $\omega(jt/n) = x_j$ for $j = 0,1,\ldots,n$. Passing to the limit we formally obtain

$$T_3(t)f(x) = c_\infty \int_{\omega \in \Omega_x} \exp\{-\beta S(\omega;t)\} f(\omega(t))\mu(d\omega) \tag{7}$$

where $\mu(d\omega) = \prod_{0 \leq s \leq t} dx_s = \lim_{n\to\infty} dx_1 \ldots dx_n$, $c_\infty = \lim_{n\to\infty} c_n$, and Ω_x is the space of paths starting at x, i.e. $\omega \in \Omega_x$ iff $\omega : [0,\infty) \to \mathbb{R}^\ell$ and

$\omega(0) = x$.

For $\beta = i$, the right hand side of (7) is the celebrated and mysterious Feynman integral representation formula for the solution u of the Schrödinger equation with potential V and initial data f. It is mysterious for at least three reasons. (i) $|c_n| \to \infty$ as $n \to \infty$. (ii) μ is not a well-defined measure on Ω_x in the conventional sense. (iii) if μ existed as a Weiner-like measure, then one would expect that the paths ω which were somewhere differentiable form a set of μ-measure zero; but in this case we would have $S(\omega;t) = \infty$ a.e. Nevertheless, we can interpret (7) as a rigorous formula for the solution (or wave function) u in the following manner. Interpret the integral in (7) as a limit as $n \to \infty$ of the right hand side of (5) where S is defined as in (6). Then, by the Trotter product formula, the right (= left) side of (5) converges in $X = L^2(\mathbb{R}^\ell)$ to $T_3(t)f(x) = u(t,x)$, the unique solution of (1) (with $\alpha = i$) and (2).

If we take $\beta = 1$ rather than i, thereby replacing the Schrödinger equation by the heat equation, then we can interpret the solution u as a Wiener integral. This was done by Kac [13] in 1951. Here is a brief explanation.

For D, $t > 0$, x, $y \in \mathbb{R}^\ell$ define

$$P_t(x,y) = (4\pi Dt)^{-\ell/2} \exp(-|x-y|^2/4Dt);$$

as a function of $x-y$, this is the probability density function of the normal distribution with mean vector zero and covariance matrix $2Dt$ times the identity. Regard $P_t(x,y)$ as the density function for a probability measure $P_t(x,\Gamma)$ defined by $P_t(x,y)dy = P_t(x,dy)$. Let $\dot{\mathbb{R}}^\ell$ be the one-point compactification of $\mathbb{R}^\ell$ and let $\Omega = \prod_{0 \le t < \infty} \dot{\mathbb{R}}^\ell$ consist of all functions from $[0,\infty)$ to $\dot{\mathbb{R}}^\ell$. Ω is a compact Hausdorff space in its product topology.

Given n, $0 \le t_1 < t_2 < \ldots < t_n$, and $\Phi \in C((\dot{\mathbb{R}}^\ell)^n, \mathbb{R})$, define a corresponding continuous function ϕ on Ω to $\mathbb{R}$ by

$$\phi(\omega) = \Phi(\omega(t_1),\ldots, \omega(t_n)).$$

let $C_0(\Omega)$ be the set of such ϕ defined by finitely many parameters. For $x \in \mathbb{R}^\ell$ and ϕ as above, define

$$Q_x(\phi) = \int_{\mathbb{R}^\ell}\cdots\int_{\mathbb{R}^\ell}\Phi(x_1,\ldots,x_n)P_{t_1}(x,dx_1)P_{t_2-t_1}(x_1,dx_2)\ldots$$

$$P_{t_n-t_{n-1}}(x_{n-1},dx_n).$$

Then Q_x is a positive linear functional on $C_0(\Omega)$ such that $Q_x(1) = 1$. Moreover, $C_0(\Omega)$ is dense in $C(\Omega)$ by the Stone-Weierstrass theorem. Thus by the Riesz representation theorem there is a probability measure W_x defined on the Borel sets of Ω such that

$$Q_x(\phi) = \int_\Omega \phi(\omega)W_x(d\omega)$$

holds for all $\phi \in C_0(\Omega)$. Moreover, the support of W_x is contained in $\Omega_x = \{\omega \in \Omega : \omega \text{ is continuous, } \omega \text{ is } \mathbb{R}^\ell\text{-valued, and } \omega(0) = x\}$. The W_x's are connected via

$$W_x(\Gamma) = W_0(\Gamma + x) = W_0(\{\omega \in \Omega : \omega(t) = \nu(t) + x \text{ for each } t, \text{ where } \nu \in \Gamma\}).$$

<u>Wiener measure</u> refers to the collection $W = \{W_x : x \in \mathbb{R}^\ell\}$ of measures on Ω (or sometimes to W_0 which determines W). W_x is called "Wiener measure starting at x".

Setting $D = 1/2m$ and $\beta = 1$, formula (5) can be rewritten as

$$(T_1(\tfrac{t}{n})T_2(\tfrac{t}{n}))^n f(x) = \underbrace{\int_{\mathbb{R}^\ell} \cdots \int_{\mathbb{R}^\ell}}_{n} P_{t/n}(x,dx_1)P_{t/n}(x_1,dx_2) \cdots$$

$$\cdots P_{t/n}(x_{n-1},dx_n)\exp[-\sum_{j=1}^{n} V(x_j)(\tfrac{t}{n})]f(x_n)$$

$$\to \int_{\Omega_x} \exp\left\{-\int_0^t V(\omega(s))ds\right\} f(\omega(t))W_x(d\omega) \tag{8}$$

as $n \to \omega$, and this is the Kac integral representation formula for the solution of the heat equation (1), (2) (with $\alpha = -1$). In case V is real-valued, (8) holds if V is continuous on $\mathbb{R}^\ell \backslash N$ where N is a closed set of (Newtonian) capacity zero [15].

Now consider the initial value problem

$$i\ \partial u/\partial t = \frac{-\Delta}{2m} u + V(x)u\ ,\ u(0,x) = f(x) \tag{9}$$

where $V \in \mathbb{C}(\mathbb{R}^\ell \backslash N, \mathbb{R})$ and $N = \overline{N}$ has capacity zero. Let $m = m_0 + im_1$ where $m_0 > 0$, $m_1 < 0$. Then (9) is a well-posed parabolic problem, and the unique solution of it is given by

$$u_m(t,x) = \lim_{n\to\infty}(T_{1,m}(t/n)T_2(t/n))^n\ f(x), \tag{10}$$

the limit being in L^p norm $(1 \le p \le \infty)$ provided $m_1 > 0$. Here T_1, T_2 are as before, but the dependence of T_1 on m has been explicitly indicated notationally. The function

$$m \longmapsto u_m(t,\cdot)$$

from $\{m \in \mathbb{C} : \text{Re } m, -\text{Im } m > 0\}$ to $L^2(\mathbb{R}^\ell)$ is an analytic function. It has radial boundary values on the positive real axis. That is, if $m_0 > 0$,

$$\lim_{m_1\to 0^-} ||u_m(t,\cdot) - u_{m_0}(t,\cdot)|| = 0$$

for each $t \geq 0$ and u_{m_0} satisfies the Schrödinger problem (1), (2), for $t \geq 0$ with $m = m_0$ and $\alpha = i$. This follows by applying a suitable version of the Fatou-Privalov theorem. Nelson [15] obtained this result for almost all positive values of m_0. It seems likely that it holds for all positive values of m_0 (cf. Jørgensen [12]).

If in (1) m is nonreal and $\alpha = -1$, the solution given by the Trotter formula is no longer given by a Wiener integral (cf. Cameron [1]). But clearly this whole circle of ideas (Feynman formula, Kac formula, Wiener integral, Feynman integral, Chernoff formula, Trotter formula) forms a closely related family.

4. A GENERALIZED CHERNOFF FORMULA

Theorem [11]. *Let* $\{F(t) : t \geq 0\}$ *be a family of contractions on* X *such that* $F(0) = I$ *and for some dense set* $D \subset X$, *there is a positive integer* k *such that for* $f \in D$, $F^{(j)}(0)f = 0$ *for* $1 \leq j < k$ *and* $F^{(k)}(0)f = k!\,Af$ *where* A, *the closure of its restriction to* D, *generates a* (C_0) *contraction semigroup* T. *Then for each* $f \in X$,

$$\lim_{n\to\infty} [F(t/n)]^{n^k} f = T(t^k)f$$

holds uniformly for t *in bounded intervals of* $[0,\infty)$.

Proof. Note that this reduces to Chernoff's formula when $k = 1$. For $f \in D$ and $t > 0$ small, Taylor's formula gives

$$F(t)f = \sum_{j=0}^{k} t^j F^{(j)}(0)f/j! + o(t^k)$$

$$= f + t^k Af + o(t^k)$$

by hypothesis. Let $s = t^k$ and $G(s) = F(t)$. Then

$$G(s)f = f + s\,Af + o(s)\ ,$$

whence by the (ungeneralized) Chernoff formula,

$$\lim_{n\to\infty} G(s/m)^m f = T(s)f$$

holds with the required uniformity. Taking $m = n^k$ and replacing s by t^k gives the advertised result. □

We shall apply this result with $k = 2$ in the context of cosine functions.

When the initial value problem

$$d^2u/dt^2 = Au\ ,\ u(0) = f_1,\ u'(0) = f_2 \tag{10}$$

is well-posed for $u : \mathbb{R} \to X$, then A generates a cosine function $C = \{C(t) : t \in \mathbb{R}\}$ and the unique solution of (1) is given by

$$u(t) = C(t)f_1 + \int_0^t C(s)f_2\,ds.$$

If B generates a group $\{T(t) : t \in \mathbb{R}\}$, then $A = B^2$ generates a cosine function given by $C(t) = (T(t) + T(-t))/2$, and the above formula for u becomes an abstract version of d'Alembert's formula for the solution of the one-dimensional wave equation.

We are interested in the following class of generators:

$$\mathbb{G} = \{A : A \text{ generates a contraction cosine function and a } (C_0) \text{ contraction semigroup}\}.$$

<u>Cosine function product formula [11]. Let</u> $A_1,\ldots,A_M \in \mathbb{G}$ <u>and for some dense set</u> $E \subset X$ <u>suppose that the closure</u> A <u>of</u> $(A_1 +\ldots+ A_M)/2$ <u>on</u> E <u>generates a</u> (C_0) <u>contraction semigroup</u> T. <u>Let</u> C_j <u>be the cosine function generated by</u> A_j. <u>Then for each</u> $f \in X$,

$$\lim_{n\to\infty} \{C_1(\tfrac{t}{n}) \ldots C_M(\tfrac{t}{n})\}^{n^2} = T(t^2)f ,$$

uniformly for t in bounded intervals.

Proof. This follows from the generalized Chernoff formula by taking $k = 2$, $D = E$, $F(t) = C_1(t) \ldots C_M(t)$, and noting that for $f \in E$ we have $C_j(0)f = f$, $C_j'(0)f = 0$, $C_j''(0)f = A_jf$. □

Let $X = L^p(\mathbb{R}^\ell)$, $1 \le p < \infty$ or $BUC(\mathbb{R}^\ell)$. Let $b_j \in C^1(\mathbb{R}, \mathbb{R})$ with $\max_{j,k}\{||b_j||_\infty, ||\partial b_j/\partial x_k||_\infty\} < \infty$, and let $a_j \in \mathbb{R}$. Then B_j defined by

$$B_j = a_j\partial/\partial x_j + ib_j(x)$$

generates a (C_0) group of isometries on X, and $A_j = B_j^2$ belongs to $\mathfrak{C}$. The group [resp. cosine function] generated by B_j [resp. A_j] is given by

$$T_j(t)f(x) = \exp\{i\int_0^t b_j(x + a_je_js)ds\}f(x + a_je_jt)$$

$$[\text{resp. } C_j(t)f(x) = \tfrac{1}{2}(T_j(t)f(x) + T_j(-t)f(x))].$$

Here e_j is the jth vector in the usual orthonormal basis for $\mathbb{R}^\ell$; e_j gives the direction in which $\partial/\partial x_j$ is computed.

Let $q \in C(\mathbb{R}^\ell, \mathbb{R})$ be nonpositive and let $B_{\ell+1}$ be the operator of multiplication by $i(-q(x))^{1/2}$. (The continuity of q is not essential if $X = L^p$ with $p < \infty$.) Set $M = \ell + 1$. Then $A_M = B_M^2 \in \mathfrak{C}$ and A_M is multiplication by $q(x)$. Moreover, the closure A of $(A_1 + \ldots + A_M)/2$ on the smooth functions with compact support is given by

$$A = \frac{1}{2}\sum_{j=1}^{\ell}(a_j\frac{\partial}{\partial x_j} + ib_j(x))^2 + \frac{1}{2}q(x)$$

(or $A = \frac{1}{2m}\Delta + \frac{1}{2}q(x)$ if $a_j = \frac{1}{m}$, $b_j = 0$ for all j) and generates a (C_0) contraction semigroup T on X. Setting $V = -\frac{1}{2}q(\ge 0)$, the cosine function product formula implies that

$$u(t,x) = \lim_{n\to\infty}\{ \prod_{j=1}^{\ell} (\tfrac{1}{2}(T_j(\tfrac{\sqrt{t}}{n}) + T_j(\tfrac{-\sqrt{t}}{n})) \cos(2tV)\}^{n^2} f(x) \qquad (11)$$

is the unique solution of

$$\frac{\partial u}{\partial t} = \frac{1}{2} \sum_{j=1}^{\ell} (a_j \frac{\partial}{\partial x_j} + ib_j(x))^2 u - V(x)u ,$$

$$u(0,x) = f(x).$$

When each $b_j = 0$, formula (11) involves only multiplication and translation operators; no integral operators as in (5) occur in (11). In particular, (11) is a particularly simple approximation formula for the Wiener integral formula (8).

5. FURTHER RESULTS AND REMARKS

The $k = 2$ case of the generalized Chernoff formula has connections with the Lie theory of skew-adjoint operators (cf. [7], [16], [4]). By this we mean the following. Let U consists of all unitary operators on an infinite dimensional complex Hilbert space X. When U is equipped with the strong operator topology it is not a Lie group, but, for the moment, act as if it were. Its Lie algebra $\mathfrak{S}$, which consists of all generators of one-parameter sub-groups of U, is, by Stone's theorem, all skew-adjoint operators on X. The (Lie algebraic) operations of addition and commutation are only partially defined on $\mathfrak{S}$. For information concerning these partially defined gadgets, see [7], [16], [3], [8], [4], [10]. For another application of the Chernoff formula see [9].

We next give an axample in which the generalized Chernoff formula is applied with a value of k greater than two, namely, with $k = 4$. Let A_1, A_2 be bounded operators on the Banach space X. Note that for

$j = 1,2$, the group T_j and cosine function C_j generated by A_j satisfy

$$||T_j(t)|| \le e^{\omega_j|t|}, \quad ||C_j(t)|| \le e^{\omega_j|t|}$$

where $\omega_j = ||A_j||$. In what follows we shall implicitly use Remark (ii) at the end of Section 2 and apply the generalized Chernoff formula to an F satisfying $||F(t)|| \le e^{\omega t}$ for some $\omega \ge 0$ and small positive t.

Let

$$D(t) = C_1(t)C_2(t),$$

$$F(t) = T_1(-t^2/2)D(t)T_2(-t^2/2).$$

Using $C_j'(0) = 0 = C_j'''(0)$, $C_j''(0) = A_j$, $C_j'''' = A_j^2$, straightforward calculations give $F(0) = I$, $F'(0) = F''(0) = F'''(0) = 0$ and

$$F''''(0) = -D''''(0) = (4!)B$$

where

$$B = -(A_1^2 + A_2^2 + 2A_1A_2)/24.$$

Hence

$$\lim_{n\to\infty}\{T_1(\tfrac{-t}{2n})C_1(\tfrac{t}{n})C_2(\tfrac{t}{n})T_2(\tfrac{-t}{2n})\}^{n^4} = \exp\{-t^4B\},$$

uniformly for t in bounded intervals. (We remark that the generalized Chernoff formula holds with the limit in the uniform generator topology when each of the generators involved is bounded, so that $F(\cdot)$ is continuous and differentiable in this topology.)

Next we generalize a recent theorem of Shaw [18].

We now recall the hypothesis of the generalized Chernoff formula and call it

(Hyp k) $\{F = F(t) : t \geq 0\}$ <u>is a family of contractions on</u> X <u>such that</u> $F(0) = I$ <u>and for some dense set</u> $D \subset X$ <u>there is a positive integer</u> k <u>such that for</u> $f \in D$ <u>there is an</u> $\varepsilon_f > 0$ <u>such that</u> $F(\cdot)f \in C^k([0,\varepsilon_f),X)$, $F^{(j)}(0)f = 0$ <u>for</u> $1 \leq j < k$ <u>and</u> $F^{(k)}(0)f = k!\ Af$ <u>where</u> A, <u>the closure of</u> $A|_D$, <u>generates a</u> (C_0) <u>contraction semigroup</u> T.

<u>Proposition.</u> <u>If</u> F <u>satisfies</u> (Hyp k), <u>then so does</u> $G = \{G(t) : t \geq 0\}$ where

$$G(t) = [2I - F(t)]^{-1}.$$

<u>Proof.</u> Shaw [18] established this result for $k = 1$ and $F(t) = (I - tA)^{-1}$.

First, $G(t)$ is a contraction since for $||L|| \leq 1$,

$$||(2I - L)^{-1}|| = \frac{1}{2}\,||(I - L/2)^{-1}||$$

$$= 2^{-1}||\sum_{n=0}^{\infty}(L/2)^n|| \leq 2^{-1}\sum_{n=0}^{\infty}||L^n||/2^n \leq 1.$$

Next, $G(0) = 2I - F(0) = I$, and for $f \in D$,

$$G'(t)f = G(t)F'(t)G(t)f,$$

$$G'(0)f = F'(0)f$$

which proves the result if $k = 1$. Next, if $k \geq 2$, $F'(0) = G'(0) = 0$ on D, and so

$$G''(t)f = G'(t)F'(t)G(t)f + G(t)F''(t)G(t)f + G(t)F'(t)G'(t)f,$$

$$G''(0)f = F''(0)f.$$

In general the proof is by induction. The above argument shows

$$G^{(k)}(t)f = \Sigma G^{(a)}(t)F^{(b)}(t)G^{(c)}(t)f$$

where the integers a, b, c in the summation satisfy $b \geq 1$, $0 \leq a \leq k-1$, $0 \leq c \leq k-1$, and $a + b + c = k$. The full result follows easily from this.□

In the context of the cosine function product formula we can deduce, as a corollary,

$$\lim_{n\to\infty}[2I - C_1(\tfrac{t}{n}) \ldots C_M(\tfrac{t}{n})]^{n^2} f = T(t^2)f,$$

$$\lim_{n\to\infty}[2I - T_1(\tfrac{t}{n}) \ldots T_M(\tfrac{t}{n})]^{n} f = T(2t)f$$

where all of the notation is as in that theorem and T_j denotes the semigroup generated by A_j.

REFERENCES

1. R.H. Cameron, A family of integrals serving to connect the Wiener and Feynman integrals, J. Math. and Phys. 39 (1960), 126-140.
2. P.R. Chernoff, Note on product formulas for operator semigroups, J. Functional Anal. 2 (1968), 238-242.
3. P.R. Chernoff, Product Formulas, Nonlinear Semigroups, and Addition of Unbounded Operators, Mem. Amer. Math. Soc. 140 (1974), 121 pp.
4. P.R. Chernoff, Universally commutable operators are scalars, Michigan Math. J. 20 (1973), 101-107.
5. R.P. Feynman, Space-time approach to non-relativistic quantum mechanics, Rev. Mod. Phys. 20 (1948), 367-387.

6. R.P. Feynman and H.P. Hibbs, Quantum Mechanics and Path Integrals, McGraw-Hill, New York, 1965.

7. J.A. Goldstein, A Lie product formula for one parameter groups of isometries on Banach spaces, Math. Ann. 186 (1970), 299-306.

8. J.A. Goldstein, Some counterexamples involving self-adjoint operators, Rocky Mtn. J. Math (1972), 142-149.

9. J.A. Goldstein, Semigroup-theoretic proofs of the central limit theorem and other theorems of analysis, Semigroup Forum 12 (1976), 189-206, 388.

10. J.A. Goldstein, The universal addability problem for generators of cosine functions and operator groups, Houston J. Math. 6 (1980), 365-373.

11. J.A. Goldstein, Cosine functions and the Feynman-Kac formula, Quart. J. Math. 33 (1982), 303-307.

12. P.E.T. Jørgensen, Analytic continuation in m (the mass) of the Schrödinger equation for highly singular potentials, preprint.

13. M. Kac, On some connections between probability theory and differential and integral equations, in Proc. Second Berkeley Symp. Math. Stat. Prob. (1951), 189-215.

14. T. Kato, Pertubation Theory for Linear Operators, Springer, Berlin, 1966.

15. E. Nelson, Feynman integrals and the Schrödinger equation, J. Math. Phys. 5 (1964), 332-343.

16. E. Nelson, Topics in Dynamics, I: Flows, Princeton U. Press, Princeton, N.J., 1969.

17. M. Reed and B. Simon, Methods of Modern Mathematical Physics II: Fourier Analysis, Self-Adjointness, Academic, New York, 1975.

18. S.-Y. Shaw, Some exponential formulas for m-parameter semigroups, Bull. Inst. Math. Acad. Sinica 9 (1981), 221-228.

19. H. F. Trotter, On the product of semi-groups of operators, Proc. Amer. Math. Soc. 10 (1959), 545-551.

Jerome A. Goldstein

Tulane University

M GOLUBITSKY, J MARSDEN & D SCHAEFFER

Bifurcation problems with hidden symmetries

§1. INTRODUCTION

Basic theory for bifurcation problems with symmetry was developed by Sattinger [1979] and Golubitsky and Schaeffer [1979b]. A symmetry group usually forces the bifurcation to be rather degenerate but simultaneously, one can take advantage of the symmetry to render some interesting problems computable.

Recent papers of Hunt [1981], [1982] make use of symmetries in what, at first sight, appears to be a nonstandard fashion. This enables him to arrive at a parabolic umbilic description for the buckling of a right circular cylinder under end loading (see also Hui and Hansen [1981]). The purpose of this paper is to establish the following points:

1. The scheme of Hunt is consistent with the general theory of Golubitsky and Schaeffer [1979].

2. There is a simple abstract procedure involving "hidden symmetries" which enables one to simplify calculations and to arrive at Hunt's procedure as a special case in a natural way.

3. The scheme proposed by Hunt for the buckling of shells can be derived by starting with, for example, the partial differential equations of Kirchhoff shell theory, and

4. The stability assignments can be computed for the bifurcation problem considered by Hunt.

A crucial $\mathbb{Z}_2$ symmetry on a subspace is used by Hunt to obtain a description of the bifurcation in terms of the parabolic umbilic. This symmetry is derived by him in a heuristic way. We show that it arises by a natural abstract construction that is verifiable for a Kirchhoff shell model.

The name "hidden" symmetry arises from two facts. First, it is a symmetry defined only on a subspace of state space. Second, this symmetry is revealed by working in a larger space that does not fix the phases of the relevant modes. This larger space is where the framework of Golubitsky and Schaeffer [1979b] holds. We shall explain these statements in more detail shortly in §2.

As Hunt notes, there are other bifurcation problems that can be dealt with by the 'hidden symmetry method', such as the buckling of stiffened structures. Another example is Schaeffer's [1980] analysis of the Taylor problem. In particular, the use of hidden symmetries enables one to see directly that certain terms in the bifurcation equation vanish. This was done by direct calculation in Schaeffer [1980]. As will be noted later, hidden symmetries also appear to play an important role in the analysis of other bifurcation problems as well, such as the Bénard problem. This is briefly discussed in Golubitsky, Swift and Knobloch [1984] and Ihrig and Golubitsky [1984].

In some physical problems, solutions of a partial differential equation on a bounded domain satisfying appropriate boundary conditions are in one-to-one correspondence with periodic solutions on an infinite domain which have additional reflection symmetries. The periodic problem is a mathematical device which helps in the understanding of the given problem in the finite domain. In particular, this device enables one to understand how hidden symmetries in the problem can be understood in the abstract formulation as symmetries on a subspace. This procedure shows why our abstract formulation includes more cases than one might at first expect.

In Section 2 we explain in more detail how the periodic extensions give rise to hidden symmetries by means of a simple example. Section 3 gives the abstract infinite dimensional formulation of bifurcation problems with hidden symmetries and Section 4 applies the methods developed to Hunt's problem of buckling cylinders. Finally in Section 5 we discuss the stability and bifurcation diagrams for the cylinder problem.

In this paper we have had to make a choice between the variational approach (based on an energy function) and the direct approach (based on the equations). In the variational approach, one is given an energy functional which is invariant under the action of the symmetry group. One then applies the splitting lemma of Gromoll and Meyer to obtain a reduced function $f: \mathbb{R}^n \to \mathbb{R}$ whose critical points are (locally) in one to one correspondence with the critical points of the energy functional. (See Golubitsky and Marsden [1983] and Buchner, Marsden, and Schecter [1983] for a general view of this approach.) Moreover, the reduced function f inherits the symmetries of the original energy function and is, itself, invariant under the group action.

The second way to obtain symmetries is to start with a differential equation whose associated differential operator has a linearization which is Fredholm of index 0. Then one may use the Liapunov-Schmidt procedure to obtain a (reduced) mapping $g: \mathbb{R}^n \to \mathbb{R}^n$ whose zeros are (locally) in one to one correspondence with the solutions to the original differential equations. Moreover, if the original differential operator is equivariant with respect to a group of symmetries, then (under suitable hypotheses), so is g.

The difference between the two approaches is significant when the unfolding (or imperfection sensitivity) problem is studied. This latter topic is discussed here only briefly. To be consistent with the spirit of

Hunts paper and with elastic buckling in general, we shall adopt the variational (or catastrophe theory) point of view.

Acknowledgements. We thank Giles Hunt for stimulating conversations which inspired this work. We also thank Stuart Antman, David Chillingworth, Ed Ihrig, and Steve Wan for several useful comments.

§2. HIDDEN SYMMETRIES AND PERIODIC BOUNDARY CONDITIONS

In this investigation of the buckling of cylindrical shells, Hunt noted that the parabolic umbilic, $\pm x^4 \pm xy^2$, appeared in a context where some less degenerate singularity (such as the elliptic or hyperbolic umbilic) seemingly should have been expected. Taking the point of view that one should attempt to explain unexpected degeneracies, Hunt looked for a context in which the parabolic umbilic would occur naturally. He found one, which he calls symmetries on a subspace. In this paper we give a context, namely that of hidden symmetries, which reveals Hunt's symmetries on a subspace in a natural way.

Let us first give a prototype example (due to Hunt) which shows how the parabolic umbilic arises from the imposition of a symmetry on a subspace. In the second half of this section we show how this situation can arise by means of a simple example.

Let $g(x,v)$ be a real-valued function satisfying

$$\begin{aligned} &\text{a)}\quad g(-x,v) = g(x,v) \quad\text{and}\quad g(0,-v) = g(0,v)\ , \text{ and} \\ &\text{(b)}\quad g(0) = 0\ ,\quad (dg)(0) = 0,\quad (d^2g)(0) = 0 \end{aligned} \tag{2.1}$$

Conditions (2.1a) state that f has a reflectional symmetry in the x-variable and a reflectional symmetry on the subspace consisting of the v-axis. Conditions (2.1b) state that g has a degenerate singularity at the origin.

Note. The function g could arise via the splitting lemma from an infinite dimensional variational problem. Conditions (2.1b) state that the kernel of the Hessian of the original variational problem is two-dimensional, while conditions (2.1a) reflect certain symmetry properties of this variational problem.

Writing the first few terms of the Taylor expansion of f consistent with (2.1) one finds

$$g(x,v) = Ax^2v + Bv^4 + Cx^2v^2 + Dx^4 + \ldots\ldots$$

The important point to note here is that x^2v is the only cubic term which can be non-zero. The symmetry on the subspace forces the coefficient of v^3 to be 0. Now if $A \cdot B \neq 0$ then f is right equivalent to the parabolic umbilic (cf. Zeeman and Trotman [1975]). More precisely, there exists a diffeomorphism $\phi\colon \mathbb{R}^2 \to \mathbb{R}^2$ such that

$$g(\phi(x,v)) = \varepsilon v^4 + x^2 v$$

where $\varepsilon = \operatorname{sgn} B$. If the coefficient of v^3 were nonzero one would obtain either the elliptic or hyperbolic umbilic.

To motivate the abstract set up in the following section, consider the following simple example. Suppose that one has a bifurcation problem in variational form which is posed on the interval $[-\pi,\pi]$ with periodic and possibly other boundary conditions assumed. Often such problems have $O(2)$ as a symmetry group; the rotations $SO(2)$ act by translation $\zeta \mapsto \zeta + \theta$ where θ is an element of $SO(2) \approx S^1$ and the orientation reversing element of $O(2)$ acts by flipping the interval $\zeta \mapsto -\zeta$. Typically, the kernel of the linearization of the bifurcation problem at an eigenvalue of this linearization will be the 2-dimensional space V_k generated by $\{\cos k\zeta, \sin k\zeta\}$. Solutions to the original bifurcation problem which correspond to V_k by a splitting lemma argument are said to have wave number k. However there is often an extra parameter in the original bifurcation problem, such as an aspect ratio, which alters the eigenvalue structure. For special values of this parameter it is possible to have an eigenvalue of multiplicity 4. Typically, in such instances, the wave numbers are consecutive. For definiteness, suppose the corresponding eigenspace is $V_k \oplus V_{k+1}$. In the example below we study the case $\mathbb{R}^4 = V_1 \oplus V_2$ with explicit coordinates given by

$$(x,y,v,w) \to x \cos \zeta + y \sin \zeta + v \cos 2\zeta + w \sin 2\zeta. \tag{2.2}$$

We now discuss why one studies bifurcation problems on an interval $[-\pi,\pi]$ with periodic boundary conditions. Often one has a physical problem posed on the finite interval which one tries to solve by solving a corresponding problem on the infinite interval and looking for periodic solutions of period 2π. This reformulation introduces $O(2)$ as a group of symmetries. However, in the original problem on the finite interval there may be additional boundary conditions besides periodicity which limit the periodic solutions allowable. For example, the only solutions to the transformed problem on the infinite interval which may be relevant are the ones which start and end symmetrically; i.e., those solutions which are invariant under the flip $\gamma_0(\zeta) = -\zeta$. See Figure 2-1 which illustrates this for waves with $k = 2$.

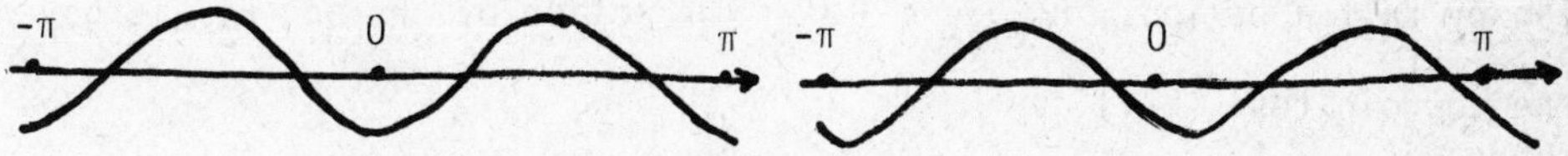

flip invariant   periodic but not flip invariant

Figure 2.1

A similar situation occurs in the analysis by Schaeffer [1981] of the Taylor problem and, as we shall see, can be used as the basis for the analysis by Hunt [1982] of the buckling of a cylindrical shell though the situation in the latter case is yet more complicated.

The context hypothesized above allows hidden symmetries. Let $\mathbb{R}^4 = V_1 \oplus V_2$. The translation $\zeta \to \zeta + \theta$ of $SO(2)$ acts on V_k by rotation through the angle $k\theta$. The flip γ_0 acts on $\mathbb{R}^4$ with the coordinates (2.2) by

$$\gamma_0(x,y,v,w) = (x,-y,v,-w).$$

Let Δ be the 2-element group generated by γ_0. Let F_Δ, the fixed point set for Δ, be defined by

$$F_\Delta = \{(x,0,v,0)\} .$$

Note that F_Δ corresponds to the periodic solutions $x \cos \zeta + v \cos 2\zeta$ which are exactly the periodic solutions which begin and end symmetrically.

We are thus interested in the elements of F_Δ. Consider the "naive symmetry group" $N(\Delta)$ consisting of those elements of $O(2)$, which leave F_Δ setwise invariant; that is, $N(\Delta)$ is generated by the flip γ_0 and the translation by half a period, $h(\zeta) = \zeta + \pi$. The action of h on F_Δ is given in the coordinates (x,v) by

$$h(x,v) = (-x,v).$$

We now ask, "Is there a hidden symmetry in this problem?", that is, is there a symmetry on a subspace other than the elements of $N(\Delta)$? The answer is yes. Let $q(\zeta) = \zeta + \frac{\pi}{2}$ be translation by a quarter period. Then q acts on $\mathbb{R}^4$ by $q(x,y,v,w) = (y,-x,-v,-w)$. In particular, on the fixed point subspace of $N(\Delta)$, namely $F_{N(\Delta)} = \{(0,v)\} \subset F_\Delta$, q acts by $q(0,v) = (0,-v)$.

Now one sees that if one were to solve the hypothetical problem above by a splitting lemma argument one would be looking for the critical point structure of a function $f\colon \mathbb{R}^4 \to \mathbb{R}$ which is invariant under the action of $O(2)$. By looking for "physically reasonable" solutions one tries to find the critical point structure of $g\colon \mathbb{R}^2 \to \mathbb{R}$ where $g = f \mid F_\Delta$ and g satisfies the symmetry conditions (2.1a); in particular, g satisfies the hypotheses of symmetry on a subspace studied by Hunt.

As noted above, the analysis of the buckling of the cylinder proposed by Hunt [1982] is somewhat more complicated though the end result is similar. The reason for this complication is that in the buckling problem two copies of $O(2)$ act as symmetries. More precisely, one copy of $O(2)$ occurs because the finite cylinder is replaced by the infinite cylinder with periodic boundary conditions imposed. The second copy of $O(2)$ acts on the problem since the cylinder itself is invariant under rotation about its axis.

§3. ABSTRACT FORMULATION

We begin with the following situation. Let Γ be a subgroup of $O(n)$ and assume $g: \mathbb{R}^n \to \mathbb{R}$ is invariant under Γ: $g(\gamma x) = g(x)$ for $x \in \mathbb{R}^n$ and $\gamma \in \Gamma$. Suppose we are interested in the critical points of g that possess a given symmetry. That is, let $\Delta \subset \Gamma$ be a given subgroup and let F_Δ be its fixed point set:

$$F_\Delta = \{y \in \mathbb{R}^n \mid \delta y = y \text{ for all } \delta \in \Delta\};$$

we are interested in critical points of g that lie in F_Δ. Let $h = g|F_\Delta$.

Lemma 1. Let $y \in F_\Delta$. Then g has a critical point at y if and only if h does.

Remark. This lemma is a special case of the "principle of symmetric criticality" due to Palais [1979]. We give a direct proof for the case at hand.

Proof. If g has a critical point at $y \in F_\Delta$, then obviously y is also a critical point for h. Conversely, let $y \in F_\Delta$ be a critical point for h. To show y is a critical point of g, we use the following remark. Let $G: \mathbb{R}^n \to \mathbb{R}^n$ be equivariant with respect to Γ: i.e. $g(\gamma x) = G(x)$ for all $\gamma \in \Gamma$. Now $G(F_\Delta) \subset F_\Delta$ since for $x \in F_\Delta$ and $\delta \in \Delta$, $G(x) = G(\delta x) = \delta G(x)$ and so $G(x) \in F_\Delta$. Now let $H = G|F_\Delta: F_\Delta \to F_\Delta$. Then for $x \in F_\Delta$, it is clear that

$$G(x) = 0 \quad \text{if and only if} \quad H(x) = 0 \tag{3.1}$$

To complete the proof of the lemma, let $G(x) = \nabla g(x)$. Then since $\Gamma \subset O(n)$, we have $G(\gamma x) = G(x)$ and $H(x) = \nabla h(x)$ for $x \in F_\Delta$. The result therefore follows by (3.1). ∎

We shall call a subgroup $\Lambda \subset \Gamma$ <u>fixed point complete</u> if

$$\Lambda = \{\gamma \in \Gamma \mid \gamma y = y \text{ for all } y \subset F_\Lambda\} \tag{3.2}$$

If our symmetry subgroup Δ is not fixed point complete, we can always enlarge Δ to symmetry subgroup $\overline{\Delta}$ that is fixed point complete and for which $F_\Delta = F_{\overline{\Delta}}$. This is reasonable since we are looking for critical points in F_Δ and augmenting Δ by group elements that pointwise fix F_Δ does not change the fixed point set.

To study the critical points of h on F_Δ, it is useful to find the symmetries of h. To locate these, we first consider the subgroup $N(\Delta) \subset \Gamma$ defined by

$$N(\Delta) = \{\gamma \in \Gamma \mid \gamma(F_\Delta) \subset F_\Delta\} \tag{3.3}$$

It is clear that h is invariant under $N(\Delta)$, so $N(\Delta)$ is a symmetry group for h. The notation $N(\Delta)$ is used because of the next lemma.

<u>Lemma 2</u>. $N(\Delta)$ <u>is the normalizer of</u> $\overline{\Delta}$ <u>in</u> Γ.

<u>Proof</u>. Recall that if H is a subgroup of a group G, its normalizer is defined by $N_H = \{g \in G \mid g^{-1}Hg \subset H\}$.

First, suppose that $\gamma \in N(\Delta)$. To show that $\gamma \in N_{\overline{\Delta}}$, let $\delta \in \overline{\Delta}$; we must prove that $\gamma^{-1}\delta\gamma \in \overline{\Delta}$. But if $y \in F_\Delta$ then $\gamma y \in F_\Delta$, so $\delta\gamma y = \gamma y$ and $\gamma^{-1}\delta\gamma y = \gamma^{-1}\gamma y = y$. Thus $\gamma^{-1}\delta\gamma \in \overline{\Delta}$ since $\overline{\Delta}$ is fixed-point complete.

Conversely, suppose $\gamma \in N_{\overline{\Delta}}$. If $y \in F_\Delta$ and $\delta \in \overline{\Delta}$, then $\gamma^{-1}\delta\gamma \in \overline{\Delta}$, so $\gamma^{-1}\delta\gamma y = y$ or $\delta\gamma y = \gamma y$ for all $\delta \in \overline{\Delta}$. Thus y is fixed by all $\delta \in \overline{\Delta}$, so by definition of $F_\Delta = F_{\overline{\Delta}}$, $\gamma y \in F_\Delta$. Thus $\gamma \in N(\Delta)$ by (3.3). ∎

Since $\overline{\Delta}$ acts trivially on F_Δ, we can "discard" it from our symmetry group of h. In view of lemma 2 one can do this by letting the group $D(\Delta)$

be defined by

$$D(\Delta) = N(\Delta)/\overline{\Delta}$$

and calling it (or $N(\Delta)$) the naive group of symmetries of h. Notice that there is a well defined action of $D(\Delta)$ on F_Δ and that h is invariant under this action.

There is a second way h can inherit symmetries from Γ. Let Σ be a proper fixed point complete subgroup of Γ and assume $\Delta \subset \Sigma$. Thus, $F_\Sigma \subset F_\Delta$. As above, $N(\Sigma)$ leaves F_Σ invariant and $h|F_\Sigma$ is invariant under $D(\Sigma)$. The new symmetries obtained this way are the hidden symmetries. Here is the formal definition.

Definition. A hidden symmetry of h is a nontrivial element of $N(\Sigma)$ for some proper subgroup Σ of Γ containing Δ, which is not in $N(\Delta)$.

Remarks 1. One could take the view that one is searching for F_Σ as much as for Σ; given F_Σ, Σ can be defined as the isotropy group of typical points in F_Σ. In the example we shall see that Σ and F_Σ are found simultaneously.

2. If we are looking for zeros of a map $H: F_\Delta \to F_\Delta$ commuting with $N(\Delta)$ (rather than critical points of an invariant function h), then the mere existence of $F_\Sigma \subsetneq F_\Delta$ can put restrictions on H. Indeed, H must map F_Σ to itself, a fact that does not in general follow just from the equivariance of H under $N(\Delta)$.

3. In Hunt's example we shall see that one can choose $\Sigma = N(\Delta)$. The group theoretic reason for this is given in remark 4 below. In other examples, one probably will need to understand the lattice of fixed point complete subgroups of Γ and the lattice of isotropy subgroups. For

examples involving convection, either the planar Bénard problem or spherical convection, these lattices can be computed and in certain cases hidden symmetries are important in understanding solutions with a given symmetry group Δ. For example, in Buzano and Golubitsky [1983] this occurs with the rectangular solutions (see Golubitsky, Swift and Knobloch [1984], §IV for a discussion). For the spherical Bénard problem, the lattice of isotropy subgroups for representations of $O(3)$ is worked out in Ihrig and Golubitsky [1984]. They find that bifurcating solutions corresponding to fixed point sets of dimension one (with maximal isotropy subgroups--these are found using a theorem of Cicogns) are often unstable. In this case one needs to go to the next level in the lattice and there hidden symmetries may be important. See Golubitsky [1983] for a general introduction to ideas and examples involving the lattice of isotropy subgroups.

4. We now ask whether there can be symmetries more subtle than hidden symmetries. In general, the answer seems to be yes. However, in Hunt's example and in the case where Δ is a maximal isotropy subgroup these subtle symmetries do not exist. This last fact was pointed out to us by Ed Ihrig, cf. Remark 5 below.

Before discussing subtle symmetries we review our description of hidden symmetries. In our discussion we have assumed the existence of a function $g: \mathbb{R}^n \to \mathbb{R}$ invariant under Γ and a fixed point complete subgroup Δ of Γ. Our interest lies in understanding the restrictions placed on $h = g|F_\Delta$. So far we have observed two types of restrictions on h. The first observation states that h is invariant under the group $N(\Delta)$ which acts naturally on F_Δ. We have called elements in $N(\Delta)$ "naive symmetries". In the second observation we have shown, by iteration, that if $\Sigma \supsetneq \Delta$ is a proper, fixed point complete subgroup of Γ then $h|F_\Sigma$ is invariant under the subgroup $N(\Sigma)$. We have called elements in $N(\Sigma)$ "hidden symmetries". Moreover, Hunt's symmetry on a subspace is just a specific instance of a hidden symmetry.

We now question whether there are any additional group theoretic restrictions placed on h by the existence of the large group Γ. Such additional symmetries we call <u>subtle symmetries</u>. The only way that subtle symmetries may arise is as follows. Suppose there is an element γ in $\Gamma \sim N(\Delta)$ for which

$$\gamma^{-1}(F_\Delta) \cap F_\Delta \supsetneq F_\Gamma . \tag{3.4}$$

(Aside: F_Γ is contained in each fixed point space and all symmetries fix vectors in F_Γ. Thus (3.4) states that the intersection in the LHS of (3.4) contains a nontrivial vector. Moreover, since $\gamma \notin N(\Delta)$ the intersection on the LHS of (3.4) is a proper subspace of F_Δ.) Observe that when (3.4) is valid, we obtain a further restriction on h. For if $w \in \gamma^{-1}(F_\Delta) \cap F_\Delta$ then both w and γw are in F_Δ; thus $h(\gamma w) = h(w)$. We call such elements γ which are not hidden symmetries, subtle symmetries.

We may clarify the issue of subtle symmetries as follows. We claim that $F_\Delta \cap \gamma^{-1}(F_\Delta)$ is itself a fixed point subspace. Observe that

$$\begin{aligned} &\text{(a)} \quad F_{\gamma^{-1}\Delta\gamma} = \gamma^{-1}(F_\Delta), \text{ and} \\ &\text{(b)} \quad F_G \cap F_H = F_{\langle G,H\rangle} \end{aligned} \tag{3.5}$$

where G, H are subgroups of Γ and $\langle G,H\rangle$ is the subgroup they generate. To prove the claim, let $T = \langle \Delta, \gamma^{-1}\Delta\gamma\rangle$. It follows from (3.5) that

$$F_T = F_\Delta \cap \gamma^{-1}(F_\Delta)$$

Moreover,

$$\Delta \subsetneq T \subsetneq \Gamma$$

since (3.4) is assumed valid and $\gamma \notin N(\Delta)$.

Thus, when searching for subtle symmetries we look for fixed point subspaces F_T satisfying

$$\begin{aligned}
&\text{(a)} \quad F_\Gamma \subsetneq F_T \subsetneq F_\Delta ,\\
&\text{(b)} \quad \gamma(F_T) \subset F_\Delta, \text{ and} \qquad\qquad (3.6)\\
&\text{(c)} \quad \gamma(F_T) \neq F_T .
\end{aligned}$$

Note that (3.6c) follows from the fact that γ is not a hidden symmetry; hence $\gamma \notin N(T)$.

Thus we see that naive symmetries correspond to elements of Γ which leave F_Δ invariant, hidden symmetries correspond to elements of Γ which leave some subspace F_Σ of F_Δ invariant, and subtle symmetries correspond to elements γ of Γ which map a subspace F_T of F_Δ onto a different subspace $\gamma(F_T)$ of F_Δ.

5. There are no hidden symmetries and there are no subtle symmetries when Δ is a maximal isotropy subgroup.(Proof due to Ed Ihrig.) In order to find a hidden symmetry or a subtle symmetry we would need to find a fixed point subspace F_T satisfying (3.6a). Now suppose that F_T is any fixed point subspace, then we claim that F_T is the union of fixed point subspaces of isotropy subgroups. First observe that if $v \in F_T$ then $F_{\Sigma_v} \subset F_T$ where Σ_v is the isotropy subgroup of v. (This follows from the fact that T fixes v and thus is contained, by definition, in Σ_v.) Since $v \in \Sigma_v$ it follows that $F_T = \bigcup_{v \in T} F_{\Sigma_v}$. Second, use (3.6a) to observe that $F_{\Sigma_v} \subset F_T \subset F_\Delta$ and hence that $\Delta \subset \Sigma_v$. By (3.6a) we can choose a vector $v \in F_T \sim F_\Gamma$ and for such a v, $\Sigma_v \neq \Gamma$. It now follows that if Δ is a maximal isotropy

subgroup we must have $\Delta = \Sigma_v$. Hence $F_\Delta = F_{\Sigma_v} = F_T$ which contradicts the second equality in (3.6a).

We now discuss how one reduces an infinite dimensional problem to the situation of looking for critical points of a function $g: \mathbb{R}^n \to \mathbb{R}$ invariant under a group $\Gamma \subset O(n)$.

Let X be a Banach space and $\langle\ ,\ \rangle$ a (not necessarily complete) inner product on X. (In many examples X is a $W^{s,p}$ Sobolev function space, and $\langle\ ,\ \rangle$ is an L^2, H^1 or H^2 inner product).* Let $f: X \to \mathbb{R}$ be a C^∞ function defined on a neighborhood of $0 \in X$ with $f(0) = 0$ and $Df(0) = 0$. Eventually f will depend on parameters but this is suppressed for the moment.

Let K be the kernel of $D^2f(0)$; i.e.

$$K = \{v \in X \mid D^2f(0)\cdot(v,w) = 0 \quad \text{for all} \quad w \in X\}$$

Assume that f admits a smooth $\langle\ ,\ \rangle$ gradient $\nabla f: X \to X$ so $\nabla f(0) = 0$ and let $T = D\nabla f(0): X \to X$. It is easy to check that $\langle D(\nabla f(0))\cdot v, w\rangle = D^2f(0)\cdot(v,w)$, so $K = \operatorname{Ker} T$ and T is $\langle\ ,\ \rangle$ - symmetric.

Assume T is Fredholm of index zero; in particular, X admits the following $\langle\ ,\ \rangle$ -orthogonal decomposition into closed subspaces.

$$X = K \oplus \operatorname{Range} T .$$

Under these hypotheses, the critical points of f are in one to one correspondence for those of a reduced function

$$g: K \to \mathbb{R}$$

---

* The generalization to the case in which X is a manifold and $\langle\ ,\ \rangle$ is a weak Riemannian metric on X is routine.

defined in a neighborhood $0 \in K$. This correspondence is by means of the graph of an implicitly defined function $\phi: K \to \text{Range } T$ defined by solving the equation $P \circ \nabla f = 0$ where P is the projection to Range T. One can also show that the problem of finding normal forms for f can be reduced to that for g, which satisfies $g(0) = 0$, $Dg(0) = 0$ and $D^2 g(0) = 0$. This is the splitting lemma, which is the variational analogue of the Liapunov-Schmidt procedure (see Golubitsky and Marsden [1983] for details and references).

Suppose that Γ is a group of isometries which act continuously on X by linear transformations and leave f invariant; that is

$$f(\gamma x) = f(x) \quad \text{for all} \quad x \in X, \quad \gamma \in \Gamma .$$

By differentiating this relation, it readily follows that Γ leaves K invariant and g is invariant as well; that is, $\gamma x \in K$ if $\gamma \in \Gamma$ and $x \in K$ and

$$g(\gamma x) = g(x) \quad \text{for} \quad \gamma \in \Gamma, \quad x \in K.$$

Aside. In the buckling problem of Hunt [1982], X corresponds to a Sobolev space of 2L-periodic displacements of a right circular cylinder, where periodic means with respect to movement along the z-axis, the axis of the cylinder, and $\Gamma = O(2) \times O(2)$ is the natural symmetry group of the problem. We take $[-L,L]$ as the fundamental interval along the cylinder's axis. Hunt is interested in solutions which are symmetric with respect to reflection about the midpoint of this interval. These solutions comprise the fixed point set for a subgroup Δ of Γ. This and the discussion in §2 are motivations for the construction of F_Δ given above. Another motivation is provided by Schaeffer [1980], in which functions satisfying a desired set of boundary conditions for the Navier-Stokes equation on a fixed domain $\Omega \times [-L,L]$, where $\Omega \subset \mathbb{R}^2$, can be characterized as periodic functions on $\Omega \times \mathbb{R}$ which are invariant under reflection with respect to the planes $z = \pm L$. Again the states satisfying these boundary conditions can be characterized as the fixed point set for a subgroup Δ.

In many examples, including Hunt's, there is a further reduction in dimension that can be done by finding a cross section for the action. The method is similar to that of Golubitsky and Schaeffer [1981] in which reduction from a five dimensional kernel to a two dimensional subspace was important (interestingly, in this example, the subspace was a fixed point set for the group D_2).

Suppose $V \subset F_\Delta$ is a subspace satisfying:

$$\begin{aligned} &\text{(a)}\quad V \text{ saturates } F_\Delta;\ \text{that is,}\ \bigcup_{\gamma \in N(\Delta)} \gamma V = F_\Delta, \text{ and} \\ &\text{(b)}\quad \text{for each } x \in V,\ T\mathcal{O}_x + \bigcup_{\gamma \in N(\Delta)_x} \gamma V = F_\Delta \end{aligned} \tag{3.7}$$

where $\mathcal{O}_x$ is the $N(\Delta)$ orbit of x in F_Δ and $N(\Delta)_x$ is the isotropy subgroup of x in $N(\Delta)$. By Lemma 1, we seek the critical points of $h = g|F_\Delta$ on F_Δ. We now show that the critical points of h are determined by $k = h|V$.

<u>Lemma 3</u>. <u>Let</u> V <u>satisfy</u> (3.7) <u>and let</u> $k = h|V$. <u>Then the critical points of</u> h <u>are the</u> $N(\Delta)$ <u>orbits of the critical points of</u> k.

<u>Proof</u>. By invariance of h under $N(\Delta)$, it is enough to show that for $x \in V$, x is a critical point of h if and only if it is one for k. From $h(\gamma x) = h(x)$ we get, in terms of differentials,

$$dh(\gamma x) \circ \gamma = dh(x).$$

[In terms of gradients, since the action is orthogonal we have $\nabla h(\gamma x) = \gamma \nabla h(x)$.] Obviously if h has a critical point at $x \in V$ so does k and the orbit of x consists of critical points. Suppose $x \in V$ is a critical point of k. Then $dh(x)|V = 0$ and so $dh(x)|\gamma V = 0$ for $\gamma \in N(\Delta)_x$ by $dh(\gamma x)\circ\gamma = dh(x)$. Since h is constant on $\mathcal{O}_x$, $dh(x)|T_x\mathcal{O}_x = 0$. Thus by (3.7b), $dh(x) = 0$. By (3.7a) we have not missed any critical points. ∎

Thus, no information is lost by restricting attention to the cross-section V. As in Golubitsky and Schaeffer [1981], we expect one can prove that no information is lost in the unfolding theory as well.

§4. BUCKLING OF A CYLINDRICAL SHELL

We now describe a context in which one can in principle rigorously arrive at Hunt's model for cylindrical shell buckling. We do not provide a complete exposition, but only indicate a framework with enough details so the symmetry groups become apparent and the hidden symmetry is revealed. We use a shell model for simplicity, but one could in principle also use a three dimensional model.

First we outline a framework for nonlinear Kirchhoff shell theory (cf. Naghdi [1972], Marsden and Hughes [1983] and references therein). Let M be a reference two manifold and C a space of deformations $\phi: M \to \mathbb{R}^3$. Each $\phi \in C$ is required to be an embedding of M into $\mathbb{R}^3$ and is to satisfy any relevant displacement boundary conditions. For each ϕ, let $F = D\phi$ be the deformation gradient, F^T its transpose, and $C = F^TF$, a positive definite symmetric two tensor on M, the Cauchy Green tensor. (Apart from the positioning of tensor indices, C is the pull-back of the Eucliden metric on $\mathbb{R}^3$ to M). Let k denote the (referential) second fundamental form of the embedding ϕ; i.e. k is the second fundamental form of the deformed surface $\phi(M)$, regarded as a symmetric two tensor on

M.

Kirchhoff shell theory deals with elastic stored energy functions of the form $W(C,k)$. We shall assume that the shell is homogeneous, i.e. W is independent of the reference point $X \in M$, is isotropic and is invariant under isometries of M. The real shell being modelled has a finite thickness which is incorporated into $W(C,k)$.

If $\underset{\sim}{\lambda}$ denotes a prescribed dead load on ∂M, then the energy function is

$$V_{\underset{\sim}{\lambda}} : \mathcal{C} \to \mathbb{R}$$

$$V_{\underset{\sim}{\lambda}} = \int_M W(C,k)\, dA - \int_{\partial M} \underset{\sim}{\lambda} \cdot \phi \, ds$$

Equilibrium configurations of the shell are the critical points of $V_{\underset{\sim}{\lambda}}$.

We choose the length scale so that the radius of the cylinder is unity. For a cylindrical shell with periodic boundary conditions, we choose

$$M = S^1 \times [0,L]$$

where S^1 is the unit circle in the plane. We consider deformations $\phi: M \to \mathbb{R}^3$ of Sobolev class H^s, $s \geq 4$ which are immersions that map $z =$ constant planes to $z =$ constant planes and which remain H^s when extended periodically in z; thus letting ϕ^x, ϕ^y and ϕ^z be the components of ϕ, the extension satisfies

$$\phi^x(\theta,z+L) = \phi^x(\theta,z), \; \phi^y(\theta,z+L) = \phi^y(\theta,z)$$

and

$$\phi^z(\theta,z+L) = \phi^z(\theta,z) + \phi^z(\theta,L)$$

where $\theta \in S^1$.

Note that the deformed cylinder has a well-defined length, say aL.

We choose X to be the space of H^s displacements u defined by $u(\theta,z) = \phi(\theta,z) - (\cos\theta, \sin\theta, z)$, so $V_{\underset{\sim}{\lambda}}$ becomes defined on a neighborhood of $u = 0$.

We assume that the linearized problem at (and hence near) $u = 0$ is strongly elliptic (so the Fredholm alternative is available -- see for example Marsden and Hughes [1983, Ch. 6]). The linearized equations are fourth order and the linearized energy is quadratic in second derivatives of u. Motivated by analogous examples for the Morse lemma (see Golubitsky and Marsden [1983] and Buchner, Marsden and Schecter [1983]), we find that $\langle\ ,\ \rangle$ may be chosen to be the H^2 inner product.

Let us assume that the linearized elastic moduli, the length L and a parallel end load of magnitude λ are chosen so that the modes described in Hunt [1982] (and references therein) comprise the kernel K of the linearized equations in X. This kernel has dimension six. It contains the two basic displacements shown in Figure 4-1 (adapted from Hunt [1982]).

The general element of the six dimensional space K has the form

$$u = \left[\alpha_1 \cos\frac{\pi z}{L}\cos 2\theta + \alpha_2 \cos\frac{\pi z}{L}\sin 2\theta,\ \alpha_3 \sin\frac{\pi z}{L}\cos 2\theta + \alpha_4 \sin\frac{\pi z}{L}\sin 2\theta,\ \beta_1 \cos\frac{2\pi z}{L} + \beta_2 \sin\frac{2\pi z}{L}\right]. \tag{4.1}$$

This function may be characterized by two amplitudes namely $q_1 = \{\Sigma\alpha_i^2\}^{1/2}$ and $q_2 = (\beta_1^2 + \beta_2^2)^{1/2}$, and various phases. The phases associated with the four dimensional subspace corresponding to the α's are a little subtle and fortunately do not matter for what follows. The case illustrated in Figure 4-1 has $\alpha_2 = \alpha_3 = \alpha_4 = 0$. In this case all displacements vanish for $z = L/2$, at which points the cross section is circular. In general there is a plane where the displacements vanish if and only if $\alpha_1\alpha_4 = \alpha_2\alpha_3$.

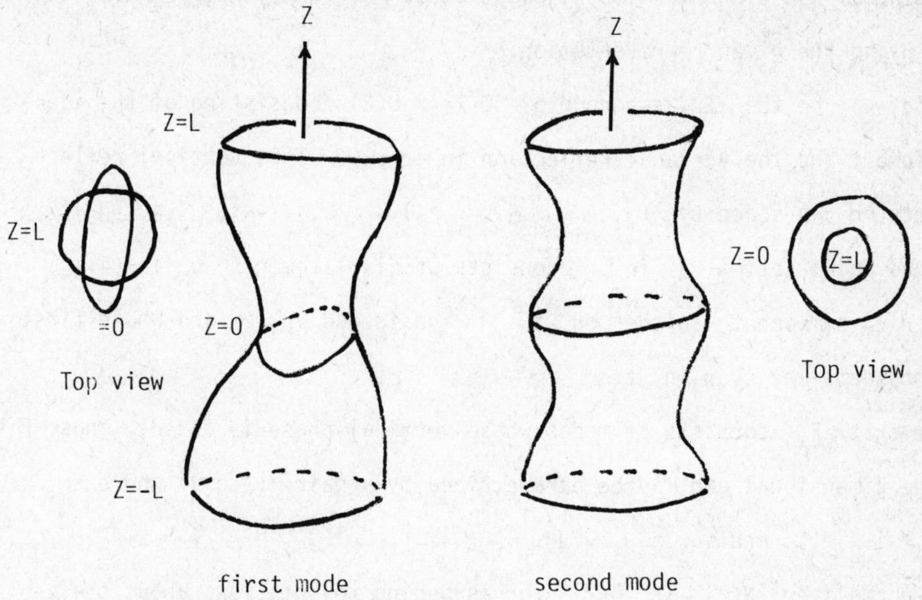

Figure 4-1

This fact will not be needed below, since we will shortly be restricting to the subspace with $\alpha_3 = \alpha_4 = 0$.

As we have hinted at in the preceding paragraph, a full bifurcation analysis directly on K is somewhat complicated. However, our interest is in F_Δ and the analysis on F_Δ is greatly simplified by symmetries and hidden symmetries.

The group Γ is taken to be $O(2) \times O(2)$; the first $O(2)$ consists of rotations about the z-axis and reflections in vertical planes. The second $O(2)$ consists of (2L-periodic) translations along the z-axis and reflections in z = constant planes. These $O(2)$ actions induce an action of Γ on M and $\mathbb{R}^3$ and hence on X. Elements $\gamma \in \Gamma$ act on configurations ϕ and displacements u by $\gamma\phi = \gamma\circ\phi\circ\gamma^{-1}$ and $\gamma u = \gamma\circ u\circ\gamma^{-1}$). It is clear that the potential function V_λ is invariant under Γ. Note that Γ also acts on K using the given representation.

We let Δ be the $\mathbb{Z}_2$ subgroup of $O(2) \times O(2)$ consisting of the identity element and the vertical reflection in $z = 0$. This vertical reflection acts on $\alpha\beta$-space by $(\alpha_1,\alpha_2,\alpha_3,\alpha_4,\beta_1,\beta_2) \to (\alpha_1,\alpha_2,-\alpha_3,-\alpha_4,\beta_1,-\beta_2)$. The fixed point set of Δ in K consists of displacements u in (4.1) whose third component vanishes on $z = 0$ and is odd in z and whose first two components are even in z.

Note that F_Δ consists of modes whose vertical phase is fixed. Thus F_Δ is three dimensional and may be parametrized by a pair (ξ,q_2) where $\xi = \alpha_1 + i\alpha_2 \in \mathbb{C}$ and $q_2 = \beta_1 \in \mathbb{R}$.

The normalizer $N(\Delta)$ is $O(2)$ corresponding to rotations about the z-axis and reflections in the vertical planes $\theta = \theta_0$. It acts on F_Δ by

$$(\xi,q_2) \mapsto (e^{i\theta}\xi,q_2) \qquad \text{(rotation by an angle } \theta)$$

and

$$(\xi,q_2) \mapsto (e^{2i\theta}\,\overline{\xi},q_2) \quad \text{(reflection in the plane } \theta = \theta_0)$$

Since $g|F_\Delta$ is invariant under $N(\Delta)$ it must be a function of $|\xi|^2$ and q_2.

To look for hidden symmetries we look at the fixed point set $F_{N(\Delta)}$ of $N(\Delta)$. This is the set of axisymmetric displacement; that is, $F_{N(\Delta)} = \{(\xi,q_2) \in F_\Delta | \xi = 0\}$. Observe using (4.1) that the quarter-period vertical translation maps elements of $F_{N(\Delta)}$ to their negatives. This is the hidden symmetry in Hunt's problem.

The set V in Lemma 3 can be chosen to be the set in F_Δ on which ξ is real, i.e. $\alpha_1 = 0$. By Lemma 3, we can restrict to V with no loss of information. Since $F_{N(\Delta)} \subset V$ the hidden symmetry still restricts $k = h|V$. Letting $q_1 = \alpha_1$ we see that k is a function of q_1^2, q_2^2 and $q_1^2 q_2$. (we note in passing that V is the fixed point set of the $\mathbb{Z}_2$ action $\xi \to \overline{\xi}$).

§5. REMARKS ON BIFURCATION AND STABILITY

Expanding k in a Taylor series, we get the form

$$k = \frac{1}{2}(aq_1^2 + bq_2^2) + cq_1^2 q_2 + dq_1^4 + eq_2^4 + fq_1^2 q_1^2 + \text{h.o.t.}$$

where "h.o.t." means "higher order terms". At a bifurcation point such as (2.1) k satisfies $a = b = 0$. Set

$$f = cq_1^2 q_2 + dq_1^4 + eq_2^4 + fq_1^2 q_2^2 + \text{h.o.t.}$$

As noted in §2, one sees that f is right equivalent to the parabolic umbilic, assuming $c, e \neq 0$. More precisely, f is right equivalent to

$$g = \frac{1}{4}\delta q_2^4 + \varepsilon q_1^2 q_2 ,$$

where $\delta = \text{sgn}\, e$ and $\varepsilon = \text{sqn}\, c$. The universal unfolding of the parabolic umbilic generally requires four parameters.

However, we consider here only those terms in the universal unfolding which are consistent with the present symmetry. This leads to an unfolding with just two free parameters, of the form

$$\tilde{G} = \frac{1}{2}(\alpha q_1^2 + \beta q_2^2) + \varepsilon q_1^2 q_2 + \frac{1}{4} q_2^4 \tag{5.1}$$

where we have chosen the $+$ sign for the q_2^4 term. We expect one can show that $\tilde{G}$ is a universal unfolding for the parabolic umbilic in the context of hidden symmetries. Formal calculations indicate this is in fact correct; a rigorous argument may be possible directly or using the work of Damon [1983].

The bifurcation diagram for (5.1) is shown in Figure 5-1 for $\varepsilon = +1$. In this figure the $\alpha\beta$-plane is divided into 6 regions by the curves $\alpha = 0$, $\beta = 0$ and $\beta = -\alpha^2$. The unfolding $\tilde{G}$ has the same number of critical points for any two pairs of (α,β) which lie in the same region. The number of these critical points, along with the signs of the eigenvalues of $d^2\tilde{G}$ at those critical points, is given in Figure 5-1. We use s to indicate a negative eigenvalue and u to indicate a positive eigenvalue.

If we now consider a bifurcation problem $H(q_1,q_2,\lambda)$ depending on the distinguished parameter λ for which $H(q_1,q_2,0) = g(q_1,q_2)$. Then the unfolding theorem, which we assume valid in the context of hidden symmetries, allows us to identify $H(q_1,q_2,\lambda)$ for fixed λ with $G(q_1,q_2,\alpha(\lambda),\beta(\lambda))$ where $\alpha(0) = \beta(0) = 0$. Therefore, the bifurcation problem $H(q_1,q_2,\lambda)$ may be identified with a smooth path $(\alpha(\lambda),\beta(\lambda))$ in the unfolding space $\tilde{G}$. Cf. Golubitsky and Schaeffer [1979a].

For the particular problem at hand we identify λ with the load as in Hunt [1982] and related bifurcation problems. Our reasoning is that the load λ should enter the potential $V_{\underset{\sim}{\lambda}}$ of §3 as the coefficient of a positive definite quadratic term plus, perhaps, higher order terms. Under this assumption we identify $V_{\underset{\sim}{\lambda}}$ with the path $\alpha(\lambda) = \beta(\lambda) = -\lambda$; that is, we consider the bifurcation problem

$$H(q_1,q_2,\lambda) = -\frac{1}{2}\lambda(q_1^2 + q_2^2) + q_1^2q_2 + \frac{1}{4}q^4 \tag{5.2}$$

Note we have taken $\varepsilon = +1$.

We now ask in which ways can the bifurcation problem H in (5.2) be perturbed. Since the curve $\alpha(\lambda) = \beta(\lambda) = -\lambda$ intersects each of the dividing curves in Figure 5-1 transversely, it is likely that the universal unfolding of H will depend on just one additional parameter σ.

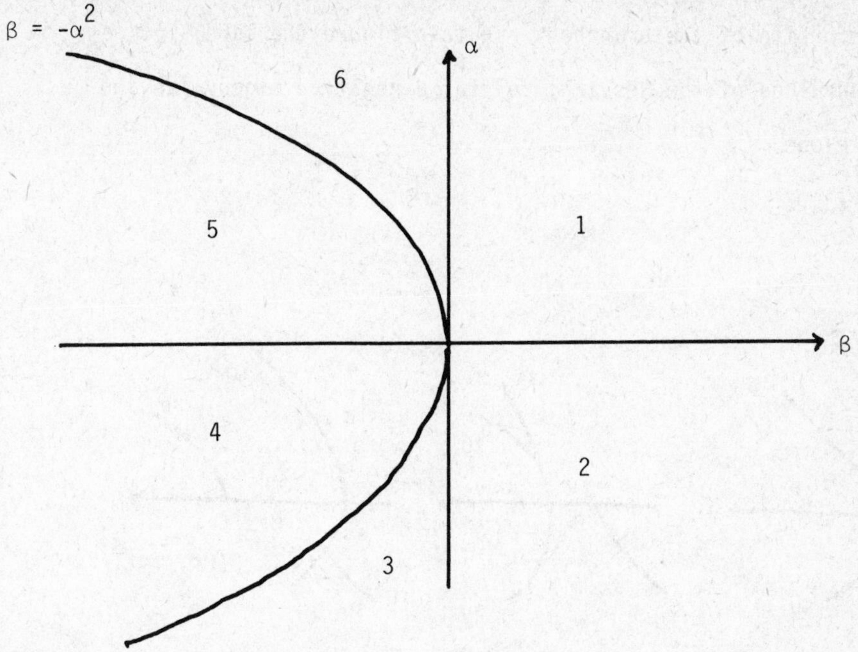

Figure 5-1.

We rewrite the unfolding (5.1) as

$$G(q_1,q_2,\lambda,\sigma) = -\tfrac{1}{2}\lambda(q_1^2 + q_2^2) + \tfrac{1}{2}\sigma(q_1^2 - q_2^2) + q_1^2 q_2 + \tfrac{1}{4}q^4 \tag{5.3}$$

where we have taken $\varepsilon = +1$. We wish to think of (5.3) as an unfolding with λ as the distinguished parameter and σ as an imperfection parameter.

We now study the bifurcation diagrams associated with G for fixed values of σ. For the reasons stated above, we feel that it is likely that G is a universal unfolding of H; however, we cannot prove this fact. In any case, G, itself, will give us sample kinds of behavior which may be found in Hunt's context.

By a bifurcation diagram we mean the set defined by $G_{q_1} = G_{q_2} = 0$ for fixed σ. These bifurcation diagrams are shown in Fig. 5-2 for $\sigma < 0$, $\sigma = 0$ and $\sigma > 0$. The stabilities may be computed directly (or recovered from Figure 5-1) and are shown schematically. Since G is even in q_1, we show only the orbits of the branches. In this figure the labels s and u refer to eigenvalues of the Hessian; s for a negative eigenvalue and u for a positive one.

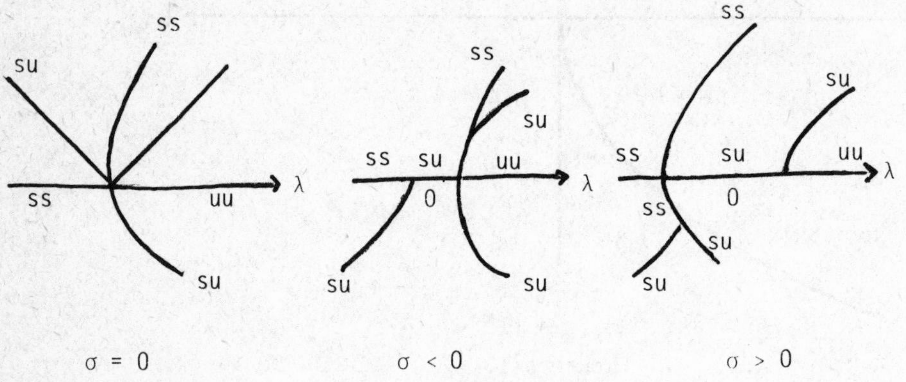

Figure 5-2

In the unperturbed problem ($\sigma = 0$) notice that as λ increases, the trivial (unbuckled) solution becomes unstable, the stability being picked up by the q_2 mode alone with $q_1 = 0$. Note that of the two axisymmetric states which are possible, concave or convex buckling, only one is stable. For the sake of argument we assume convex as in the first mode of Fig. 4-1. An interesting perturbation occurs for $\sigma > 0$. Here the shell buckles into one of the two asixymmetric states either convex or concave. Should the buckling be concave then the bifurcation diagram predicts that this solution will loose stability with a snap-through bifurcation to a convex axisymmetric state. For $\sigma < 0$ the existence of an interval in λ with no stable solutions suggests that the shell buckles to a state involving other modes, a situation not covered by this analysis.

As should be clear from the discussion above, we have not attempted to develop the theory of universal unfoldings for potentials with hidden symmetries either with or without a distinguished parameter. What we have tried to indicate is that such a theory might generate new and intersting behavior.

REFERENCES

1. M. Buchner, J. Marsden and S. Schecter [1983]. Examples for the infinite dimensional Morse lemma, SIAM J. Math. Anal. 14 (1983) 1045-1055.

2. E. Buzano and M. Golubitsky [1983]. Bifurcation on the hexagonal lattice and the planar Bénard problem, Phil. Trans. Roy. Soc. Lon. A308, 617-667.

3. J. Damon [1983]. The unfolding and deferminary theorems for subgroups of A and K insingularities, Proc. Symp. Pure Math. 40 pp. 233-254.

4. M. Golubitsky [1983]. The Bénard problem, symmetry and the lattice of isotropy subgroups. Bifurcation Theory, Mechanics and Physics. C.P. Bruter etal (eds). D. Reidel Publishing Co., 1983. 225-256.

5. M. Golubitsky and J. Marsden [1983]. The Morse lemma in infinite dimensions via singularity theory, SIAM J. Math. Anal. 14 (1983) 1037-1044.

6. M. Golubitsky and D. Schaeffer [1979a]. A theory for imperfect bifurcation via singularity theory. Comm. Pure. Appl. Math. 32, 21-98.

a situation not covered by this analysis.

As should be clear from the discussion above, we have not attempted to develop the theory of universal unfoldings for potentials with hidden symmetries either with or without a distinguished parameter. What we have tried to indicate is that such a theory might generate new and interesting behavior.

Acknowledgement Research supported in part by ARO contract DAAG-29-79-C-0086 (J.M., M.G.), DOE contract DE-AT03-82ER12097 (J.M.), NSF grant MCS-8101580 (MG) and NSF grant MCS-79-02010 (D.S.).

REFERENCES

1. M. Buchner, J. Marsden and S. Schecter [1983]. Examples for the infinite dimensional Morse lemma, SIAM J. Math. Anal. 14, 1045-1055.

2. E. Buzano and M. Golubitsky [1983]. Bifurcation on the hexagonal lattice and the planer Bénard problem, Phil. Trans. Roy. Soc. Lon. A308, 617-667.

3. J. Damon [1983]. The unfolding and deferminary theorems for subgroups of A and K. Singularities, Proc. Symp. Pure Math. 40, 233-254.

4. M. Golubitsky [1983]. The Bénard problem, symmetry and the lattice of isotropy subgroups. Bifurcation Theory, Mechanics and Physics. C.P. Bruter et al (eds). D. Reidel Publishing Co., 225-256.

5. M. Golubitsky and J. Marsden [1983]. The Morse lemma in infinite dimensions via singularity theory, SIAM J. Math. Anal. 14, 1037-1044.

6. M. Golubitsky and D. Schaeffer [1979a]. A theory for imperfect bifurcation via singularity theory. Commun. Pure. Appl. Math. 32, 21-98.

7. M. Golubitsky and D. Schaeffer [1979b]. Imperfect bifurcation in the presence of symmetries. Commun. Math. Phys. 69, 205-232.

8. M. Golubitsky and D. Schaeffer [1981]. Bifurcation with O(3) symmetry including applications to the Bénard problem. Commun. Pure Appl. Math. 35, 81-109.

9. M. Golubitsky, J. Swift and E. Knobloch [1984]. Symmetries and pattern selection in Rayleigh-Bénard convection. Physica D. To appear.

10. D. Hui and J.S. Hansen [1981]. The parabolic umbilic catastrophe and its application in the theory of elastic stability. Quart. Appl. Math. 39, 201-220.

11. G.W. Hunt [1981]. An algorithm for the nonlinear analysis of compound bifurcation. Phil. Trans. Roy. Soc. Lon. 300, 443-471.

12. G.W. Hunt [1982]. Symmetries of elastic buckling, Eng. Struct. 4, 21-28.

13. E. Ihrig and M. Golubitsky [1984]. Pattern selection with O(3) symmetry. Physica D. Nonlinear Phenomena. To appear.

14. J. Marsden and T.J.R. Hughes [1983]. Mathematical foundations of elasticity, Prentice-Hall.

15. P. Naghdi [1972]. The theory of Shells, in Handbuch der Physik (C. Truesdell, ed.) VIa/2, Springer.

16. R.S. Palais [1979]. The principle of symmetric criticality, Commun. Math. Phys. 69, 19-30.

17. D. Sattinger [1979]. Group theoretic methods in bifurcation theory, Springer Lecture Notes in Math. #762.

18. D. Schaeffer [1980]. Qualitative analysis of a model for boundary effects in the Taylor problem, Math. Proc. Camb. Phil. Soc. 87, 307-337.

19. E.C. Zeeman and D.J.A. Trotman [1975]. The classification of elementary catastrophes of codimension ≤ 5, Springer Lect. Notes Math. 525, 263-327.

Martin Golubitsky, Jerrold Marsden and David Schaeffer
University of Houston, University of California - Berkeley and Duke University

P HOLMES & D WHITLEY

On the attracting set for Duffing's equation I: Analytical methods for small force and damping

1. INTRODUCTION.

In this paper we review a number of analytical results on the periodically forced Duffing equation

$$\ddot{x} + \delta\dot{x} - x + x^3 = \gamma\cos\omega t, \tag{1.1}$$

where $\delta \geq 0$ is the dissipation parameter, $\gamma \geq 0$ the excitation force amplitude and $\omega > 0$ the forcing frequency. Rewriting (1.1) as a first order system with $u = x$, $v = \dot{x}$, we have

$$\dot{u} = v$$

$$\dot{v} = u - u^3 - \delta v + \gamma\cos\omega t . \tag{1.2}$$

For most of our analysis we will assume that $\gamma = \varepsilon\bar{\gamma}$ and $\delta = \varepsilon\bar{\delta}$ are both (small) $O(\varepsilon)$, so that (1.2) is a perturbation of a Hamiltonian system with energy:

$$H(u,v) = \frac{v^2}{2} - \frac{u^2}{2} + \frac{u^4}{4}, \tag{1.3}$$

We sketch the level curves of this function in Figure 1. In particular, we note the presence of a double homoclinic orbit (saddle connection) and three families of periodic orbits.

Systems of the form (1.1) occur in many problems in mechanics and physics. Holmes [1979] and Moon and Holmes [1979] showed that (1.1) arises as the simplest reasonable model for a periodically forced beam buckled either mechanically or by a nonuniform magnetic field. Holmes and Marsden [1981]

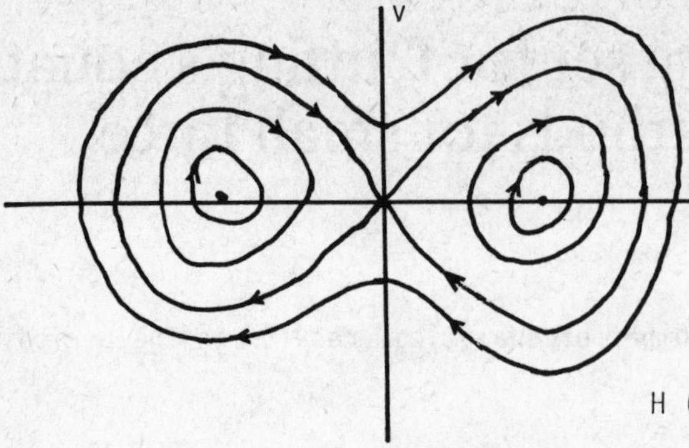

Figure 1. Level curves of $H(u,v)$

studied a partial differential version of equation (1.1) describing the behavior of such a continuous structure and were able to obtain results similar to some given in the first part of this paper. In these and other studies it soon became apparent that the solutions of (1.1-2) can behave in a rich and chaotic fashion, and it was conjectured that, for large sets of parameter values, Duffing's equation might have a strange attractor. We will define attracting sets and attractors in section two.

While we concentrate on the particular example of Duffing's equation in this paper, we stress that the general ideas developed here, and many of the specific methods, are applicable to other problems in nonlinear oscillations. As we proceed, we will indicate where such generalizations have been developed.

In section two we show that all solutions of (1.2) are attracted to a compact invariant set A, the <u>attracting set</u>. The rest of the paper is devoted to a study of the geometrical structure of this set. In section 3 we outline the global perturbation calculations of Greenspan and Holmes [1982] (cf. Greenspan [1981]), using Melnikov theory (Melnikov [1963]). These methods allow us to prove that, for open sets of $|\gamma|, |\delta|$ sufficiently small, equation (1.1) has infinite invariant hyperbolic sets (Smale horseshoes) coexisting with arbitrarily large finite collections of stable periodic orbits. They also allow us to give a characterization of the attracting set as the closure of a certain unstable manifold. Moreover, Newhouse's results [1979, 1980] then imply that our attracting set probably contains countably many stable periodic orbits for a residual set of parameter values near those at which the homoclinic tangencies occur.

The Poincaré map for (1.2) is an important tool in our analysis. Regarding (1.2) as an autonomous flow on $\mathbb{R}^2 \times S^1$ generated by the three dimensional vector field

$$\begin{aligned} \dot{u} &= v \\ \dot{v} &= u-u^3-\delta v+\gamma\cos\omega\theta; \quad (u,v;\theta) \in \mathbb{R}^2 \times S^1 , \\ \dot{\theta} &= 1 \end{aligned} \tag{1.5}$$

we take a cross section $\Sigma^{t_o} = \{(u,v;\theta) | \theta = t \in [0,\frac{2\pi}{\omega}]\}$ and consider the time $\frac{2\pi}{\omega}$ first-return or <u>Poincaré map</u>

$$P(u,v): U \to \Sigma^{t_o}; \; U \subseteq \Sigma^{t_o}, \tag{1.6}$$

induced by the flow ϕ_t of (1.5). Letting $\phi_{t-t_o}(u,v,t_o)$ denote the solution based at $(u(t_o),v(t_o)) = (u,v)$ at $t = t_o$, we have

$$P^{t_o}(u,v) = \pi.\phi_{\frac{2\pi}{\omega}}(u,v;t_o) , \tag{1.7}$$

where π denotes projection onto the first factor ($\mathbb{R}^2$). The existence of a trapping region, proven in 2, guarantees that the Poincaré map is globally defined.

In what follows we shall often regard the dissipation δ and frequency as fixed parameters and vary only the force amplitude γ. We will therefore denote the Poincare map as $P_\gamma^{t_o}$ or simply P_γ, when the particular cross section is unimportant. (We note that any two maps $P_\gamma^{t_1}, P_\gamma^{t_2}$ are equivalent, cf. Chillingworth [1976].) We also note that any cross section Σ^{t_o} is just a copy of the (u,v) Euclidean plane, and the Poincare map P_γ can therefore be regarded as acting on $\mathbb{R}^2$.

The space available here prevents us from giving a full background to the differentiable dynamical and global analytical methods used in this paper. For general background we recommend the books by Chillingworth [1976], the notes of Newhouse [1980] and Bowen [1978] and the forthcoming text by Guckenheimer and Holmes [1983].

In particular, some familiarity with the horseshoe construction due to Smale [1963, 1967], cf. Moser [1973], Newhouse [1980], will be assumed in this paper. A nice elementary discussion of the horseshoe can be found in Chillingworth [1976]. We feel that a proper understanding of this example is an essential prerequisite to attempts to understand chaos and strange attractors in dynamical systems.

The reader may also find it helpful to consult a number of earlier papers on the Duffing equation and variants of it, especially Holmes [1979], Moon and Holmes [1979], and Greenspan and Holmes [1982, 1983]. Andronov, Vitt and Khaiken [1966] continues to provide the best background in nonlinear oscillations and planar systems.

2. A TRAPPING REGION AND AN ATTRACTING SET.

We define the open disc D $\mathbb{R}^2$ with boundary D given by the level curve

$$L(u,v) = \frac{v^2}{2} + \nu uv - \frac{u^2}{2} + \frac{u^4}{4} = \ell \tag{2.1}$$

for some (large) ℓ and $0 < \nu < \delta$. Note that this is a slight modification of the Hamiltonian (1.3). For large ℓ the terms v^2 and u^4 dominate and ∂D is not only diffeomorphic to a circle, but any ray $v = \alpha u$ intersects D in just two, diametrically opposite points. We will show that, for suitably large ℓ, depending on δ and γ, the vectorfield (1.2) is directed inward at all points on D, and hence that solutions based on D enter its interior in arbitrarily small time. Thus the Poincare map P takes D into its interior.

Differentiating (2.1) along solution curves we have

$$\begin{aligned}\nabla L\cdot(\dot{u},\dot{v}) &= v\dot{v} + \nu\dot{u}v + \nu u\dot{v} - u\dot{u} + u^3\dot{u} \\ &= -(\delta-\nu)v^2 - \nu u^2(u^2-1) - \nu\delta uv + \gamma(v+\nu u)\cos\omega t \\ &\le -(\delta-\nu)v^2 - \nu u^2(u^2-1) + \nu\delta|uv| + \gamma|v+\nu u| \ .\end{aligned} \tag{2.2}$$

It is easy to check that, for sufficiently large u and v, this expression is negative on the v axis and on any ray $v = \alpha u$. In fact, on such rays it becomes

$$\nabla L\cdot(\dot{x},\dot{y}) \le -(\delta-\nu)\alpha^2u^2 - \nu u^4 + \nu(1+\delta\alpha)u^2 + \gamma(\nu+\alpha)|u| \ , \tag{2.3}$$

which is clearly negative for large u and/or large α. We conclude that the vector field behaves as claimed for ℓ sufficiently large.

We can now define the <u>attracting set</u> for the Duffing equation as

$$A_\gamma = \overline{\bigcap_{n\ge 0} P_\gamma^n(D)} \ , \tag{2.4}$$

where the overbar denotes closure. We can immediately see that A has zero area and hence no interior, since the vector field (1.5) has negative trace:

$$\text{trace } Df = -\delta < 0 , \tag{2.5}$$

and using the relationship

$$DP_\gamma = e^{\frac{2\pi}{\omega}Df} ,$$

we have $\det(DP_\gamma) = \exp((\frac{2\pi}{\omega})\text{trace}(Df)) = e^{\frac{-2\pi\delta}{\omega}} < 1$. The properties of two dimensional diffeomorphisms with constant Jacobian less than one therefore play an important rôle in our analysis. We note that the famous quadratic map of Hénon [1976] and the cubic map studied earlier by Holmes [1979] in connection with Diffing's equation are examples of such diffeomorphisms.

We summarize the results of this section in:

<u>Lemma 2.1</u>. <u>There is an open disc</u> $D \subset \mathbb{R}^2$ <u>such that</u> $\overline{P_\gamma(D)} \subset D$. <u>The attracting set</u> $A_\gamma = \bigcap_{n\geq 0} \overline{P_\gamma(D)}$ <u>is closed and has empty interior</u>.

In the rest of this paper we will study the structure of the attracting set A_γ . We will show that, for certain (open sets of) parameter values, A_γ is extremely complicated, containing countable sets of periodic orbits and uncountable sets of bounded non-periodic "chaotic" orbits. However, we shall also see that A_γ frequently contains finite or countable sets of <u>stable</u> periodic motions along with the unstable chaotic motions, and thus that it does not necessarily qualify as an indecomposable <u>strange attractor</u>. There is still no generally accepted definition of a strange attractor, but a reasonable working definition is that:

Definition 2.2. A strange attractor is a closed, invariant, attracting set which contains a dense orbit and in which solutions exhibit sensitive dependence upon initial conditions.

The attracting sets we find tend to have everything except a dense orbit!

3. BIFURCATIONS TO HOMOCLINIC ORBITS AND SINKS FOR SMALL γ, δ.

In this section we sketch the results of global perturbation methods. Full details are given in Greenspan and Holmes 1982 and here, after a brief summary of the method, we only give the main results. The procedure is based on the fact that, for $\gamma = \varepsilon\bar{\gamma}$, $\delta = \varepsilon\bar{\delta}$, $0 < \varepsilon \ll 1$, (1.2) is a small perturbation of the integrable Hamiltonian system

$$\dot{u} = v$$
$$\dot{v} = u - u^3 . \qquad (3.1)$$

The general theory developed by Greenspan and Holmes [1982], following Melnikov [1963], app;ies to systems of the form

$$\dot{x} = f(x) + \varepsilon g_\mu(x,t), \; x = \binom{u}{v} \in \mathbb{R}^2 \qquad (3.2)$$

where $f(x) = \binom{\partial H/\partial v}{-\partial H/\partial u}$ is a Hamiltonian vectorfield and $g_\mu(\cdot,t) = g_\mu(\cdot,t+T)$ is T-periodic in t and depends upon parameters $\mu \in \mathbb{R}^k$. The functions f and g_μ are assumed sufficiently smooth (jointly) in their arguments and in addition we require that:

(A1) $\dot{x} = f(x)$ has a homoclinic orbit $q^0(t)$ to a hyperbolic saddle point p_0 (i.e., the compact piece of the level curve $H(q^0)$ contains a saddle point).

(A2) $\Gamma^0 = \{q^0(t) | t \in \mathbb{R} \cup \{p_0\}$. The interior of Γ_0 is filled with a one parameter family $\{q^\alpha(t) | \alpha \in (-1,0)\}$ of periodic orbits converging on Γ^0

as $\alpha \to 0$.

(A3) Let $H(q^\alpha) = h^\alpha$ and the period of $q^\alpha(t)$ be T^α . Then $T^\alpha = T(h^\alpha)$ is a differentiable function and $dT^\alpha/dh^\alpha > 0$ inside Γ^0 .

(A2-3) imply that $T^\alpha, dT^\alpha/dh^\alpha \to \infty$ as $\alpha \to 0^-$. Our assumptions imply that the unperturbed phase portrait looks like "one half" of Figure 1. Under assumption (A1) it can be shown that, for $\varepsilon \neq 0$, small, the Poincare map P_ε of (3.2) has a hyperbolic saddle point $p_\varepsilon = p_o + O(\varepsilon)$ with local stable and unstable manifolds $W^s_{loc}(p_\varepsilon)$, $W^u_{loc}(p_\varepsilon)$ which are ε-C^r close to the unperturbed saddle-loop Γ_o near p_ε . In fact the saddle p_o perturbs to a periodic orbit $\gamma_\varepsilon(t)$ whose <u>global</u> stable and unstable manifolds $W^s(\gamma_\varepsilon)$, $W^u(\gamma_\varepsilon)$ are filled with one parameter families of solutions $q^s_\varepsilon(t,t_o)$, $q^u_\varepsilon(t,t_o)$ which each can be expanded in a power series on a <u>semi-infinite interval</u>:

$$\begin{aligned} W^s&: q^s_\varepsilon(t,t_o) = q^o(t-t_o) + \varepsilon q^s_1(t,t_o) + O(\varepsilon^2) \ , \ t \in [t_o,\infty) \\ W^u&: q^u_\varepsilon(t,t_o) = q^o(t-t_o) + \varepsilon q^u_1(t,t_o) + O(\varepsilon^2) \ , \ t \in (-\infty,t_o] \ . \end{aligned} \tag{3.3}$$

Here the "t_o" refers to the fact that the solution is based at $t = t_o$ on the cross section

$$\Sigma^{t_o} = \{(x,\theta) | \theta = t^o \in [0,T]\} \ . \tag{3.4}$$

Thus the first variational equations,

$$\dot{q}^{s,u}_1(t,t_o) = Df(q^o(t-t_o))q^{s,u}_1(t,t_o) + g_\mu(q^o(t-t_o),t) \ , \tag{3.5}$$

can be integrated on semi-infinite integrals. A delicate analysis of solutions passing near the saddle (Greenspan [1981]) shows that similar results apply to periodic orbits and one can use expansions of the form

$$q_t^\alpha(t,t_o) = q^\alpha(t-t_o) + \varepsilon q_1^\alpha(t,t_o) + O(\varepsilon^2) , \tag{3.6}$$

on arbitrarily long intervals $t \in [t_o, t_o+T^\alpha]$, where $0 < \varepsilon \le \varepsilon_o$ and ε_o is independent of α.

The next important idea is that of the distance, $d(t_o)$, at the point $q_t(0)$, between the stable and unstable manifolds $W^u(\gamma_\varepsilon)$, $W^s(\gamma_\varepsilon)$ on the section Z^{t_o}, measured normal to the unperturbed vector $f(q\ (0))$. We have

$$d(t_o) = \frac{f(q^o(0))\wedge(q_\varepsilon^u(t_o)-q_\varepsilon^s(t_o))}{|f(q^o(0))|}$$

$$= \frac{\varepsilon f(q^o(0))\wedge(q_1^u(t_o)-q_1^s(t_o))}{|f(q^o(0))|} + O(\varepsilon^2) , \tag{3.7}$$

where the wedge product is $a\wedge b = a_1b_2 - a_2b_1$. We also introduce the time dependent "distance"

$$\begin{aligned}\Delta(t,t_o) &= f(q^o(t-t_o))\wedge(q^u(t,t_o) - q_1^s(t,t_o)) \\ &\stackrel{\text{def}}{=} \Delta^u(t,t_o) - \Delta^s(t,t_o) .\end{aligned} \tag{3.8}$$

Note that

$$d(t_o) = \varepsilon\Delta(t_o,t_o)/|f(q^o(0))| + O(\varepsilon^2) , \tag{3.9}$$

from (3.7). We now compute the time derivative of Δ, taking Δ^s first (and supressing the arguments):

$$\dot{\Delta}^s = Df(q^o)\dot{q}_o\wedge q_1^s + f(q_o)\wedge\dot{q}_1^s . \tag{3.10}$$

From (3.5) and the fact that $\dot{q} = f(q_o)$, this yields

$$\dot{\Delta}^s = Df(q^o)f(q^o)\wedge q_1^s + f(q_o)\wedge[Df(q^o)q_1^s + g_\mu(q^o,t)]$$

$$= \text{trace } Df(q^o)\Delta^s + f(q^o)\wedge g_\mu(q^o,t) \ ,$$

$$= f(q_o)\wedge g_\mu(q^o,t) \ . \tag{3.11}$$

(Recall that f is Hamiltonian and hence trace $Df \equiv 0$.) Integrating (3.11) from t_o to $+\infty$, we then have

$$\Delta^s(\infty,t_o) - \Delta^s(t_o,t_o) = \int_{t_o}^{\infty} f(q^o(t-t_o)\wedge g_\mu(q^o(t-t_o),t)dt \tag{3.12}$$

However $\Delta^s(\infty,t_o) = \lim_{t\to\infty} f(q^o(t-t_o))\wedge q_1^s(\infty,t_o)$ and $\lim_{t\to\infty} q^o(t-t_o) = p_o$,

so that $\lim_{t\to\infty} f(q^o(t-t_o)) = 0$ while $q_1^s(\infty,t_o)$ is bounded; thus $\Delta^s(\infty,t_o) = 0$

and (3.12) gives us a formula for $\Delta^s(t_o,t_o)$. A similar calculation gives

$$\Delta^u(t_o,t_o) = \int_{-\infty}^{t_o} f(q_o(t-t_o))\wedge g_\mu(q^o(t-t_o),t)dt \ , \tag{3.13}$$

and addition of (3.12) and (3.13) and use of (3.9) yields

$$d(t_o) = \varepsilon \int_{-\infty}^{\infty} f(q^o(t))\wedge g_\mu(q^o(t),t+t_o)dt/|f(q^o(0))| + O(\varepsilon^2) \ . \tag{3.14}$$

(Note that we have changed variables in the integral for convenience.) We now define the <u>Melnikov function</u> (Melnikov [1963]):

$$M_\mu^\infty(t_o) = \int_{-\infty}^{\infty} f(q^o(t))\wedge g(q^o(t),t+t_o))dt \tag{3.15}$$

Since $|f(q^o(0))| = O(1)$, $M^\infty(t_o)$ provides a good measure of separation of the manifolds at $q^o(0)$ on Σ^{t_o} . We recall that the vector $f(q^o(0))$ and its base point $q^o(0)$ are fixed on the section Σ^{t_o} and that, as t^o varies, Σ^{t_o} sweeps around $\mathbb{R}^2 \times S^1$. Thus, if $M_\mu^\infty(t_o)$ oscillates about zero with maxima and minima independent of ε , then $q_\varepsilon^u(t_o)$ and $q_\varepsilon^s(t_o)$

must change their orientation with respect to $f^{\perp}(q^{o}(0))$ as t^{o} varies. This implies that there must be a time $t_o = \tau$ such that $q_{\varepsilon}^{s}(\tau) = q_{\varepsilon}^{u}(\tau)$ and we have a homoclinic point $q \in W^{s}(p_{\varepsilon}^{\tau}) \cap W^{u}(p_{\varepsilon}^{\tau})$. But since all the Poincare maps $P_{\varepsilon}^{t_o}$ are equivalent, $W^{s}(p_{\varepsilon}^{t_o})$ and $W^{u}(p_{\varepsilon}^{t_o})$ must intersect for all $t_o \in [0,T]$. Moreover, if the zeros are simple $\frac{dM_{\mu}^{\infty}}{dt_o} \neq 0$, then it follows that the intersections are transversal. Conversely, if no zeros exist, then $q_{\varepsilon}^{u}(t_o)$ and $q_{\varepsilon}^{s}(t_p)$ retain the same orientation and hence the manifolds do not interse-t. We summarize as follows:

Theorem 3.1. Let assumption (A.1) hold and suppose that $M_{\mu}^{\infty}(t_o)$ has simple zeros. Then, for $\varepsilon \neq 0$ sufficiently small, $W^{s}(p_{\varepsilon})$ and $W^{u}(p_{\varepsilon})$ intersect transversally. If $M_{\mu}^{\infty}(t_o)$ is bounded away from zero then $W^{s}(p_{\varepsilon}) \cap W^{u}(p_{\varepsilon}) = \phi$. Moreover, if $M_{\mu}^{\infty}(\tau) = \frac{\partial M_{\mu}^{\infty}}{\partial t_o}(\tau) = 0$ but $\frac{\partial^2 M_{\mu}^{\infty}}{t^2}(\tau) \neq 0$ and $\frac{\partial M^{\infty}}{\partial \mu}(\tau) \neq 0$ at $\mu = \mu^{\infty}$, then for $\varepsilon \neq 0$ sufficiently small and μ near μ^{∞} $W^{s}(p_{\varepsilon})$ and $W^{u}(p_{\varepsilon})$ have a homoclinic orbit of quadratic tangency.

This result will be illustrated in a moment. A similar theorem describes how the infinite family of periodic orbits within Γ^{o} breaks up. We select an orbit $q^{\alpha(m)}$ of period $T^{\alpha} = mT$ (or more generally $\frac{mT}{n}$) in resonance with the forcing function g, and define the subharmonic Melnikov function

$$M_{\mu}^{m}(t_o) = \int_{0}^{mT} f(q^{\alpha(m)}t) \wedge g(q^{\alpha(m)}(t), t+t_o)dt \,. \tag{3.16}$$

We then have

Theorem 3.2. Let assumptions (A1-3) hold and suppose that $M_{\mu}^{m}(t)$ has simple zeros. Then for $\varepsilon \neq 0$ sufficiently small, near the level curve $H(q^{\alpha(m)})$ there is an isolated subharmonic orbit of period mT. If $M_{\mu}^{m}(t_o)$ is bounded away from zero there is no subharmonic near $q^{\alpha(m)}$. If $M_{\mu}^{m}(\tau) =$

$\frac{\partial M_\mu^m}{\partial t_o}(\tau) = 0$ but $\frac{\partial^2 M_\mu^m}{\partial t_o^2}(\tau)$, $\frac{\partial M_\mu^m}{\partial t_o^2}(\tau) \neq 0$ at $\mu = \mu^m$, then, for $\varepsilon \neq 0$ sufficiently small there is a saddle node bifurcation of periodic orbits near $\mu = \mu^m$. Moreover, the bifurcation values μ^m accumulate on μ^∞ as $m \to \infty$.

Similar results were obtained independently, using different methods, by Chow et. al. [1980]. The Melnikov machinery has been generalized to perturbations of multidegree and even infinite degree of freedom Hamiltonian systems in a series of papers by Holmes and Marsden [1981, 1982a,b, 1983].

We now apply these theorems to Duffing's equation, in which the force and damping $\gamma = \varepsilon\bar{\gamma}$, $\delta = \varepsilon\bar{\delta}$ are assumed to be both small and of the same order. For our problem the wedge $f \wedge g_\mu$ takes the form (cf. eq. (1.2))

$$v^\alpha(t)(\bar{\gamma}\cos\omega(t+t_o) - \bar{\delta}v^\alpha(t)) \tag{3.17}$$

and integration of (3.15) and (3.16) between the appropriate limits yields

$$M_\mu^o(t_o) = -\frac{4\bar{\delta}}{3} - \sqrt{2}\pi\omega\bar{\gamma}\ \mathrm{sech}\left(\frac{\pi\omega}{2}\right)\sin\omega t_o \tag{3.18}$$

for the (double) homoclinic orbit, and

$$M_\mu^m(t_o) = -\bar{\delta}J_1(m,\omega) - \bar{\gamma}J_2(m,\omega)\sin\omega t \tag{3.19}$$

for the periodic orbits within the homoclinic loop. Similar computations yield the subharmonic Melnikov function

$$\hat{M}_\mu^m(t_o) = -\bar{\delta}\hat{J}_1(m,\omega) - \bar{\gamma}\hat{J}_2(m,\omega)\sin\omega t_o \ , \ (m \text{ odd}) \tag{3.20}$$

for the 'large' periodic orbits which encircle both loops. The symmetry of the unperturbed flow leads to the excitation of only odd subharmonics in (3.20). In (3.19) and (3.20) $J_1(m,\omega)$ and $J_2(m,\omega)$, etc. are complicated positive functions of elliptic integrals arising from the unperturbed periodic solutions of (3.1). For details, see Greenspan and Holmes [1982]. Here we merely need note that for ω fixed, as $m \to \infty$, the ratios

$$R^m(\omega) = J_1(m,\omega)/J_2(m,\omega) \tag{3.21}$$

and

$$\hat{R}^m(\omega) = \hat{J}_1(m,\omega)/\hat{J}_2(m,\omega)$$

rapidly approach

$$R^\infty(\omega) = \frac{4}{3\sqrt{2}\pi\omega} \cosh\left(\frac{\pi\omega}{2}\right) \tag{3.22}$$

from below and above respectively. We deduce that the subharmonic bifurcation curves $\ldots, \gamma^m, \gamma^{m+2}, \ldots, \hat{\gamma}^m, \ldots$ are tangent at $\gamma = \delta = 0$ to a family of rays

$$\gamma = R^m(\omega)\delta, \ldots \quad \text{and} \quad \gamma = \hat{R}^m(\omega)\delta \tag{3.23}$$

which accumulate on the homoclinic bifurcation ray $\gamma = R^\infty(\omega)\delta$ from below and above respectively. The uniform validity of our perturbation results implies that, for any pair γ, δ, with $|\gamma|, |\delta| < \varepsilon_0$ and $\gamma > R^M(\omega)\delta$, "inner" subharmonics of all (admissible) orders $m \leq M$ coexist. For $\gamma > \hat{R}^M(\omega)\delta$, "outer" subharmonics of all odd orders $m \geq M$ coexist with a countable set of inner subharmonics and transverse homoclinic points. See figure 4, below.

We next consider the stability of these subharmonic orbits and the global structures associated with them. To do this, we develop an idea of Melnikov [1963] and write the general system (3.2) in action angle coordinates $I = I(u,v)$, $\theta = \theta(u,v)$ in the interior of Γ^0; so that it becomes

$$\dot{I} = \varepsilon\left(\frac{\partial I}{\partial u} g_1 + \frac{\partial I}{\partial v} g_2\right) \overset{\text{def}}{=} F(i,\theta,t)$$

$$\dot{\theta} = \frac{\partial H(I)}{\partial I} + \varepsilon\left(\frac{\partial \theta}{\partial u} g_1 + \frac{\partial \theta}{\partial v} g_2\right) \overset{\text{def}}{=} \Omega(I) + G(I,\theta,t) \; . \tag{3.24}$$

We then make a small perturbation from a resonant torus $I = I^m$ of frequency $\Omega(I^m) = \frac{2\pi}{mT}$,*letting

$$I + I^m + \sqrt{\varepsilon}h$$

$$\theta = \Omega(I^m)t + \phi \tag{3.25}$$

After some simple calculations, and the application of the averaging theorem (Hale [1969]) we obtain an autonomous system of equations describing the dynamics in a "resonant band" near $I = I^\alpha$ up to $O(\varepsilon)$:

$$\dot{\phi} = \sqrt{\varepsilon}\Omega'(I^m)h + \varepsilon\tilde{G}(\phi,h)$$

$$+ O(\varepsilon^{3/2}) \; . \tag{3.26}$$

$$\dot{h} = \sqrt{\varepsilon}\frac{1}{2}M_\mu^m \; (mT\phi/2\pi) + \varepsilon\tilde{F}(\phi,h)$$

Here $\tilde{F}$ and $\tilde{G}$ are the averages (with respect to t) of transformed versions of the functions F, Ω and G of (3.24) and their derivatives, and $\Omega'(I^m) = \partial^2 H/\partial I^2|_{I=I^m}$. For full details, see Greenspan and Holmes [1982, 1983]. We note that, at $O(\sqrt{\varepsilon})$, (3.26) is Hamiltonian and that the zeros of the Melnikov function correspond to fixed points $(\phi,h) = (\overline{\phi},0)$. Moreover via the transformation (3.25), such fixed points correspond, as they should, to subharmonics of period MT . However, with the additional $O(\varepsilon)$ terms present in (3.26) we can now determine the stability types of these subharmonics and the global structure of their stable and unstable manifolds.

*This can be generalized to all rational tori of frequency $2_\pi n/mT$.

If this structure is stable, then the averaging theorem permits us to conclude that the stable and unstable manifolds of the (subharmonic) periodic points of the full Poincare map will have the same qualitative structure as the corresponding manifolds of the averaged system (cf. Guckenheimer and Holmes [1983], Chapter 4.)

For the periodic orbits of Duffing's equation inside Γ^o, (3.26) takes the form (cf. 3.19)

$$\begin{aligned}
\dot{\phi} &= \sqrt{\varepsilon}\;\Omega'(I^m)h \\
& \qquad\qquad\qquad\qquad\qquad\qquad + O(\varepsilon)\,, \\
\dot{h} &= \sqrt{\varepsilon}\,\frac{1}{2\pi}\,[-\bar{\delta}J_1(m,\omega)-\bar{\gamma}J_2(m,\omega)\sin(m\phi)]
\end{aligned} \tag{3.27}$$

and the $O(\varepsilon)$ terms have a Jacobian with constant negative trace, implying that the fixed points are either sinks, saddles or (at bifurcation) saddle-nodes. We therefore conclude that the structure of solutions in each resonance band near $I + I^m$ is as shown in Figure 2. We note that Morosov [1973] obtained similar pictures for a different version of Duffing's equation.

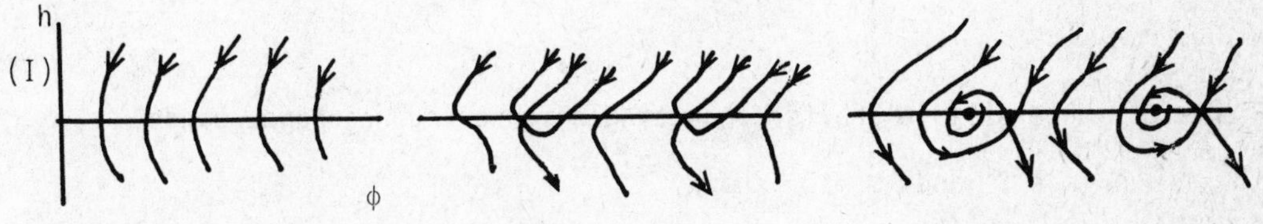

(a) $\gamma < R^m(\omega)\delta$, (b) $\gamma \quad R^m(\omega)\delta$ (bifurcation) (c) $\gamma > R^m(\omega)\delta$
Figure 2. Bifurcation to subharmonics of order M (M>2 shown)

Performing similar calculations near all resonant and non-resonant tori between any two neighboring fundamental subharmonics of periods mT and $(m+1)T$, we find that all such tori are destroyed by the presence of damping, and that no periodic points exist between such fundamental resonances. We therefore conclude that the unstable manifolds of the order (m+1) subharmonic saddle must intersect the stable manifolds of the order m subharmonics. (It is clear that no invariant tori can exist for $\delta > 0$, since the Poincare map contracts areas and hence cannot admit an invariant closed curve with non-empty interior.) See Figure 3.

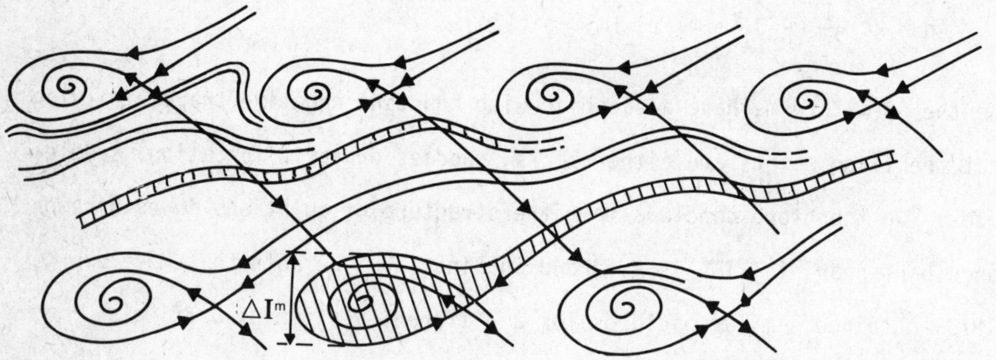

Figure 3. Heteroclinic orbits connecting neighboring resonance bands ($m = 3$ and $m = 2$ shown).

This structure is repeated between each pair of resonance bands and, assuming that the manifolds intersect transversely,* it is possible to prove, using the lambda lemma (Palis [1969], cf. Newhouse [1980]), that in

*Even if the 'primary' intersections are not transversal, 'secondary' intersections will be, and the manifolds still accumulate on each other.

any neighborhood of the stable manifold of a sink or saddle of order M there are pieces of the stable manifolds of all sinks and saddles of order $m < M$. The resultant packing of alternate domains of attractions of the stable subharmonics leads to sensitive dependence on initial conditions, especially for higher order subharmonics, since from the first integral of (3.27) it can be shown that the width ΔI^m (Figure 3) of the domains of attraction goes like

$$\Delta I^m \sim \sqrt{\varepsilon \frac{m^3}{4\pi^3} e^{-\pi m/\omega}} \, . \tag{3.28}$$

Assembling these results and performing similar computations for the outer subharmonics, we obtain the global picture of Figure 4 for the fixed points of P and their invariant manifolds. We have selected a force level $\gamma \in (\hat{R}^3(\omega)\delta, \hat{R}^1(\omega)\delta)$, for which all inner and outer subharmonics above and including order 3 are excited.

Note that, in Figure 4 we have not shown stable subharmonics near the transversal homoclinic orbits associated with the 'main' saddle point near $(u,v) = (0,0)$. This is not merely a result of our poor draughtmanship, or the finite size of our pencil. We note that, for application of averaging, the right hand side of (3.26) (before averaging) must be uniformly bounded, so that, taking $\sqrt{\varepsilon}$ sufficiently small, we can invert a certain transformation (Hale [1969]). Unfortunately, we have

$$\Omega'(I) \sim -\omega^3 e^{2\pi m/\omega}/m^3 \tag{3.29}$$

for the Duffing example, and thus, unlike our existence results, <u>our stability results are not uniformly valid</u> in ε. In fact it is almost certain (cf. Greenspan and Holmes [1982]) that, near the homoclinic points, all the subharmonic sinks lose stability repeatedly in period doubling bifurcations and that secondary transverse homoclinic and heteroclinic orbits are created. Specifically, we strongly suspect, but cannot prove by perturbation methods, that as γ,δ increase (ε is no longer small) the stable and

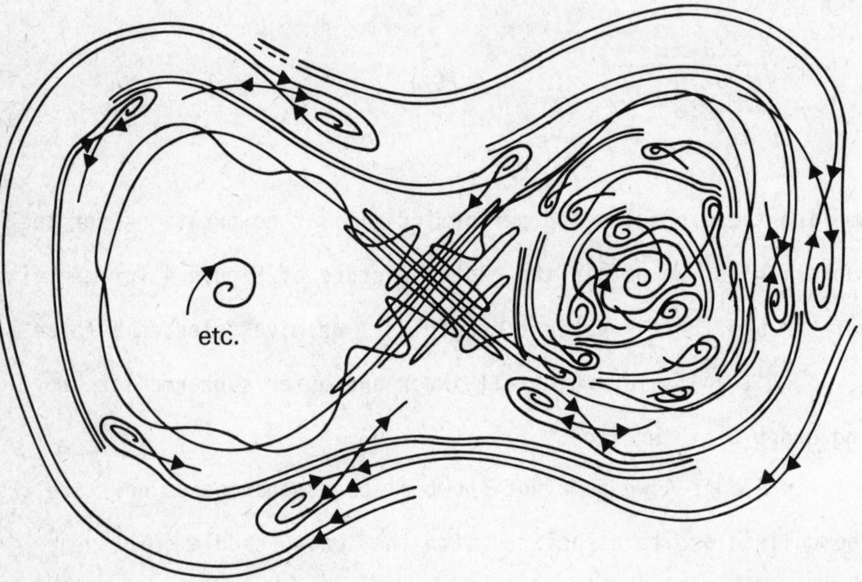

Figure 4. The Poincaré map P_γ, $\gamma \in (\hat{R}^3(\omega)\delta, \hat{R}^1(\omega)\delta)$.

unstable manifolds of saddles within a given resonant band intersect each other and ultimately that the unstable manifold of the order m saddle intersects the stable manifold of the order $(m+1)$st, yielding a homoclinic cycle. Such behavior would be analogous to the destruction of invariant (KAM) tori and development of stochastic seas in non-integrable Hamiltonian systems (cf. Chirikov [1979], Greene [1979]). It is known that in that case, stable elliptic islands can coexist with large scale stochasticity and we believe that in this case also stable sinks can still exist 'within' the homoclinic tangles.

Before leaving these analytical results we note that the presence of transverse homoclinic orbits implies via the Smale-Birkhoff homoclinic theorem (Smale [1963, 1967], cf. Moser [1973] that, near any transverse homoclinic point there is a compact invariant Cantor set Λ^N for some iterate P_γ^N of the Poincare map, on which P_γ^N is homeomorphic to a shift on two symbols: i.e., P_γ^N has a chaotic invariant set, containing countably many unstable periodic orbits, uncountably many irregular bounded non-periodic motions and a dense orbit. Such invariant sets are called horse-shoes and are examples of strange invariant sets which are not however attractors. Moreover, the proven presence of quadratic homoclinic tangencies near the ray $\gamma = R^\infty(\omega)\delta$ implies, via Newhouse's [1979, 1980] results, that P_γ has invariant 'wild hyperbolic sets' containing persistent tangencies of stable and unstable manifolds near these bifurcation points. This leads us to suspect that P_γ not only has the finite sets of periodic (subharmonic) sinks found above, but that it can also have infinite sets of sinks for a residual subset of parameter values.

4. THE STRUCTURE AND CONTENTS OF THE ATTRACTING SET FOR SMALL γ,δ.

We now assemble the perturbation results of the previous section and the global stability results of §2 in a grand finale. First we give a result which will establish the global structure of our attracting set, namely that it is the closure of a certain unstable manifold.

Using the results of §2, pick $\gamma \in (\hat{R}^M(\omega)\delta, \hat{R}^{M-2}(\omega)\delta)$ so that the outer band of order M (odd) subharmonics is excited but the next further out band (of order $M-2$) is not. Choose a saddle p in the band of order M and a saddle q in the band of order $M+2$. Recall from Lemma 2.1 that the Poincaré map P_γ has an open, simply connected trapping region D with $\overline{P_\gamma(D)} \subset D$. Henceforth we drop the subscript γ. The perturbation results show that D can be chosen such that both components of the stable manifold of p under P^M, $W^s(p,P^M)/\{p\}$, intersect the boundary of D exactly once; hence $D/W^s(p,P^M)$ consists of two sets S and T. We may assume that $q \in S$ and, by the stability results for γ,δ small, that T contains a single sink of period M. The Melnikov theory shows that both components of $W^s(q,P^{M+2})/\{q\}$ intersect the unstable manifold $W^u(p,P^M)$, (cf. figure 4), and under these conditions we have the following result.

<u>Proposition 4.1</u>. $A = \overline{\bigcap_{n\geq 0} P^n(D)} = \overline{W^u(p)}$

<u>Proof</u>. First note that, since the local unstable manifold $W^u(p)$ lies in D and $\overline{P(D)} \subset D$, we clearly have $W^u(p) \subset A$. But A is closed, so $\overline{W^u(p)} \subseteq A$. It remains to show $A \subseteq \overline{W^u(p)}$.

Let x and y lie in the intersection of $W^s(q,P^{M+2})$ and $W^u(p.P^M)$, one in each component of $W^s(q,P^{M+2})/\{q\}$, and take the order of $\{p,x,y\}$ along $W^u(p,P^M)$ in the direction away from p to be p,x,y. Let $x' = P^{-M(M+2)}(x)$, $y' = P^{-M(M+2)}(y)$. Then p,x',y',x,y lie along $W^u(p,P^M)$ in that order and the arcs $\widehat{xx'}$ of $W^s(q,P^{M+2})$ and $W^u(p,P^M)$ bound a closed region R which lies inside D somewhat as shown in Figure 5. Let $R = \bigcap_{n\geq 0} P^n(R)$.

We now claim that $R \subset \overline{W^u(p)}$. To establish the claim, take a point $r \notin \overline{W^u(p)}$. Then we can find an $\varepsilon > 0$ so that $B_\varepsilon(r)$, the ε-ball centered at r, is disjoint from $W^u(p)$. By our choice of R, $B_\varepsilon(r)$ is also dis-

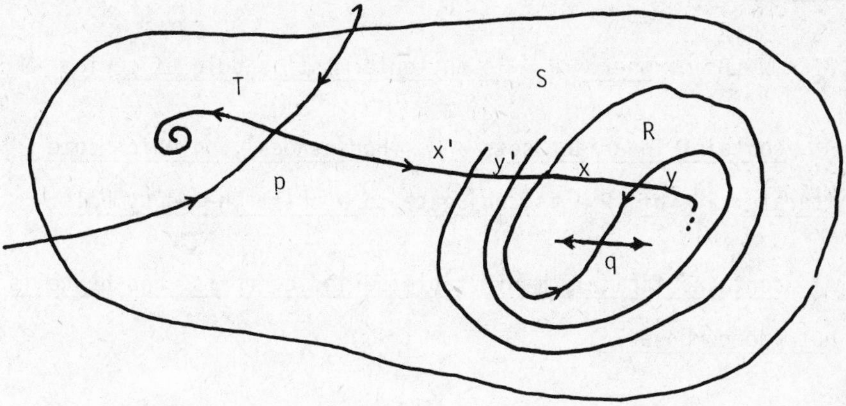

Figure 5. Proof of Proposition 4.1.

joint from the boundary of $P^k(R)$, for sufficiently large k. Since for $x \in D$, $\det DP(x) < 1$, the Lebesgue measure of the sets $P^k(R)$ tends to zero as $k \to \infty$. Thus for large k, $B_\varepsilon(r) \cap P^k(R) = \emptyset$, so that $r \notin R$ and therefore $R \subset \overline{W^u(p)}$ as claimed.

Now for small γ, δ points not trapped in the outermost resonance band eventually pass through to the next lower band, i.e., all points in $S' = S \setminus \bigcup_{k=0} P^k(T)$ are mapped into R by some iterate of P, and so $\bigcap_{n \geq 0} P^n(S') = R$. Since $\bigcap_{n \geq 0} P^{Mn}(W^s(p, P^M)) = \{p\}$ and $\bigcap_{n \geq 0} P^{Mn}(T)$ is the component of $W^u(p, P^M)/\{p\}$ lying in T, it follows that $A \subseteq \overline{W^u(p)}$ and the proof is complete.

Taking the proposition together with the existence and stability results of §3 we have our grand finale:

<u>Theorem 4.2</u>. <u>For</u> $\gamma, \delta \neq 0$ <u>sufficiently small and</u> $\gamma \in (\hat{R}^M(\delta), \hat{R}^{M-2}(\delta))$,

the attracting set $A = \bigcap_{n=0} P_\gamma^n(D)$ _for the Duffing equation satisfies the following conditions_:

(i) $A_\gamma = \overline{W^u(P^M)}$, _where_ P^M _is an 'outermost' saddle of period_ M.

(ii) A_γ _contains invariant sets_ Λ^N (_horseshoes_), _on which some iterate_ P_γ^N _of_ P_γ _is conjugate to a shift on two symbols_.

(iii) A_γ _contains finitely many stable periodic orbits, and hence is not indecomposable_.

We note that, if the unstable manifold $W^u(p_1)$ of the 'central' saddle point p_1 near $(u,v) = (0,0)$ intersects the stable manifold of one of the outer saddles p (and hence of all of them) then we can equally well characterize A_γ as $\overline{W^u(p_1)}$. As noted by Greenspan and Holmes [1982] this appears to be the situation actually observed with moderate γ, δ.

We end this section, and the paper, with some additional description of the attracting set and we illustrate our results with some numerical computations due to Ueda [1981].

We first remark that the presence of 'double' transverse homoclinic orbits, due to the symmetry of the equation, implies that we can find strips S_+, S_- 'parallel' to the stable manifold $W^s(p_1)$ of the period one saddle near $(u,v) = (0,0)$ such that their images under some iterate of P, $P^N(S_\pm)$ are as shown in Figure 6. Thus, since the intersections $P^N(S_\pm) \cap S_\pm$ are non-empty for all choices of + and -, we can find orbits which circulate arbitrarily many times around the right hand loop (..++++...) followed by arbitrarily many times around the left hand loop (...----...), etc. Such orbits provide a specific realization of the shift on two symbols (+,-)

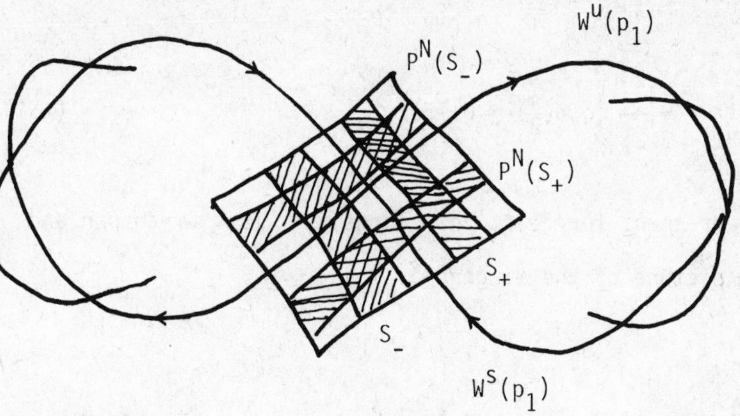

Figure 6. The double homoclinic structure.

and represent chaotic motions similar to some of those observed in numerical integrations (cf. Figure 7). However, all the chaotic orbits found in this way are of unstable 'saddle' type and coexist with stable subharmonic sinks, and thus we do not expect to see sustained motions like those of Figure 7. (for low γ,δ). In fact for low γ,δ , numerical work shows that a transient period of chaos is generally followed by stable low order subharmonic motions (cf. Greenspan and Holmes [1982, fig 6.4]) and that higher values of γ and δ are necessary if sustained chaos is to be generated.

In Figure 8 we show the results of numerical integrations due to Ueda [1981]. Figures 7a and 7b show the stable and unstable invariant manifolds $W^s(p_1)$ and $W^u(p_1)$ of the saddle point p_1 for the Poincare map. In 8(a) the force level is low, $\bar{\gamma} < R^1(\omega)|_{\omega=1}^{\bar{\delta}}$, and the map is very simple, possessing only the saddle p_1 and two sinks near the unperturbed centers at $(u,v) = (\pm 1,0)$. In Figure 8(b) we show the manifolds for a higher force level, $\bar{\gamma} \in (\hat{R}^3(1)\bar{\delta},\hat{R}^1(1)\bar{\delta})$, for which all subharmonics but the outermost (period 1) are excited. We note that the manifolds are observed to touch first at $\overline{\gamma}/\delta$ 0.76, in comparison with the theoretical value from equation (3.22) of

$$\overline{\gamma}/\delta = R^{\infty}(\omega)\Big|_{\omega=1} = \frac{4}{3\sqrt{2\pi}}\cosh\left(\frac{\pi}{2}\right) \simeq 0.753; \tag{4.1}$$

quite a creditable agreement for relatively large ε! (see Greenspan and Holmes [1982] for a picture of the tangency).

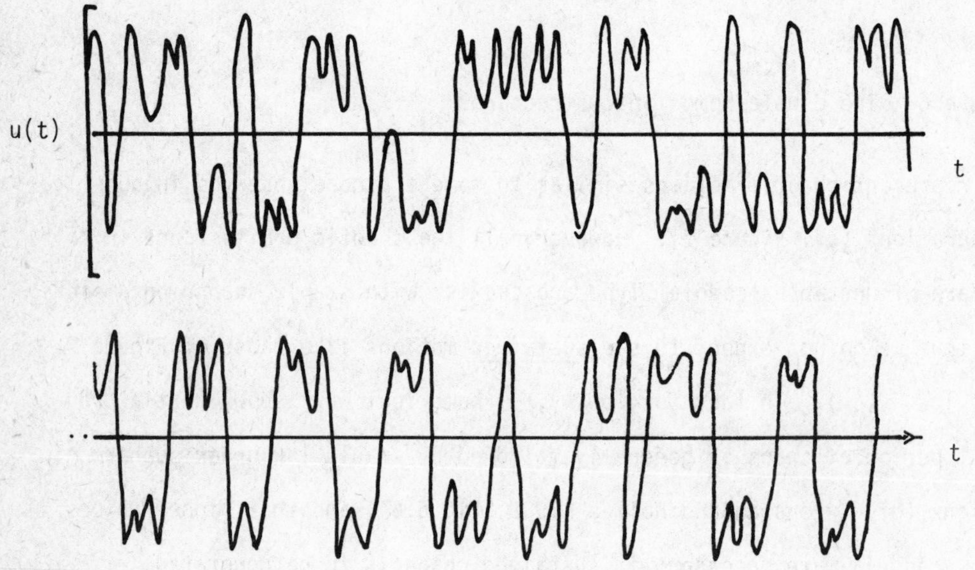

Figure 7. A typical time series of $u(t)$ in the 'chaotic' region.

(Facing page)

Figure 8. Poincare maps for Duffing's equation, $\delta = 0.25$, $\omega = 1.0$. (a), (b) invariant manifolds of p_1. (c) A single orbit.

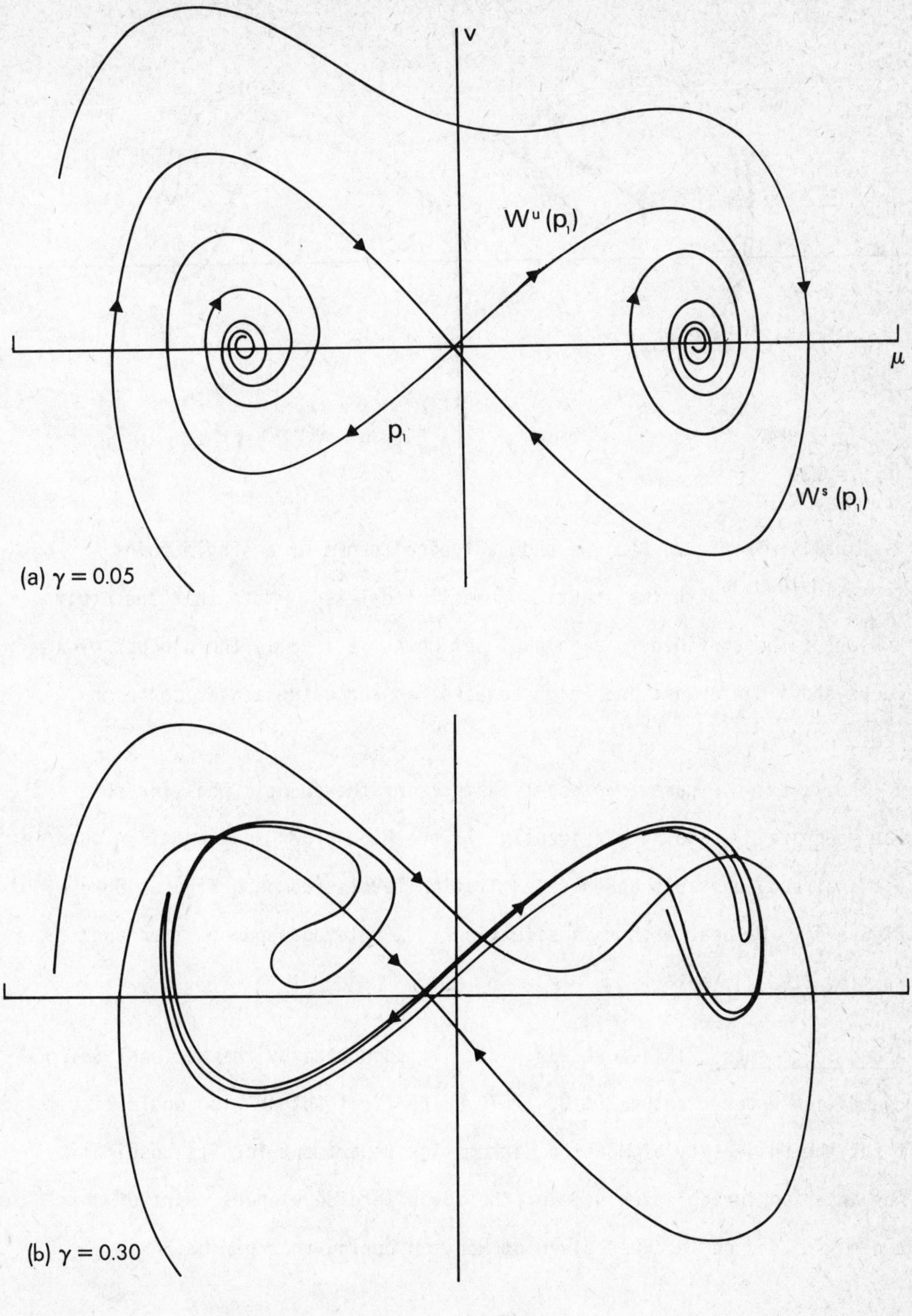
v
$W^u(p_1)$
μ
p_1
$W^s(p_1)$
(a) $\gamma = 0.05$
(b) $\gamma = 0.30$

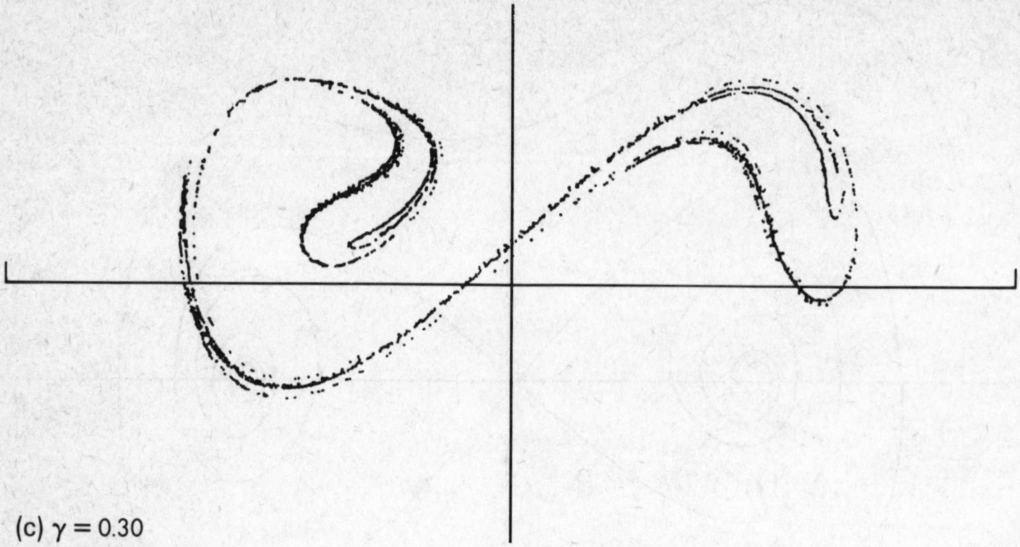

(c) $\gamma = 0.30$

Finally, in Figure 8(c) we show a typical orbit of a single point $\{P^n(x)\}_{n=50}^{n=1000}$, with the starting transient deleted. Note that the orbit appears to be confined to a compact set which is locally the product of a curve and a Cantor set and which appears, as suggested above, to be the closure $\overline{W^u(p_1)}$.

The last two Figures represent examples of the chaotic behavior observed for moderate γ and δ; evidently all the high order subharmonics found in 3 are already unstable at these excitation levels (compare Figures 8b-c with Figure 4). We deal with this situation in a related paper (Holmes and Whitley [1983]).

Acknowledgements. This work was partially supported by the National Science Foundation under grant # MEA 80-17570. The first author also would like to thank the University of Houston Mathematics Department for its hospitality for a period during which some of the ideas were developed. Part of this paper is based on lectures given at Houston during that period.

REFERENCES

1. A.A. Andronov, E.A. Vitt and S.E. Khaiken, Theory of Oscillators, Pergamon Press, Oxford, 1966.
2. R. Bowen, On Axiom A Diffeomorphisms, Amer. Math. Soc. Regional Conference Series in Math #35, 1978.
3. D.R.J. Chillingworth, Differential Topology with a View to Applications, Pitman, London, 1976.
4. B.V. Chirikiv, A Universal Instability of Many Fimensional Oscillator Systems, Physics Reports 52 (1979), 263-379.
5. S.N. Chow, J.K. Hale and J. Mallet-Paret, An Example of Bifurcation to Homoclinic Orbits. J. Diff. Eqns. 37 (1980), 351-373.
6. J.M. Greene, Method for Determining a Stochasitc Transition. J. Math. Phys. 20 (1979), 1183-1201.
7. B.D. Greenspan, Bifurcations in Periodically Forced Oscillations: Subharmonics and Homoclinic Orbits. Ph.D. Thesis, Center for Applied Mathematics, Cornell University, 1981.
8. B.D. Greenspan and P.J. Holmes, Homoclinic Orbits, Subharmonics and Global Bifurcation in Forced Oscillators. To appear in Nonlinear Dynamics and Turbulence, eds. G. Barenblatt, G. Iooss and D.D. Joseph, Pitman, London, 1982.
9. B.D. Greenspan and P.J. Holmes, Repeated Resonance and Homoclinic Bifurcation in a Periodically Forced Family of Oscillators (submitted for publication), 1983.

10. J. Guckenheimer and P.J. Holmes, Nonlinear Oscillations, Dynamical Systems and Bifurcations of Vectorfields, Addison-Wesley (to appear), 1983.

11. J.K. Hale, Ordinary Differential Equations, Wiley, New York, 1969.

12. M. Henon, A Two Dimensional Mapping with a Strange Attractor, Comm. Math. Phys. 50 (1976), 69-77.

13. P.J. Holmes, A Nonlinear Oscillator with a Strange Attractor, Phil. Trans. Roy. Soc. A292 (1979), 419-448.

14. P.J. Holmes and J.E. Marsden, A Partial Differential Equation with Infinitely Many Periodic Orbits: Chaotic Oscillations of a Forced Beam. Archive for Rational Mechanics and Analysis 76 (1981), 135-166.

15. P.J. Holmes and J.E. Marsden, Horseshoes in Perturbations of Hamiltonian Systems with Two Degrees of Freedom. Comm. Math. Phys. 82 (1982a), 533-544.

16. P.J. Holmes and J.E. Marsden, Melnikov's Method and Arnold Diffusion for Perturbations of Integrable Hamiltonian Systems. J. Math. Phys. 23 (1982b), 689-675.

17. P.J. Holmes and J.E. Marsden, Horseshoes and Arnold Diffusion for Hamiltonian Systems of Lie Groups. Indiana U. Math. J., 1983 (in press).

18. P.J. Holmes and D.C. Whitley, On the Attracting Set for Duffing's Equation, II: A Geometrical Model for Moderate Force and Damping. Proc. 'Order in Chaos,' Los Alamos National Laboratory, May 1982 (1983)--to appear in Physica D.

19. V.K. Melnikov, On the Stability of the Center for Periodic Perturbations. Trans. Moscow Math. Soc., 12 (1963), 1-57.

20. F.C. Moon and P.J. Holmes, A Magnetoelastic Strange Attractor. J. Sound and Vibration, 65 (1979), 275-296.

21. A.D. Morosiv, Approach to a Complete Qualitative Study of Duffing's Equation. USSR J. Comp. Math and Math. Phys., 13 1134-1152.

22. J. Moser, Stable and Rando- Motions in Dynamical Systems, Princeton University Press, 1973.

23. S.E. Newhouse, The Abundance of Wild Hyperbolic Sets and Non-Smooth Stable Sets for Diffeomorphisms. Publ. IHES 50 (1979), 101-151.

24. S.E. Newhouse, Lectures on Dynamical Systems, in Dynamical Systems, C.I.M.E. Lectures Bressanone, Italy, June 1978, Progress in Mathematics #8, Birkhauser, Boston, 1980.

25. J. Palis, On Morse-Smale Dynamical Systems, Topology 8, (1969), 385-405.

26. S. Smale, Diffeomorphisms with Many Periodic Points. In Differential and Combinatorial Topology, ed. S.S. Cairns, Princeton Univ. Press, 1963.

27. S. Smale, Differentiable Dynamical Systems. Bull. Amer. Math. Soc. 73 (1967), 747-817.

28. Y. Ueda, Personal Communication; also see Explosion of Strange Attractors Exhibited by Duffing's Equation. Ann. N.Y. Acad. Sci. 357 (1981), 422-434.

Philip J. Holmes and David C. Whitley

Cornell University

L MARKUS

Generic conservative dynamics

1. PHILOSOPHICAL AND MATHEMATICAL MOTIVATIONS.

Genericity is a fundamental philosophical principle that has often provided an intellectual focus in diverse branches of mathematics, and that has served as a central guide in much of the research in theoretical dynamics during the past few decades. Very roughly speaking, a mathematical model of a physical phenomenon is generic in case "unnatural irregularities" and "unlikely coincidences" have been ruled out of the considerations. Thus this principle reflects our attempts to deal with the typical or essential nature of the phenomenon. Of course, what is deemed "unnatural" or "unlikely" depends on the attitude or approach that we bring to our theoretical analysis, and on which mathematical - physical aspects we intend to isolate or emphasize. It is an important aspect of the science of applied mathematics to define these terms in a manner that is agreeable to the appropriate engineer or scientist, in each particular area.

In this sense a basic activity of the applied mathematician, or theoretical engineer, or scientist is to pose significant questions - as well as to seek relevant answers. In particular, much of mathematical research can be viewed as an attempt to construct conceptual frameworks with specialized technical vocabularies so as to facilitate the discussion about physical reality in a revealing way.

As an illustration of these ideas let us consider a free (unforced) oscillator: say, a mass m with position coordinate x along a line of motion, at time t. Then Newton's law asserts

$$m \frac{d^2x}{dt^2} = F(x, \frac{dx}{dt}) ,$$

where the force $F(x,y)$ depends continuously and smoothly on the position x and the velocity $y = \frac{dx}{dt}$. We display this dynamical system graphically as a steady-state (time-independent) vector field $(y, F(x,y))$ at each point (x,y) in the state-space (2-dimensional real phase-plane): say $m = 1$, and use the notation $\dot{x}$ for $\frac{dx}{dt}$ as usual)

$$\dot{x} = y$$

$$\dot{y} = F(x,y)$$

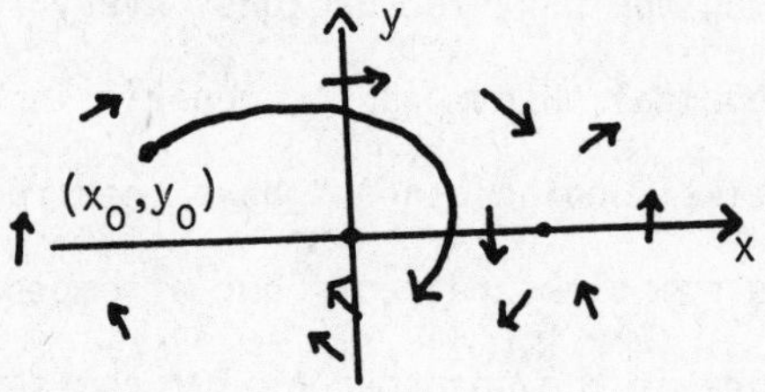

Figure 1 Vector Field in $\mathbb{R}^2$

Then for each initial state (x_0,y_0), there exists a unique solution curve $x(t)$, $y(t)$ that starts at (x_0,y_0) when $t = 0$. In this situation we can introduce the analogy of the solution curves or trajectories with the streamlines of a fluid flow along the vector field $(y, F(x,y))$. This graphical technique, while failing to display the t-dependence explicitly, has the great advantage of displaying the total "phase-portrait" of all solution curves in a single over-view. In particular we can immediately recognize the equilibrium states as the critical points or zeros of the given vector field, and we observe periodic orbits as simple closed curves in the state-space.

For further definiteness let us consider the case of a damped linear oscillator,

$$\ddot{x} = F(x,\dot{x}) = -k^2x - 2b\dot{x} \qquad \text{(constants } b \geq 0 \text{ , } k > 0\text{).}$$

The corresponding first order differential system in the $(x, y=\dot{x})$ phase plane is given by

$$\dot{x} = y$$

$$\dot{y} = -k^2x - 2by \ .$$

Possible qualitative behaviors for the solutions are indicated in Figure 2.

To specify a generic set of such linear oscillators we could exclude both the critically damped $(b = k)$, and the conservative $(b = 0)$ since these excluded cases are defined by "unlikely coincidences" for the value of the frictional coefficient $b \geq 0$. On the other hand, if our theory concerns general nonlinear oscillations, then all linear cases could be excluded from the study of generic dynamics.

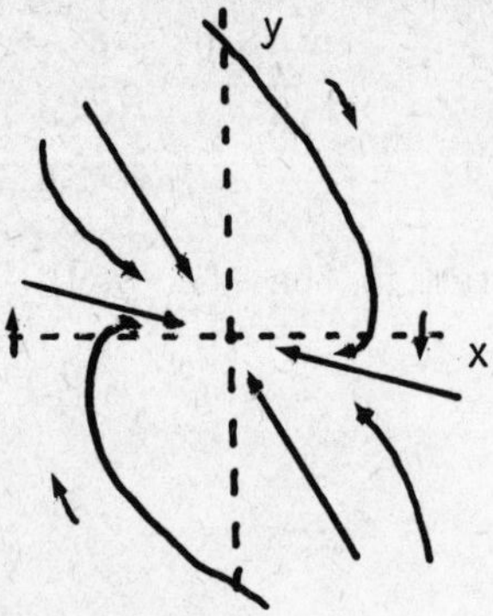

Overdamped (Node)

$b > k$

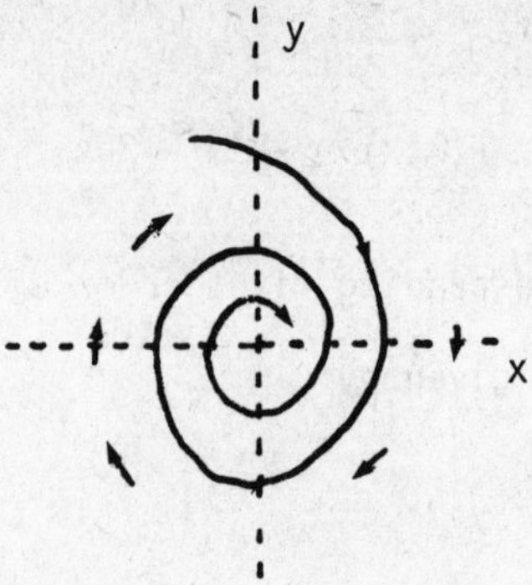

Underdamped (Spiral Focus)

$k > b > 0$

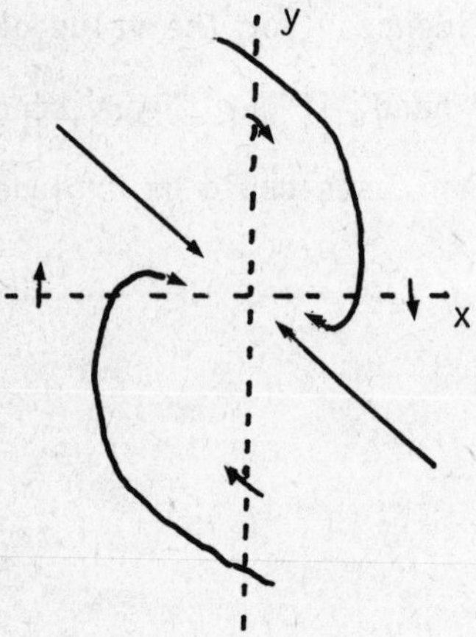

Critically Damped

(One-Branch Node)

$b = k$

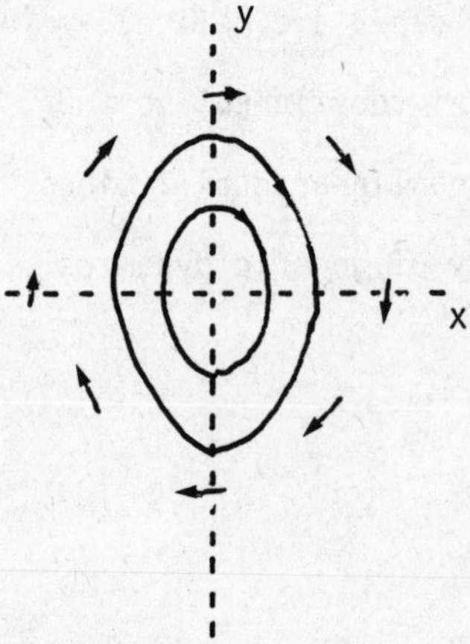

Conservative (Center)

$b = 0$

Figure 2

Damped Linear Oscillator

$$\ddot{x} + 2b\dot{x} + k^2 x = 0 \quad (\text{constants } b \geq 0 \text{ , } k > 0)$$

2. GENERICITY - RELATIVE TO A CLASS OF SYSTEMS.

Consider a first-order differential system

$$S)\quad \dot{x}^i = f^i(x^1, x^2, \ldots, x^n) \qquad \text{for} \quad i = 1, 2, \ldots, n$$

or in vector notation with the state n-vector x ,

$$\dot{x} = f(x) \, .$$

Here the real vector function $f(x)$ is defined in the real n-dimensional state space $\mathbb{R}^n$ and is everywhere continuous and smooth (all partial derivatives of all orders continuous everywhere).

We define another smooth differential system

$$S_\varepsilon)\quad \dot{x} = f(x) + \varepsilon(x)$$

to be in the ε_0-neighborhood of S) in case:

$$\max_{i,j}\left[|\varepsilon^i(x)| + \left|\frac{\partial\varepsilon^i}{\partial x^j}\right| + \ldots + |D^k\varepsilon^i|\right] < \varepsilon_0 \, .$$

Here the neighborhood is determined by $\varepsilon_0 > 0$ and the total order k of partial derivatives specified. Furthermore, in order to demand that the approximation be especially close in delicate regions, say as $x \to \infty$, we allow $\varepsilon_0 = \varepsilon_0(x)$ to be any continuous positive function for $x \in \mathbb{R}^n$. Thus the neighborhood of S) can be made more restrictive by increasing the order k , and decreasing the error tolerance $\varepsilon_0(x)$. With this concept of neighborhood we can define a <u>topology (Whitney C^∞-topology) on the collection $\mathcal{S}$</u> of all such differential systems. In particular we can consider an open subset Σ of $\mathcal{S}$ (small perturbations remain within Σ) or a dense subset Σ of $\mathcal{S}$ (each system of $\mathcal{S}$ can be approximated arbitrarily by elements of Σ). An open and dense subset Σ of $\mathcal{S}$ is called generic. Usually we also allow finite, or even countable intersections, of such open-dense subsets as generic subsets of $\mathcal{S}$.

Consider the linear differential system in $\mathbb{R}^n$

L) $\dot{x} = Ax$

where A is a real constant $n \times n$ matrix. Of course, within the space $\mathcal{S}$ of all differential system in $\mathbb{R}^n$, the set $\mathcal{L}$ of all linear systems is closed and nowhere dense (its complement is open-dense in $\mathcal{S}$). Hence generically, there are no linear differential systems in $\mathcal{S}$.

In order to remedy this unfortunate situation we introduce the concept of genericity relative to a class $\mathcal{C}$ of differential systems.

Definition. Let $\mathcal{C}$ be a subset of the space $\mathcal{S}$ of all smooth differential systems in $\mathbb{R}^n$. A subset Σ of $\mathcal{C}$ is called generic relative to $\mathcal{C}$ in case:

i) Σ is relatively open in $\mathcal{C}$
(small perturbations of $S \in \Sigma$ remain in Σ - provided they are in the class $\mathcal{C}$)

ii) Σ is dense in $\mathcal{C}$
(each system in $\mathcal{C}$ can be arbitrarily closely approximated by systems of Σ).

Also if Σ is a countable intersection of such relatively open-dense subsets of $\mathcal{C}$, then Σ is $\mathcal{C}$-generic.

Example 1. Let $\mathcal{L}$ be the collection of all linear differential systems in $\mathbb{R}^n$. Thus $\mathcal{L}$ is a closed subset of $\mathcal{S}$, and moreover the topology on $\mathcal{L}$ collapses to the usual topology on the coefficient matrices.

We can specify the generic set $\Sigma_0 \subset \mathcal{L}$ of (nondegenerate) linear systems with

$$\det A \neq 0 ,$$

that is, the eigenvalues $\lambda_1, \lambda_2, \ldots, \lambda_n$ of A are all nonzero. Another generic subset Σ_H consists of all (hyperbolic) linear systems with $\mathrm{Re}\,\lambda_i \neq 0$ for $i = 1, \ldots, n$. However, the stable linear systems, say, specified by

$$\mathrm{Re}\,\lambda_i < 0 \qquad i = 1, \ldots, n$$

fail to constitute a generic set in $\mathcal{L}$ - because, while forming an open set in $\mathcal{L}$, they fail to be dense in $\mathcal{L}$.

Example 2. Consider smooth differential systems in the (x,y) phase plane $\mathbb{R}^2$

$$S) \quad \dot{x} = f(x,y) , \quad \dot{y} = g(x,y) .$$

We restrict attention to the class $\mathcal{D} \subset \mathcal{S}$ of dissipative systems, say those systems for which the point at ∞ is a repellor - that is, ∞ is a (nondegenerate) unstable source. Then $\mathcal{D}$ is an open subset of $\mathcal{S}$ - hence $\mathcal{D}$ is a Baire space in which each generic subset is dense. We now specify generic conditions, that is, properties specifying a $\mathcal{D}$-generic set.

1) Each critical point in $\mathbb{R}^2$ is isolated and hyperbolic (No eigenvalue λ is pure imaginary).
2) Each periodic orbit is isolated and hyperbolic (No multiplier μ has unit modulus; in the plane, $\mu \neq 1$)
3) No two saddle points are joined along separatrices.

Pontryagin [2] found this generic subset of $\mathcal{D}$, and moreover showed that it was open and dense in $\mathcal{D}$, when he first formulated the concept of structural stability.

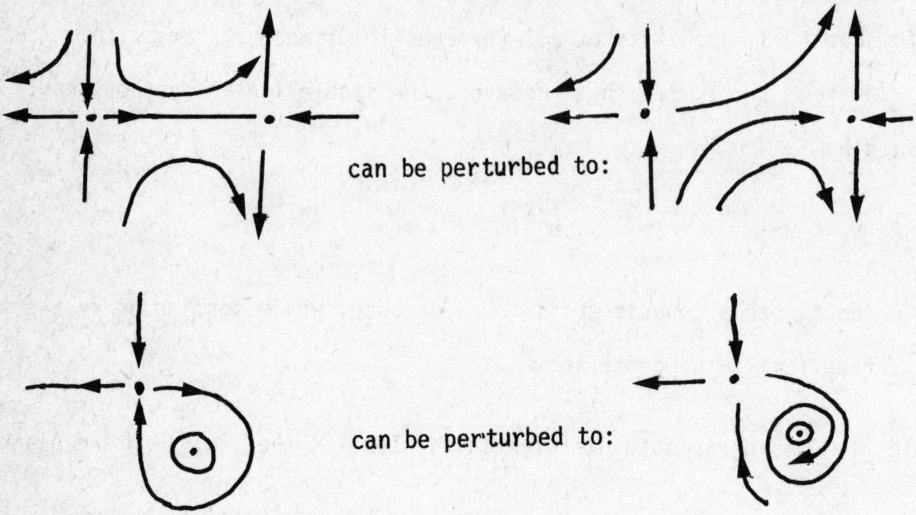

Figure 3

Perturbations to $\mathcal{D}$-genericity in $\mathbb{R}^2$

3. GENERIC CONSERVATIVE (HAMILTONIAN) DYNAMICAL SYSTEMS.

The frictionless multi-oscillator gives a prototype model for conservative dynamics, which we interpret in the mode of Hamilton. Here, for n degrees of positional freedom $x \in \mathbb{R}^n$,

$$\ddot{x}^i = -\text{grad } G(x)$$

or

$$\begin{cases} \dot{x}^i = \dfrac{\partial H}{\partial y^i} & (\dot{x}^i = y_i) \\ \dot{y}_i = -\dfrac{\partial H}{\partial x^i} & (\dot{y}_i = -\dfrac{\partial G}{\partial x^i}) \quad \text{for } i = 1, 2, \ldots, n . \end{cases}$$

Here the Hamiltonian function is

$$H(x,y) = \sum_{i=1}^{n} \frac{(y_i)^2}{2} + G(x) .$$

For our more general study we consider any smooth Hamiltonian function $H(x^1, \ldots, x^n, y_1, \ldots, y_n)$ in $\mathbb{R}^{2n}$, which is a real function of the coordinates $x \in \mathbb{R}^n$, $y \in \mathbb{R}^n$. Then the class $\mathcal{H}$ of Hamiltonian differential systems

$$\begin{cases} \dot{x} = \dfrac{\partial H}{\partial y} \\ \dot{y} = -\dfrac{\partial H}{\partial x} \end{cases}$$

is a suitable closed subset of $\mathcal{S}$ (for $\mathbb{R}^{2n}$) and we can consider generic subsets relative to $\mathcal{H}$, see [4,5].

We recall that the flow along the trajectories

$$(x_0,y_0) \to (x(t,x_0,y_0), y(t,x_0,y_0))$$

of a Hamiltonian system in $\mathbb{R}^{2n}$ has important special geometric properties[1]:
i) Energy is conserved under $(x_0,y_0) \to (x(t),y(t))$: Namely,

$$\frac{dH}{dt}(x(t),y(t)) = \frac{\partial H}{\partial x}\dot{x} + \frac{\partial H}{\partial y}\dot{y} = \frac{\partial H}{\partial x}\frac{\partial H}{\partial y} - \frac{\partial H}{\partial y}\frac{\partial H}{\partial x} \equiv 0 .$$

Thus the equi-energy hypersurfaces, $H(x,y)$ = constant, are each invariant sets under the flow in $\mathbb{R}^{2n}$.

ii) 2n - volume is conserved under $(x_0,y_0) \to (x(t),y(t))$: Namely,

$$\int\int\ldots\int dx\,dy = \int\int\ldots\int \frac{\partial(x,y)}{\partial(x_0,y_0)}\,dx_0\,dy_0 = \int\int\ldots\int dx_0\,dy_0\ .$$

This follows from calculations with

$$\operatorname{div}(H_y, -H_x) = H_{yx} - H_{xy} \equiv 0\ .$$

iii) (grad H) is normal to the energy hypersurfaces and we can further write the Hamiltonian system

$$\begin{pmatrix} \dot{x} \\ \\ \dot{y} \end{pmatrix} = J(\text{grad } H)^T\ , \text{ where } J = \begin{pmatrix} 0 & E \\ \\ -E_n & 0 \end{pmatrix} \quad \text{is a fixed matrix.}$$

Near a critical point, say the origin $x_0 = y_0 = 0$, we have

$H_x(0,0) = H_y(0,0) = 0$ and then

$$H(x,y) = H(0,0) + (x,y)\ S\binom{x}{y} + \ldots$$

where $S = S^T$ is a real symmetric matrix. Thus the linearized Hamiltonian system near the critical point is

$$\begin{pmatrix} \dot{x} \\ \\ \dot{y} \end{pmatrix} = JS\binom{x}{y}\ .$$

It is proved in linear algebra that every Hamiltonian matrix, that is a matrix of the form JS , has eigenvalues $\pm\lambda_1, \pm\lambda_2, \ldots, \pm\lambda_n$. That is, whenever λ is an eigenvalue of JS , then so is $-\lambda$ (and with the same multiplicity), see [1].

Let us examine the class $\mathcal{H}$ of Hamiltonian systems in $\mathbb{R}^2$, that is, for one positional degree of freedom:

$$\dot{x} = \frac{\partial H}{\partial y} \quad , \quad \dot{y} = -\frac{\partial H}{\partial x}$$

for a smooth Hamiltonian real function $H(x,y)$ in $\mathbb{R}^2$.

First study the Hamiltonian flow in $\mathbb{R}^2$ near a critical point, say the origin $x_0 = 0$, $y_0 = 0$, where

$$H_y(0,0) = H_x(0,0) = 0 \ ,$$

and

$$H(x,y) = (x,y) \quad S\binom{x}{y} + \ldots \qquad\qquad (\text{taking } H(0,0) = 0) \ .$$

In the linear approximation the differential system becomes

$$\begin{pmatrix} \dot{x} \\ \dot{y} \end{pmatrix} = JS\binom{x}{y}$$

with eigenvalues $\pm\lambda$. If λ is real, then the critical point is a saddle; whereas if $\lambda = ik$ (for $k > 0$) then the critical point is a (nondegenerate) center. Hence asymptotically stable critical points are not possible. This fact is also evident from the area-preservation of the Hamiltonian flow in $\mathbb{R}^2$. Similar geometric considerations show that no periodic orbit can be asymptotically stable (or strictly unstable) and each periodic orbit must be embedded in an annular family of periodic orbits-consequently the multiplier μ must equal one.

A useful model is provided by the conservative pendulum

$$\ddot{\theta} + \sin\theta = 0 \ .$$

By writing $\theta = x \pmod{2\pi}$, $\dot{\theta} = y$ we obtain

$$\dot{x} = y$$

$$\dot{y} = -\sin x$$

in the (x,y)-plane.

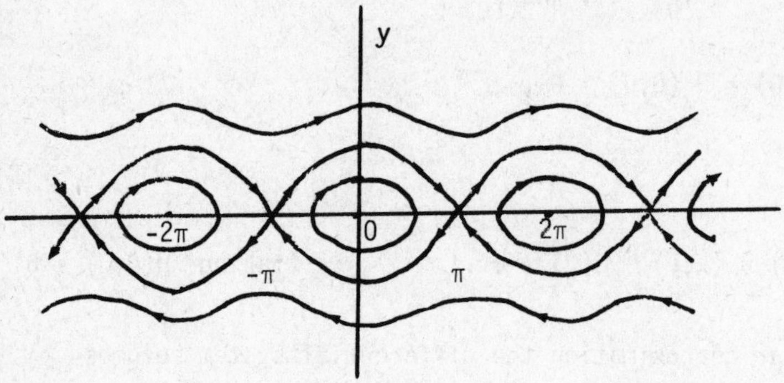

Figure 4

Pendulum Dynamics in (x,y) plane

It is clear that generic properties within $\mathcal{H}$ (for $\mathbb{R}^2$) include the following:

i) Each critical point is isolated and is either a nondegenerate center or saddle - so no source or sink can occur.

ii) Each periodic orbit is emmbedded in an annular band of periodic orbits - so each periodic orbit displays neutral stability and no limit cycle can occur.

iii) Less evident is the generic condition prohibiting the junction of different saddle points - so each separatrix of a saddle must go to

∞ , or else return to the same saddle (homoclinic case). (Note that a separatrix cannot spiral towards a critical point, as in $\mathcal{S}$.)

iv) Also the period T of the periodic orbit, at energy (or amplitude h in the annular band, cannot be constant - so $\frac{dT}{dh} \neq 0$ (except at isolated values of h). This requirement excludes the linear conservative oscillator and asserts that true nonlinearity is generic.

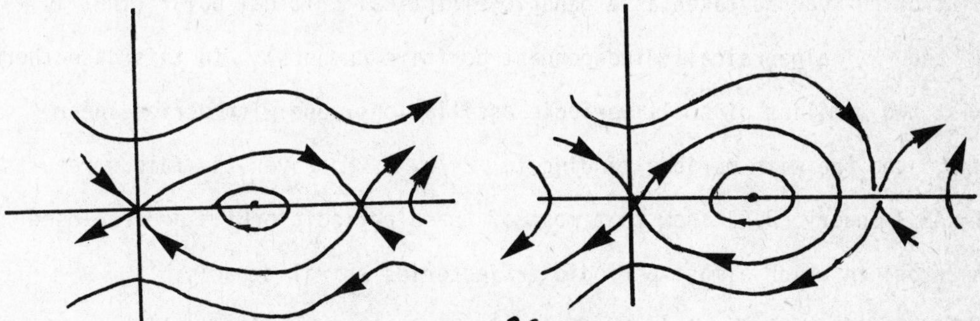

can be perturbed within $\mathcal{H}$ to:

Figure 5

Generic Hamiltonian Systems in $\mathbb{R}^2$

As a final topic we turn to the study of generic Hamiltonian systems with two or more positional degrees of freedom - that is, $\mathcal{H}$ -genericity for $\mathbb{R}^{2n}$ with $n \geq 2$. In this generality little is known about the global behavior of such systems and we concentrate on the local behavior in the vicinity of a critical point or a periodic orbit.

Consider a Hamiltonian system in $\mathbb{R}^4$

$$\text{H)} \quad \begin{cases} \dot{x}^i = \dfrac{\partial H}{\partial y_i} \\[2ex] \dot{y}_i = \dfrac{-\partial H}{\partial x^i} \end{cases} \qquad \text{with} \quad i = 1, 2$$

Here $H(x^1, x^2, y_1, y_2)$ is the given Hamiltonian function in $\mathbb{R}^4$. Near a critical point, say the origin of $\mathbb{R}^4$, the eigenvalues are $\pm\lambda_1$, $\pm\lambda_2$. If these are all real numbers, genericity demands that $\lambda_1 \neq \lambda_2$ and that these are algebraically independent. In this case the qualitative behavior of the solutions of the nonlinear system H) is precisely that of the corresponding hyperbolic linear system - namely a 4-saddle.

If $\lambda_1 = ik_1$, and $\lambda_2 = ik_2$ are both pure imaginary, then the critical point of H can be taken as a generic elliptical critical point (that is, k_1 and k_2 algebraically independent positive numbers). In this case there exist two families of small periodic oscillations, one with periods near $2\pi/k_1$ and one with periods tending to $2\pi/k_2$. Moreover the famous results of KAM - Theory [3,6] show that most of these periodic orbits are enclosed in 2-tori on which almost-periodic trajectories spiral densely.

If λ_1 is pure imaginary but λ_2 is real, then a more complicated pattern exists that incorporates parts of the elliptic and the hyperbolic structures - rather little seems to be known beyond this.

Now let us turn to generic behavior near a periodic orbit σ. First, σ occurs in a 2-band or annulus of periodic orbits, as parametrized by the energy level along which the period varies smoothly. Thus the characteristic multipliers for the "first return" or Poincaré map P are $\mu_1 = 1$, μ_2 and μ_2^{-1} - see figure 6 and subsequent clarifications. That is, within each energy hypersurface the (non-trivial) multipliers are μ_2 and μ_2^{-1}.

Following trajectories around the circuit designated by the periodic orbit σ in $\mathbb{R}^4$, we define the Poincaré map P of a 3 - transversal into itself. A fixed point of P yields a periodic orbit with a period near that of σ, while a fixed point of some iterate of P yields a longer duration periodic orbit. The eigenvalues of the map P, at the fixed point corresponding to σ are complex numbers 1, μ_2 and μ_2^{-1} (where the trivial

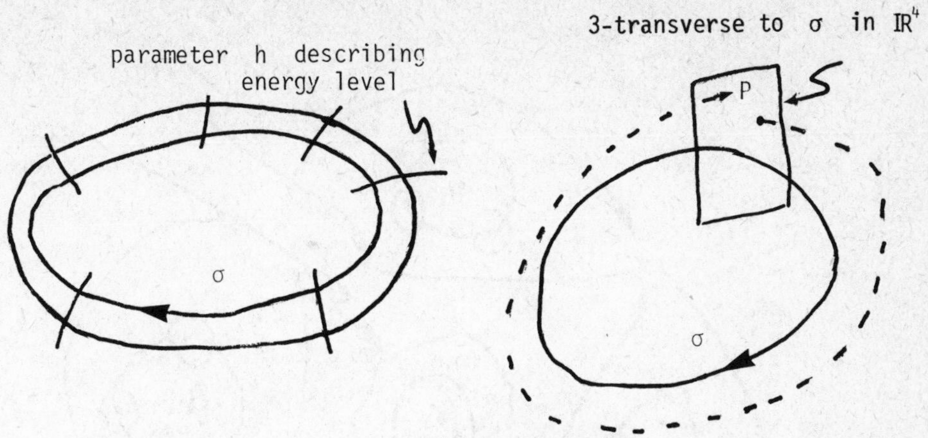

2-band or annulus through σ

Poincarré mapping around σ with eigenvalues 1, μ_2, μ_2^{-1}

Figure 6

1 corresponds to the tangent to the invariant 2-band). Hence if we restrict the map P to the energy hypersurface through σ, then the Poincaré map can be recognized as an area-preserving map of a 2-plane into itself (possibly with domain only a neighborhood of the origin), and the eigenvalues at the origin are μ_2, μ_2^{-1}. Under generic conditions we know that $\mu_2 \neq \pm 1$ and the multipliers are distinct.

In the hyperbolic case where μ_2 is real, say $\mu_2 > 1$, $\mu_2^{-1} < 1$, the topological structure of the Poincaré map - and the corresponding structure of solution trajectories near the periodic orbit - are just as is indicated by the linearized map.

The elliptic case, where $\mu_2 = e^{2\pi i f}$ and $\mu_2^{-1} = e^{-2\pi i f}$, for some real frequency f, is a famous puzzle. Any attempted sketch, see figure 7, breaks down into chaos.

The figure shows the central periodic orbit σ and a nearby periodic orbit σ_1 with about four times the original period of σ. But σ_1 is also generic elliptic and is encircled by still longer periodic orbits, etc. This sequence of long periodic elliptic orbits may converge to a complicated

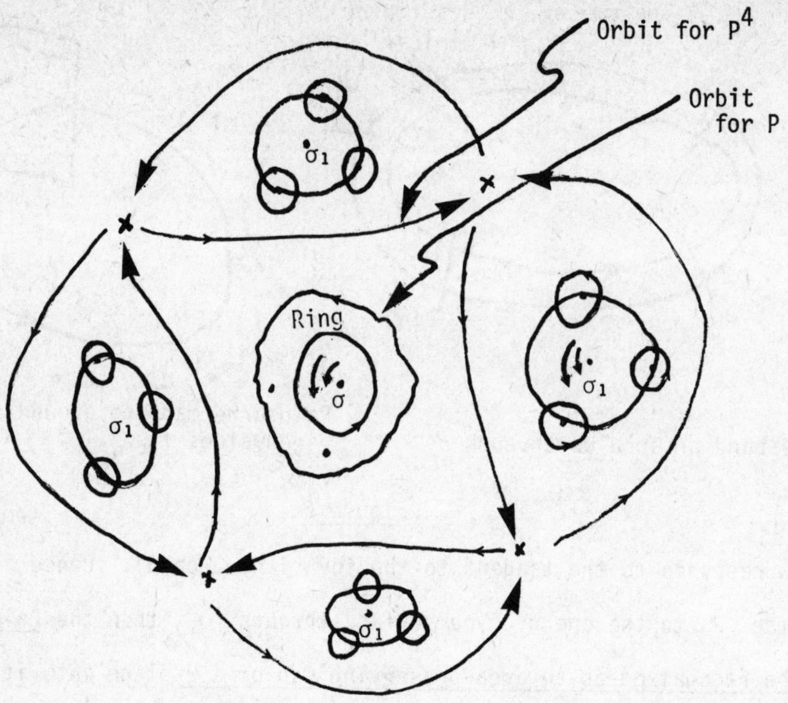

Figure 7

Poincaré Map P around Elliptic Orbit σ.

invariant set that is a topological solenoid [5].

In addition to these elliptic periodic orbits, there are alternatingly positioned hyperbolic long periodic orbits. Moreover each elliptic orbit is enclosed in noncountably many 2-tori (invariant closed curves bounding rings in the 2-plane section). The geometry is so confusing and mysterious that the more one analyses, the less one seems to understand. One could hope to prove that certain of these qualitative configurations are guaranteed occurrences, within a generic subset of the space $\mathcal{H}$. Indeed this has been accomplished with a certain degree of success [3, 4, 5].

We close this discussion of generic Hamiltonian dynamics with a few remarks on possible physical interpretations of the mathematical theory.

This is the dawning of the Age of Aquarius. The precession of the equinix through the astronomical Zodiac is caused by a wobble of the Earth's axis - an oscillation with a period of about 25,000 years. The Earth also undergoes higher order precessions, nutations and wobbles (e.g. Chandler wobble) in a complicated hierarchy of superimposed oscillations. The physical explanation for this hierarchy of oscillations is the asymmetry of the figure of the Earth, and the resulting perturbations arising from various gravitational source in the conservative dynamics of our Solar System.

Similar wobbles and hierarchies of oscillations occur in the spinning motion of an asymmetrical, frictionless gyroscopic top. The principal mathematical result that applies to these conservative dynamical problems is: <u>under generic conditions, such hierarchical oscillations are logical consequences of the axioms of Hamiltonian mechanics</u>.

REFERENCES

1. R. Abraham and J. Marsden, Foundations of Mechanics 2nd ed. N.Y. (1978).
2. A. Andronov and L. Pontryagin, Systems Grossiers, C.R. (Dokl.) Acad USSR 14 (1937) p. 247-251.
3. L. Markus, Lectures in Differentiable Dynamics, CBMS Series, No. 3 (Revised Ed.) (1980).
4. L. Markus and K. Meyer, Generic Hamiltonian Dynamical Systems, Memoir 144 AMS (1974).
5. L. Markus and K. Meyer, Periodic Orbits and Solenoids in Generic Hamiltonian Dynamical Systems, Am. J. Math. vol 102, (1980), p. 25-92.

6. J. Moser, On Invariant Curves of Area Preserving Mappings of an Annulus, Nachr. Akad. Wiss. Göttingen Math. Phy. (1962), p. 1-20.

Lawrence Markus
University of Minnesota

J E MARSDEN

Hamiltonian structures for the heavy top plasmas

Our purpose is to discuss the use of symmetry groups in a study of the chaotic dynamics of a heavy top and the Hamiltonian structure of the equations of plasma physics.

THE FREE RIGID BODY

We begin by presenting the equations of a free rigid body. The body is assumed to rotate freely about its center of mass and have a fixed angular velocity vector $\vec{\omega}$ seen by an observer fixed on the body. The body angular momentum $\vec{m}$ is defined by

$$\vec{m} = I\vec{\omega}$$

where $I = \mathrm{diag}(I_1, I_2, I_3)$ is a diagonal matrix coming from the moment of inertia tensor. Assuming that $I_1 > I_2 > I_3$, Euler's equations written in terms of $\vec{m}$ are:

$$\begin{aligned}\dot{m}_1 &= \frac{I_2 - I_3}{I_2 I_3} m_2 m_3 = a_1 m_2 m_3 \\ \dot{m}_2 &= \frac{I_3 - I_1}{I_1 I_3} m_1 m_3 = a_2 m_1 m_3 \\ \dot{m}_3 &= \frac{I_1 - I_2}{I_1 I_2} m_1 m_2 = a_3 m_1 m_2\end{aligned} \tag{1}$$

where $a_1 = (I_1 - I_3)/I_2 I_3$ etc. Two basic constants of motion are

<u>Total Angular Momemtum</u> ℓ defined by $\ell^2 = m_1{}^2 + m_2{}^2 + m_3{}^2$ (2)

<u>Energy</u> $H(m) = \frac{1}{2} \sum_j m_j{}^2/I_j$ (3)

Invariance of ℓ^2 in time follows directly from the identity $a_1 + a_2 + a_3 = 0$ and invariance of H follows from the identity $\frac{a_1}{I_1} + \frac{a_2}{I_2} + \frac{a_3}{I_3} = 0$.

The trajectories of (1) are given by intersecting the ellipsoids $H =$ constant from (3) with spheres $\ell =$ constant from (2). For distinct moments of intertia the flow on the sphere has saddle points at $(0,\pm\ell,0)$ and centers at $(\pm\ell,0,0)$, $(0,0,\pm\ell)$. The saddles are connected by four heteroclinic orbits, as indicated in Figure 1.

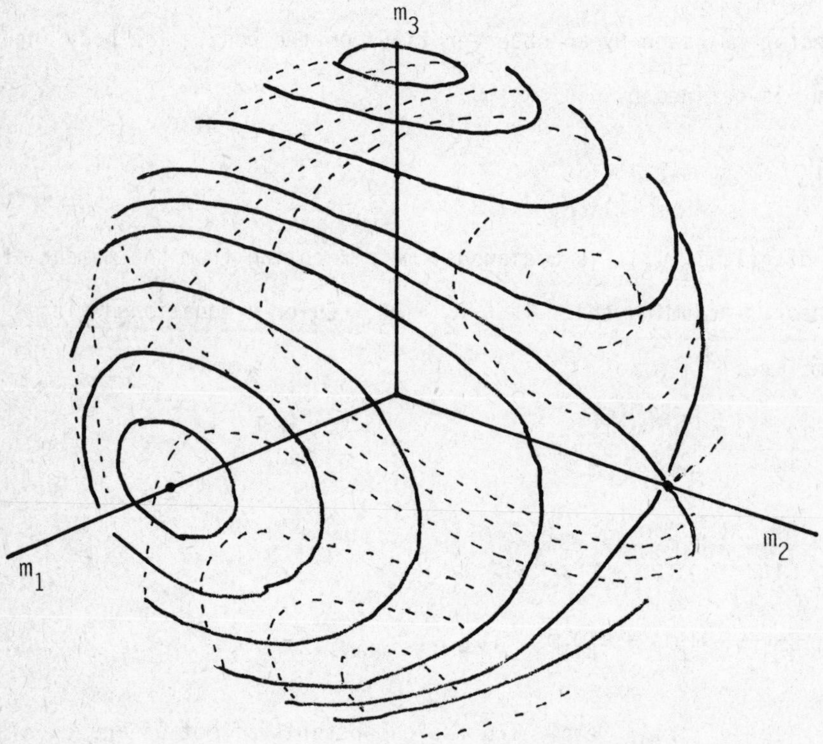

Figure 1. The spherical phase space of the rigid body for fixed total angular momentum $\ell = \sqrt{m_1^2 + m_2^2 + m_3^2}$; $I_1 > I_2 > I_3$

The orbits are explicitly known in terms of elliptic and hyperbolic functions. For example, the heteroclinic orbits lie in the invariant planes.

$$m_3 = \pm \sqrt{\frac{a_3}{a_1}}\; m_1$$

and are given by

$$m_1^+(t) = \pm \ell \sqrt{\frac{a_1}{-a_2}}\; \text{sech}\,(-\sqrt{a_1 a_3}\; \ell t)$$

$$m_2^+(t) = \pm \ell\; \tanh\,(-\sqrt{a_1 a_3}\; \ell t)$$

$$m_3^+(t) = \pm \ell \sqrt{\frac{a_3}{-a_1}}\; \text{sech}\,(-\sqrt{a_1 a_3}\; \ell t)$$

for $m_3 = \sqrt{\frac{a_3}{a_1}}\, m_1$ and by

$$m_1^-(t) = m_1^+(-t)$$

$$m_2^-(t) = m_2^+(-t)$$

$$m_3^-(t) = -m_3^+(-t)$$

for $m = -\sqrt{\frac{a_3}{a_1}}\, m_1$

(Note that $a_1 > 0$, $a_3 > 0$ and $a_2 < 0$). This may be checked by direct computation or by consulting one of the classical texts.

We now introduce a Poisson Bracket for equation (1). Given functions $F,G:R^3 \to R$ we define

$$\{F,G\}(m) = -m \cdot (\nabla F \times \nabla G) \tag{4}$$

This makes R^3 into a <u>Poisson manifold</u>. This means that on $C^\infty(\mathbb{R}^3)$, the smooth functions on $\mathbb{R}^3$, the bracket $\{\ ,\ \}$ is a Lie algebra $(\{\ ,\ \}$ is real bilinear, skew symmetric and verifies Jacobi's identity) and is a derivation in each of F and G. Later we shall explain where this bracket comes from in group theoretic terms: Here the group is SO(3) and m lives in $SO(3)^* \cong \mathbb{R}^3$, the dual of the Lie algebra of SO(3). By a straightforward calculation one can verify that (1) is equivalent to

$$\dot{F} = \{F,G\} \tag{5}$$

THE HEAVY TOP

We now turn our attention to the heavy top, i.e., a rigid body moving above a fixed point and under the influence of gravity. We let A be a given rotation in SO(3) with corresponding Euler angles denoted (ϕ,ψ,θ). The conjugate momenta are denoted p_ϕ, p_ψ, p_θ so that $(\phi,\psi,\theta,p_\phi,p_\psi,p_\theta)$ coordinatize $T^* SO(3)$. We let m denote the body angular momentum and let $v = A^{-1} k$ where k is the unit vector along the spatial z-axis. We assume that the center of mass is at $(0,0,\ell)$ when A is identity. Coordinates for the vectors (m,v) are most conveniently expressed in the body coordinate system; see Figure 2.

The phase space for the heavy top is $T^*SO(3)$. The system has, however an S^1 symmetry corresponding to rotations about the z-axis. A classical process called reduction enables one to eliminate two of the six variables. (Reduction is described in Arnold [1978] and in Abraham and Marsden [1978].) One gets a reduced space for each value of the angular momentum about the z-axis. One can show (see Marsden, Ratiu and Weinstein [1982] and references therein) that the reduced spaces for the heavy top are sympletically diffeomorphic to T^*S^2 and to a coadjoint orbits for the semi-direct product $SO(3) \times \mathbb{R}^3$; i.e., for the Euclidean group E_3. The Lie algebra of E_3 is denoted $e_3 = SO(3) \times \mathbb{R}^3$. The mapping giving this

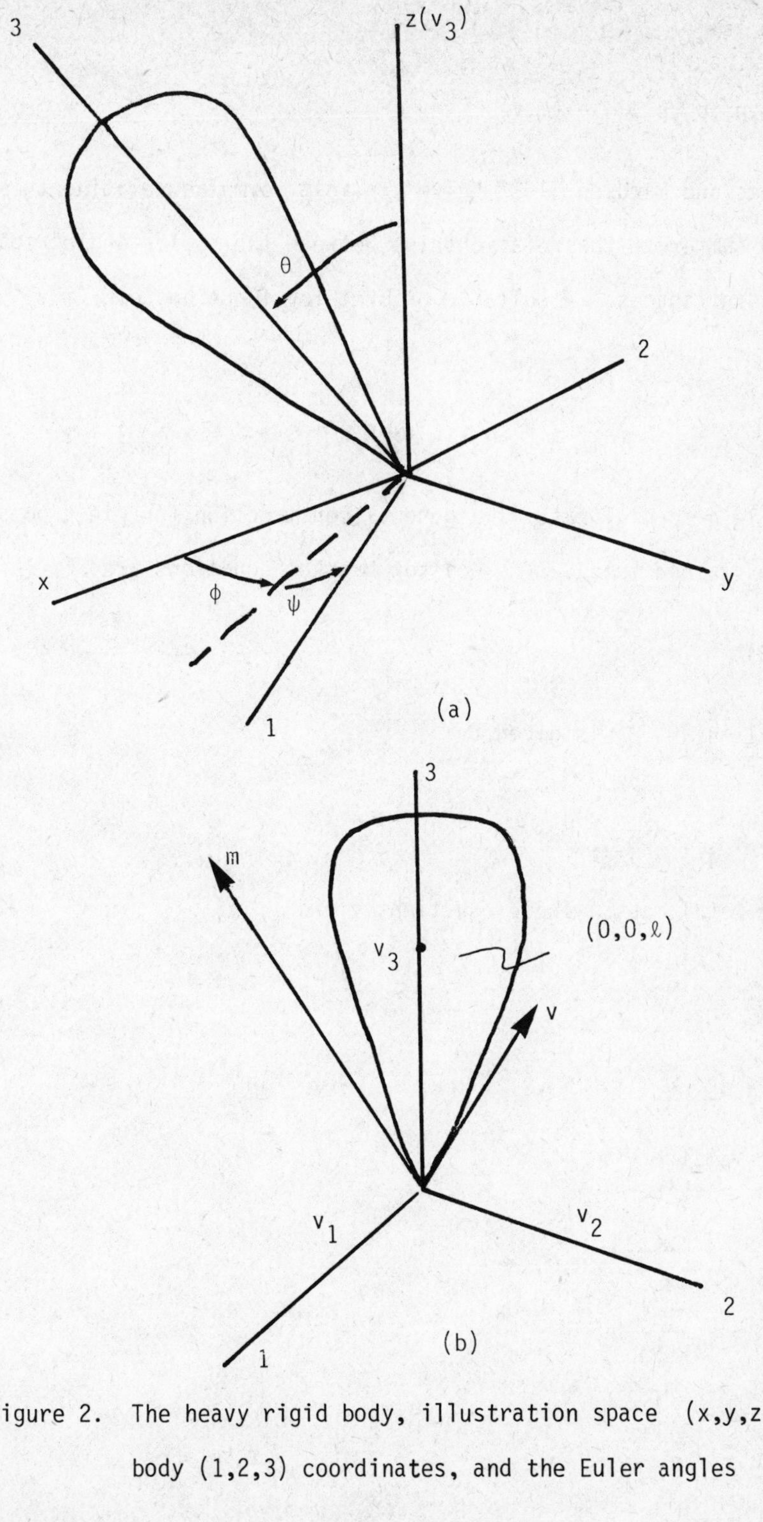

Figure 2. The heavy rigid body, illustration space (x,y,z) and body (1,2,3) coordinates, and the Euler angles (ϕ,ψ,θ).

diffeomorphism is

$$\Lambda : (\phi, \psi, \theta, p_\phi, p_\psi, p_\theta) \;\longmapsto\; (m, v)$$

Tables in Holmes and Mardsen [1983] give explicit formulae relating these quantities and summarize the relationships between the "Euler angle" spaces and the co-adjoint spaces. A suitable bracket for functions of (m,v) is given by

$$\{F,G\}(m,v) = -m\cdot(\nabla_m F \times \nabla_m G) - v\cdot(\nabla_m F \times \nabla_v G + \nabla_v F \times \nabla_m G) \tag{6}$$

Like (4), (6) is a special case of a general construction for Lie groups that will be explained below. The Poisson bracket equations are

$$\dot{F} = \{F,G\}$$

where the Hamiltonian H is given by

$$H(m,v) = \tfrac{1}{2} \sum_{j=1}^{3} \frac{m_j^2}{I_j} + Mg\ell v_3$$

and M is the total mass. These equations yield

$$\begin{aligned}
\dot{m}_1 &= a_1 m_2 m_3 - Mg\ell v_2 \\
\dot{m}_2 &= a_2 m_1 m_3 + Mg\ell v_1 \\
\dot{m}_3 &= a_3 m_1 m_2 \\
\dot{v}_1 &= \frac{m_3 v_2}{I_3} - \frac{m_2 v_3}{I_2} \\
\dot{v}_2 &= \frac{m_1 v_3}{I_1} - \frac{m_3 v_1}{I_3} \\
\dot{v}_3 &= \frac{m_2 v_1}{I_2} - \frac{m_1 v_2}{I_1}
\end{aligned} \tag{7}$$

One can check that these are equivalent to the classical equations of a heavy top; see Holmes and Marsden [1982]. This description of the heavy top in also found in Guillemin and Sternberg [1980].

The foregoing system has $||v||$ and $p_\phi = m \cdot v$ as constants of motion. This reflects the conservation law p_ϕ = constant and the preservation of the co-adjoint orbits by the equations. The conditions $||v|| = 1$ and $m \cdot v = p_\phi$ = constant also provide the identification of the co-adjoint orbit with T^*S^2. Indeed, $||v|| = 1$ describes the unit sphere S^2 and $m \cdot v = p_\phi$ specifies m as a linear functional on the unit sphere normal to S^2 leaving M restricted to $T_v S^2$ free. Thus m determines by restriction an element of $T_v^* S^2$. The coordinates θ, ψ are essentially spherical coordinates on S^2.

We discuss the Lagrange top in this framework. For $I_1 = I_2$ we get an additional S^1 symmetry, namely invariance under rotations about the 3-axis. This S^1 action corresponds to the S^1 action of rotation through ψ in the Euler angle picture. Also the momentum map can be readily checked to be just m_3. As with the free rigid body, the Lagrange top has a homoclinic orbit that has an explicit expression in terms of hyperbolic functions.

For the case of a nearly symmetric top, we have:

Theorem (Homes and Mardsen [1983]). *If* I_1/I_3 *is sufficiently large,* $I_2 = I_1 + \varepsilon$ *and* $\varepsilon > 0$ *is sufficiently small, then the Hamiltonian system for heavy top has transverse homoclinic orbits (close to the homoclinic orbit for* $\varepsilon = 0$) *in the Poincaré map for the* ψ *variable on each energy surface for* H = *constant in a certain open interval.*

The proof of this theorem involves integrating the Poisson bracket $\{H_0, H_1\}$ where $H = H_0 + \varepsilon H_1 + O(\varepsilon^2)$ around the homoclinic orbit for the Lagrange top. These techniques are based on Melnikov [1963].

One concludes that the heavy top close to the symmetric top has no analytic integrals other than the energy and angular momentum about the vertical axis.

Transverse homoclinic orbits implies the presence of Smale horseshoes. Thus, in the motion of a nearly symmetric heavy top, the dynamics is complex, having periodic orbits of arbitrarily high periods and aperiodic orbits embedded in an invariant Cantor set and so the system admits no additional analytical integrals.

Further examples of Hamiltonian systems are provided by Euler's equations for perfect fluids (see Marsden and Weinstein [1982b] and Marsden, Ratiu and Weinstein [1982] for details). In the incompressible case the analogue of m for the free rigid body is the vorticity ω and the bracket is

$$\{F,G\} = \int_\Omega \omega \cdot \left[\frac{\partial F}{\partial \omega}, \frac{\partial G}{\partial \omega}\right] \tag{8}$$

where $\delta F/\delta\omega$ is the functional derivative and $[\ ,\ ]$ is the Lie bracket of vector fields. In the compressible case the bracket on functions of the momentum density $m = \rho u$ and the density ρ is

$$\{F,G\} = \int_\Omega m \left[\frac{\partial F}{\partial m}, \frac{\partial G}{\partial m}\right] + \int \rho\left[\frac{\partial F}{\partial m} \cdot \frac{\nabla \delta G}{\delta \rho} - \frac{\delta G}{\delta m} \frac{\nabla \delta F}{\delta \rho}\right]. \tag{9}$$

The dynamics of the incompressible case is analogous to that in the rigid body and that of the compressible case parallels that of the heavy top. Other examples are provided by the MHD equations and the plasma equations described below.

LIE-POISSON BRACKETS

We now discuss generalities on bracket structures associated to Lie groups. Let G be a Lie group and $\mathfrak{g}$ be its Lie algebra. For $\xi, \eta \in \mathfrak{g}$, $[\xi,\eta]$ denotes the Lie bracket of ξ and η. Let $\mathfrak{g}^*$ denote the dual space of of $\mathfrak{g}$. For $F: \mathfrak{g}^* \to \mathbb{R}$ and the variable in $\mathfrak{g}^*$ denoted by μ, define $\delta F/\delta\mu: \mathfrak{g}^* \to \mathfrak{g}$ by

$$DF(\mu)\cdot\nu = \langle\nu, \frac{\partial F}{\partial\mu}\rangle$$

where $\langle\ ,\ \rangle$ denotes the pairing between $\mathfrak{g}^*$ and $\mathfrak{g}$ and $DF(\mu): \mathfrak{g} \to \mathbb{R}$ is the usual Frechet derivative. It is understood that $\partial F/\partial\mu$ is evaluated at the point μ. The Lie-Poisson bracket of two functions $F,G: \mathfrak{g}^* \to \mathbb{R}$ is defined by

$$\{F,G\} = -\langle\mu, [\frac{\delta F}{\delta\mu}, \frac{\delta G}{\delta\mu}]\rangle \tag{10}$$

The bracket defines a Poisson structure. This can proved directly or by understanding the relationship of (10) with canonical brackets described below -- see formula (12). From the latter, it is obvious that one obtains a Poisson structure. The bracket (10) is due to S. Lie [1890], p. 235, 294.

The Kirilov-Kostant-Souriau theorem asserts that orbits of the co-adjoint representation in $\mathfrak{g}^*$ are symplectic manifolds. See Arnold [1978] or Abraham and Marsden [1978] for the proof. Thus, $\mathfrak{g}^*$ is a disjoint union of symplectic manifolds. For $F,G: \mathfrak{g}^* \to \mathbb{R}$, a Poisson bracket is thus defined by

$$\{F,G\}(\mu) = \{F|\mathcal{O}_\mu, G|_\mu\}(u) \tag{11}$$

where $\mu\ \mathfrak{g}^*$, $\mathcal{O}_\mu$ is the orbit through μ, $F|\mathcal{O}_\mu$ is the restriction of F to $\mathcal{O}_\mu$, and $\{\ ,\ \}$ on the right hand side of (11) is the bracket on $\mathcal{O}_\mu$. This method shows that the bracket $\{F,G\}$ is degenerate; however, it determines a symplectic foliation on each leaf of which it is nondegenerate. The leaves are just the co-adjoint orbits.

Another method of ascertaining the Poisson structure is by extension. Given, $F, G: \mathfrak{g}^* \to \mathbb{R}$ extend them to maps $\hat{F}, \hat{G}: T^*G \to \mathbb{R}$ by left invariance. Then using the canonical bracket structure on T^*G, form $\{\hat{F}, \hat{G}\}$. Finally, regarding $\mathfrak{g}^*$ as $T_e^*G \subset T^*G$ restrict to $\mathfrak{g}^*$:

$$\{F,G\} = \{\hat{F},\hat{G}\}|_{\mathfrak{g}^*} \tag{12}$$

Formulas (11) and (12) both give (10). (If left invariance is replaced by right invariance, "-" in (10) is replaced by "+".)

Both of the foregoing methods are related by reduction; i.e., the reduced symplectic manifolds for the action of G on T^*G by left translation are the co-adjoint orbits (Marsden and Weinstein [1974]).

MAXWELL's EQUATIONS

In order to review and motivate reduction in more detail we shall now consider the Hamiltonian description of Maxwell's equations. As the configuration space for Maxwell's equations, we take the space $\mathfrak{A}$ of vector fields A on $\mathbb{R}^3$. These are vector potentials related to the magnetic field B by $B = \nabla \times A$. (In the more general situations of Yang-Mills fields, one should replace $\mathfrak{A}$ by the set of connections on a principal bundle). The corresponding phase space is the cotangent bundle $T^*\mathfrak{A}$ with its canonical symplectic structure and a suitable function space topology. Elements of $T^*\mathfrak{A}$ may be identified with pairs (A,Y) where Y is a vector field density on $\mathbb{R}^3$. (We shall not distinguish Y and Ydx). The pairing between A's and Y's is given by integration so that the canonical symplectic structure ω on T^* is given by

$$\omega((A_1,Y_1),(A_2,Y_2)) = \int_m (Y_2 \cdot A_1 - Y_1 \cdot A_2)dx \ , \tag{13}$$

with the associated canonical Poisson bracket

$$\{F,G\} = \int[\frac{\delta F}{\delta A}\frac{\delta G}{\delta Y} - \frac{\delta F}{\delta Y}\frac{\delta G}{\delta A}]dx \ . \tag{14}$$

Choosing the Hamiltonian

$$H(A,Y) = \tfrac{1}{2}\int|Y|^2 dx + \tfrac{1}{2}\int|\mathrm{curl}\ A|^2 \ , \tag{15}$$

Hamilton's equations are easily computed to be

$$\frac{\delta E}{\delta t} = \mathrm{curl}\ \mathrm{curl}\ A \quad \text{and} \quad \frac{\delta A}{\delta t} = Y \ . \tag{16}$$

If we write B for curl A and E for $-Y$, the Hamiltonian becomes the usual field energy

$$\tfrac{1}{2}\int|E|^2 dx + \tfrac{1}{2}\int|B|^2\ dx \tag{17}$$

and the equations (16) imply two of Maxwell's vacuum equations

$$\frac{\delta E}{\delta t} = \mathrm{curl}\ B \quad \text{and} \quad \frac{\delta B}{\delta t} = -\mathrm{curl}\ E \ . \tag{18}$$

The two remaining Maxwell equations will appear as a consequence of gauge invariance. The gauge group G consists of real valued functions on $\mathbb{R}^3$; the group operation is addition. An element $\psi \in G$ acts on $\mathcal{A}$ by the rule

$$A \to A + \nabla\psi \tag{19}$$

The translation (19) of A extends in a standard way to a canonical transformation ("extended point transformation") of $T^*\mathcal{A}$ given by

$$(A,Y) \to (A + \nabla\psi, Y) \tag{20}$$

We notice that our Hamiltonian (15) is invariant under the transformation (20). This allows us to use the gauge symmetries to reduce the number of degrees of freedom of our system. The action (20) of G on $T^*\mathfrak{A}$ has a momentum map $J: T^*\mathfrak{A} \to \mathfrak{A}^*$ where the Lie algebra $\mathfrak{g}$ of G is identified with the real valued functions on $\mathbb{R}^3$. This map J is associated to the symmetry group G in a way that generalizes the way conserved quantities are related to symmetries by Noether's theorem in classical mechanics. We may determine J by a standard formula (Abraham and Marsden [1978]): for $\psi \in \mathfrak{g}$,

$$< J\,(A,Y), \psi > = \int (Y \cdot \nabla\phi) dx = -\int (\text{div } Y)\phi dx$$

Thus we may write

$$J(A,Y) = -\text{div } Y \tag{21}$$

If ρ is an element of $\mathfrak{g}^*$ (the densities on $\mathbb{R}^3$), $J^{-1}(\rho) = \{(A,Y) \in T^*\mathfrak{g} \mid \text{div } Y = -\rho\}$. In terms of E, the condition $\text{div } Y = -\rho$ becomes the Maxwell equation $\text{div } E = \rho$ so we may interpret the elements of $\mathfrak{g}^*$ as charge densities. By a general theorem of Marsden and Weinstein [1974], the reduced manifold $J^{-1}(\rho)/G$ has a naturally induced symplectic structure. Computation leads to the following.

<u>Proposition</u>. <u>The space</u> $J^{-1}(\rho)/G$ <u>can be identified with</u> $\text{Max} = \{(E,B) \mid \text{div } E = \rho, \text{div } B = 0\}$ <u>and the Poisson bracket on</u> Max <u>is given in terms of</u> E <u>and</u> B <u>by</u>

$$\{F,G\} = \int \left(\frac{\delta F}{\delta E} \text{ curl } \frac{\delta G}{\delta B} - \frac{\delta G}{\delta E} \text{ curl } \frac{\delta F}{\delta B}\right) dx \tag{22}$$

<u>Maxwell's equations with an ambient charge density</u> ρ <u>are Hamilton's equations</u> for

$$H(E,B) = \tfrac{1}{2}\int [|E|^2 + |B|^2] \; dx$$

on the space/Max.

The bracket (22) was first introduced, using a different argument, by Born and Infeld [1934].

THE MAXWELL-VLASOV EQUATIONS

We consider a plasma consisting of particles with charge e and mass m moving in Euclidean space $\mathbb{R}^3$ with positions x and velocities v. (for simplicity we consider only one species of particle - the case of several species of particles can be treated in an analogous fashion). Let $f(x,v,t)$ be the plasma density at time t, $E(x,t)$ and $B(x,t)$ the electric and magnetic fields. The Maxwell-Vlasov equations are:

$$\begin{aligned}
&(a)\quad \frac{\partial f}{\partial t} + v\cdot\frac{\partial f}{\partial x} + \frac{e}{m}\left(E + \frac{v \times B}{c}\right)\cdot\frac{\partial f}{\partial v} = 0\\
&(b)\quad \frac{1}{c}\frac{\partial B}{\partial t} = -\operatorname{curl} E\\
&(c)\quad \frac{1}{c}\frac{\partial E}{\partial t} = \operatorname{curl} B - j, \text{ where } j = \frac{e}{c}\int v f(x,v,t)\,dv\\
&(d)\quad \operatorname{div} E = \rho_f, \text{ where } \rho_f = e\int f(x,v,t)\,dv\\
&(e)\quad \operatorname{div} B = 0
\end{aligned} \tag{23}$$

Letting $c \to \infty$ leads to the Poisson-Vlasov equation:

$$\frac{\partial f}{\partial t} + v\cdot\frac{\partial f}{\partial x} - \frac{e}{m}\,\frac{\partial \phi_f}{\partial x}\cdot\frac{\partial f}{\partial v} = 0$$

where

$$\nabla^2 \phi_f = \rho_f \tag{24}$$

In what follows we shall set $e = m = c = 1$.

The Hamiltonian for the Maxwell-Vlasov system is

$$H(f,E,B) = \int \tfrac{1}{2}|v|^2 f(x,v,t)\,dx\,dv + \int \tfrac{1}{2}[\,|E(x,t)|^2 + |B(x,t)|^2\,]dx \qquad (25)$$

while that for the Poisson-Vlasov equation is

$$H(f) = \int \tfrac{1}{2}|v|^2 f(x,v,t)\,dx\,dv + \tfrac{1}{2}\int \phi_f(x)\rho_f(x)\,dx \qquad (26)$$

The Poisson bracket for the Maxwell-Vlasov equation is as follows:

$$\begin{aligned}\{F,G\}(f,E,B) = &\int f\left\{\frac{\delta F}{\delta f},\frac{\delta G}{\delta f}\right\}dx\,dv \\ &+\int\left\{\frac{\delta F}{\delta E}\,\mathrm{curl}\,\frac{\delta G}{\delta B} - \frac{\delta G}{\delta E}\,\mathrm{curl}\,\frac{\delta F}{\delta B}\right\}dx \\ &+\int\left\{\frac{\delta F}{\delta E}\cdot\frac{\delta f}{\delta v}\frac{\delta G}{\delta f} - \frac{\delta G}{\delta E}\frac{\delta f}{\delta v}\cdot\frac{\delta F}{\delta f}\right\}dx\,dv \\ &+\int B\cdot\left[\frac{\partial}{\partial v}\frac{\delta F}{\delta f}\times\frac{\partial}{\partial v}\frac{\delta G}{\delta f}\right]dx\,dv\end{aligned} \qquad (27)$$

The bracket (27) is due to Marsden and Weinstein [1982a] and is based on reduction and an earlier attempt "by hand" by Morrison [1980].

We are now ready to discuss the meaning of the first term $\int f\left\{\frac{\delta F}{\delta f},\frac{\delta G}{\delta f}\right\}dxdv$ in (27). In the absence of a magnetic field and normalizing the mass, we can identify velocity with momentum. Thus we let $\mathbb{R}^6$ denote the usual position - momentum phase space with co-ordinates $(x_1,x_2,x_3,p_1,p_2,p_3)$ and symplectic structure $\Sigma dx_i \wedge dp_i$. Let $\mathcal{S}$ denote the group of canonical transformations of $\mathbb{R}^6$ which have polynomial growth at infinity in the momentum directions. The Lie algebra $\mathfrak{s}$ of $\mathcal{S}$ consists of the Hamiltonian vector fileds on $\mathbb{R}^6$ with polynomial growth in the momentum directions. We shall identify elements of $\mathfrak{s}$ with their generating functions so that consists of C^∞ functions on $\mathbb{R}^6$ and the left Lie algebra structure is given by $[f,g] = \{f,g\}$, the usual Poisson bracket on phase space. (See Abraham and Marsden [1978]).

The dual space $\mathfrak{s}^*$ can be identified with the distribution densities on R^6 which are rapidly decreasing in the momentum directions. The pairing between $h \in \mathfrak{s}$ and $f \in \mathfrak{s}^*$ is obtained by integration

$$\langle h,f\rangle = \int hf \, dxdp$$

As with any Lie algebra, the dual space $\mathfrak{s}^*$ carries a natural Lie-Poisson structure. In (10) we change "-" to "+" since our system is right invariant. Then with $\mu = f$, (10) becomes

$$\{F,G\}(f) = \langle f, [\frac{\delta F}{\delta f}, \frac{\delta G}{\delta f}]\rangle = \int f \left\{\frac{\delta F}{\delta f}, \frac{\delta F}{\delta f}\right\} dxdp$$

are required.

The second term in (27) is the bracket for Maxwell's equations which has been previously discussed -- see (22). We consequently turn our attention to the last two terms in (27) which represent coupling or interaction. The Hamiltonian structure for the Maxwell-Vlasov system becomes simple if we choose our variables to be densities on (x,p) space rather than (x,v) space and elements (A,Y) of $T^*\mathfrak{A}$. To avoid confusion with densities on (x,v) space, we utilize the notation f_{mom} for densities on (x,p) space.

The Poisson structure on $\mathfrak{s}^* \times T^*\mathfrak{A}$ is just the sum of those on $\mathfrak{s}^*$ and $T^*\mathfrak{A}$: for functions $\overline{F}$ and $\overline{G}$ of f_{mom}, A and Y, set

$$\{\overline{F},\overline{G}\}(f_{mom},A,Y) = \int f_{mom} \left\{\frac{\partial \overline{F}}{\partial f_{mom}}, \frac{\partial \overline{G}}{\partial f_{mom}}\right\} dxdp \; + \tag{28}$$

$$\int [\frac{\partial \overline{F}}{\partial A}\frac{\partial G}{\partial Y} - \frac{\partial \overline{G}}{\partial A}\frac{\partial \overline{F}}{\partial Y}] \, dx$$

and the Hamiltonian is just (25) written in terms of these variables. Using the classical relation between momemtum and velocity, $p = v + A$, we have

$$H(f_{mom},A,Y) = \frac{1}{2}\int |p - A(x)|^2 f_{mom}(x,p)\,dxdp$$
$$+ \frac{1}{2}\int (|Y|^2 + |\mathrm{curl}\ A|^2)dx$$

We observe that there is no coupling in the symplectic structure but there is coupling between f_{mom} and A in the first term of (29).

<u>Theorem</u>. <u>The evolution equations</u> $\dot{F} = \{F,H\}$ <u>for a function</u> F <u>on</u> $\mathfrak{s}^* \times T^*$ <u>with</u> H <u>given by (29) and</u> $\{\ \}$ <u>by (28) are the equations</u> (23a,b,c) <u>with (23b) replaced by</u> $\frac{\partial A}{\partial t} = Y$.

The proof of this theorem is a straightforward verification. The constraints can, as in Morrison [1980], be regarded as subsidiary equations which are consistent with the evolution equations. Equations (23b and e) hold since $B = \mathrm{curl}\ A$. We will now show that equation (23d) expresses the fact that we are on the zero level of the momentum map generated by the gauge transformations. The corresponding reduced space <u>decouples</u> the energy, while <u>coupling</u> the symplectic structure.

The work of Weinstein [1978] on the equations of motion for a particle in a Yang-Mills field uses the following general set-up. Let $\pi:P \to M$ be a principle G-bundle and Q a Hamiltonian G-space (or a Poisson manifold which is a union of hamiltonian G-spaces). Then G acts on T^*P and on Q, so it acts on $Q \times T^*P$ (with the product symplectic structure). This action has a momentum map J and so may be reduced at 0:

$$(Q \times T^*P)_0 = J^{-1}(0)/G \tag{30}$$

The reduced manifold (30) carries a symplectic (or Poisson, if Q was a Poisson manifold) structure naturally induced from those of Q and T^*P.

To obtain the phase space for an elementary particle in a Yang-Mills field one chooses P to be a G-bundle over 3-space M and Q a co-adjoint orbit for G (the internal variables). The Hamiltonian is constructed using a connection (i.e., a Yang-Mills field) for P. In the special case of electromagnetism, $G = S^1$ and $Q = \{e\}$ is a point.

For the Vlasov-Maxwell system we choose our gauge bundle to be

$$P = \mathfrak{A} \to M$$

where $M = \{B \mid \operatorname{div} B = 0\}$, with G the gauge group described in the previous section. As in §3, let $\mathfrak{S}$ denote the group of canonical transformations of $T^*M (= \mathbb{R}^6)$. We can let Q be either the symplectic manifold $T^*\mathfrak{S}$ or the Poisson manifold $\mathfrak{s}^*$. It is a little more direct work with $\mathfrak{s}^*$, so we shall do this.

We wish to specify an action of G on $\mathfrak{s}^*$ which, when combined with the action (20) on $T^*\mathfrak{A}$, will leave the Hamiltonian (29) invariant. A natural choice is to let $\psi \in G$ act by the (linear) map

$$f_{mom} \to f_{mom} \circ \tau_{-\nabla\psi} \tag{31}$$

where $\tau_{-\nabla\psi}: \mathbb{R}^6 \to \mathbb{R}^6$ is the "momentum translation map" defined by

$$\tau_{-\nabla\psi}(x,p) = (x, p - \nabla\psi(x)). \tag{32}$$

It is easy to verify that $\tau_{-\nabla\psi}$ is a canonical transformation, so it preserves the ordinary Poisson bracket on $\mathbb{R}^6$. It follows that the map (31) preserves the Poisson structure on $\mathfrak{s}^*$. A simple calculation gives:

<u>Lemma</u>. <u>The action of</u> G <u>on</u> $\mathfrak{s}^*$ <u>defined by</u> (31) <u>and</u> (32) <u>has a momentum map</u> $J: \mathfrak{s}^* \to \mathfrak{g}^*$ <u>given by</u>

$$\langle J(f_{mom}), \phi \rangle = -\int f_{mom}(x,p)\phi(x)\, dx dp \tag{33}$$

i.e.

$$J(f_{mom}) = -\int f_{mom}(x,p)\, dp \tag{34}$$

The right hand side of (34) is a density on $\mathbb{R}^3$ which we may denote by ρ_{mom}.

Now we define the action of G on the product $\mathfrak{s}^* \times T^*\mathfrak{A}$ by combining (31) and (20), i.e. $\psi \in G$ maps

$$(f_{mom}, A, Y) \to (f_{mom} \circ \tau_{-\nabla\psi},\ A + \nabla\psi, Y). \tag{35}$$

Combining equations (21) and (34) gives:

__Lemma__. __The momentum map__ $J: \mathfrak{s}^* \times T^*\mathfrak{A} \to \mathfrak{g}^*$ __for the action__ (35) __is given by__:

$$J(f_{mom}, A, Y) = -\int f_{mom}(x,p)\, dp - \operatorname{div} Y. \tag{36}$$

We may now describe the reduced Poisson manifold in terms of densities $f(x,v)$ defined on position-__velocity__ space.

__Proposition__. __The reduced manifold__ $(\mathfrak{s}^* \times T^*\mathfrak{A})_0 = J^{-1}(0)/G$ __may be identified with the Maxwell-Vlasov phase space__

$$MV = \{(f,B,E) \mid \operatorname{div} B = 0 \text{ and } \operatorname{div} E = \int f(x,v)\, dv.\}$$

__Proof__. To each (f_{mom}, A, Y) in $J^{-1}(0)$ we associate the triple (f,B,E) in MV where

$$f(x,v) = f_{mom}(x, v + A(x)),\ B = \operatorname{curl} A, \text{ and } E = -Y.$$

The condition $J(f_{mom}, A, Y) = 0$ is equivalent, by (36), to the Maxwell equation $\operatorname{div} E = \int f(x,v)\, dv$ in the definition of MV. It is easy to check that elements of $J^{-1}(0)$ are associated to the same (f,B,E) if and only if they are related by a gauge transformation (35), so our association gives a 1-1 correspondence between $J^{-1}(0)/G$ and MV. ■

By the general theory of reduction, MV inherits a Poisson structure from the one on $s^* \times T^*$. Since the Hamiltonian (29) is invariant under G, it follows that the Maxwell-Vlasov equations are a Hamiltonian system on MV with respect to this structure. We can compute the explicit form of the inherited Poisson structure in the variables (f,B,E). In fact, a direct calculation using the chain rule shows that (28) becomes (27). This is how one arrives at the following result.

Theorem. The bracket (27) makes (MV) into a Poisson manifold. The Maxwell-Vlasov equations are equivalent to the evolution equations $\dot{F} = \{F,H\}$ *on* (MV), *where* H *is given by* (25).

*Lectures presented at the University of Houston, November, 1981. Thanks are due to the editor for help in the preparation of the manuscript. The results here are based on joint work with Phil Holmes and Alan Weinstein.

REFERENCES.

1. R. Abraham and J. Marsden [1978]. Foundations of Mechanics, Second Edition, Addition-Wesley.

2. V. Arnold [1978]. Mathematical methods of classical mechanics. Graduate Texts in Math. No. 60 Springer Verlag, New York.

3. H. Born and L. Infeld [1935]. On the quantization of the new field theory, Proc. Roy. Soc. A. 150, 141.

4. P.J. Holmes and J.E. Marsden [1983]. Horseshoes and Arnold diffusion for Hamiltonian systems on Lie groups. Ind. Univ. Math. J. (to appear).

5. J. Marsden, T. Ratiu and A. Weinstein [1982]. Semi-direct products and reduction in mechanics (preprint).

6. J. Marsden and A. Weinstein [1974]. Reduction of symplectic manifolds with symmetry, Rep. Math. Phys., 5, pp. 121-130.

7. J. Marsden and A. Weinstein [1982a]. The Hamiltonian structure of the Maxwell-Vlasov equations, Physica D4, 394-406.

8. J. Marsden and A. Weinstein [1982b]. Coadjoint orbits, vortices and Clebsch variables for incompressible fluids, Physica D. (to appear).

9. P.J. Morrison [1980]. The Maxwell-Vlasov equations as a continuous Hamiltonian system, Phys. Lett. 80A, 383-386.

10. V.K. Melnikov [1963]. On the stability of the center for time period perturbations, Trans. Moscov Math. Soc. 12, 1-57.

11. A. Weinstein [1978]. A universal phase space for particles in Yang-Mills fields, Lett. Math. Phys. 2, 417-420.

Jerrold E. Marsden
University of California, Berkeley

J W NEUBERGER

Use of steepest descent for systems of conservation equations

Many systems of conservation equations have the form

$$u_t = \nabla \cdot F(u, \nabla u) \ , \qquad (1)$$

$u: [0,T] \times \Omega \to R^k$, $\Omega \subset R^n$. Often, however, a more complicated form is found:

$$Q(u)_t = \nabla \cdot F(u, \nabla u) \qquad (2)$$

$$S(u) = 0,$$

$u: [0,T] \times \Omega \to R^{k+q}$, $\Omega \subset R^n$, $Q: R^{k+q} \to R^k$, $S: R^{k+q} \to R^q$, $F: R^{k+q} \times R^{n(k+q)} \to R^{nk}$. Examples are found in many places (cf [1], [6]).

Some background for the present work is found in [2], [3].

Often problem (2) can be changed into (1) by using the condition $S(u) = 0$ to eliminate q of the unknowns and by using some change of variables to convert the term $(Q(u))_t$ into the form u_t. However since applications often seem to lead to problems naturally expressed in form (2), it seems worthwhile to study (2) directly. We consider homogeneous boundary conditions on $\delta\Omega$ together with initial conditions.

We introduce a strategy for a time-stepping procedure. We at first discretize in time only. Take w for a time-slice of a solution at a time t_0. We seek an estimate v at time $t_0 + \delta$ by seeking to minimize

$$\phi(v) \equiv [||Q(v) - Q(w) - \delta\nabla\cdot F((v+w)/2, \nabla((v+w)/2)||^2 \qquad (3)$$

$$+ ||S((v+w)/2||^2]/2$$

over all appropriate v satisfying given homogeneous boundary conditions. The norms in (3) are L_2 norms. We assume that Q, F and S are such that there is an integer m so that if $D(\phi) = H_0^m(\Omega)$ (the m-th order Sobolev space of R^{k+q} valued functions satisfying the given boundary conditions), then ϕ is a $C^{(2)}$ function.

If v is found so that $\phi(v) = 0$, then v is likely to be a reasonable approximation to the time-slice of a solution to (2) at $t_0 + \delta$ given that w is such an approximation at t_0. If $\phi(v) \neq 0$ but $\phi(v)$ is minimum for some v close to w, then v is still likely to be a reasonable approximation at $t_0 + \delta$.

The remainder of this note is concerned with a numerical implementation of the above.

Denote by G a regular grid which approximates Ω. Denote by K the vector space of all R^{k+q}-valued functions on G and denote by K_0 those members of K which satisfy boundary conditions appropriate to the problem at hand. Denote by $\hat{G}$ the grid composed of centers of rectangular solids generated by G. Denote by $\hat{K}$ the vector space of R^{k+q}-valued functions on $\hat{G}$ and denote by $\hat{I}: K \to \hat{K}$ the transformation which associates with an element $u \in K$ its interpolated counterpart on $\hat{K}$. Denote by $\hat{\nabla} : K_0 \to \hat{K}$ the operator which associates with each element u of K the n-tuple $u_1, \ldots, u_n$ such that u_i is the element of $\hat{K}$ obtained by taking central differences of u in its i-th argument.

Denote by $\hat{\nabla}\cdot: \hat{K}^n \to K_0$ the operator which associates with an element $z \in \hat{K}^n$ the numerical counterpart of the divergence of z.

We are now ready to state a numerical time-stepping procedure for a finite difference version of (2):

Suppose $w \in K_o$ and $\delta > 0$. Define

$$\alpha(v) = [||Q(v) - Q(w) - \delta\hat{\nabla}\cdot F(\hat{I}(v+w))/2,\ \hat{\nabla}(v+w)/2||^2 \tag{4}$$

$$+ ||S(\hat{I}(v+w)/2||^2]/2, \quad v \in K_o .$$

We seek to find $v \in K_o$ such that $\alpha(v)$ is minimum. Once again there is a question of appropriate norm. At first we consider an ℓ_2 norm for K_o and then later consider a better choice. Rewrite (4) as

$$\alpha(v) = [||L(v)||^2 + ||M(v)||^2]/2, \quad v \in K_o,$$

where

$$L(v) = Q(v) - Q(w) - \delta\hat{\nabla}\cdot F(\hat{I}(v+w)/2,\ \hat{\nabla}(v+w)/2)$$

$$M(v) = S(\hat{I}(v+w)/2), \quad v \in K_o .$$

Then the gradient of α is given by

$$(\nabla\alpha)(v) = L'(v)^*L(v) + M'(v)^*M(v) , \quad v \in K_o .$$

A steepest descent scheme for finding a minimum of α near w is started by picking $v_o = w$ and defining

$$v_{j+1} = v_j - \delta_j(\nabla\alpha)(v_j), \quad j = 0,1,2,..., \tag{5}$$

where for each j, δ_j is chosen so that if

$$g(s) = \phi(v_j - s(\nabla\alpha)(v_j)) , \quad s > 0,$$

then δ_j is the first zero of g'.

Evidence presented in [4] indicates that steepest descent relative to such an ℓ_2 norm of K_0 is substantially inferior to steepest descent relative to the appropriate Sobolev norm. In our present setting ϕ in (3) is continuous (even $C^{(2)}$) as a function for $H_0^{(m)}(\Omega) \to R$ whereas ϕ as a function on $L_2(\Omega)$ is discontinuous and only densely defined. It is not so surprising that analytical difficulties of using $L_2(\Omega)$ instead of $H_0^M(\Omega)$ should correspond to numerical difficulties in the discrete setting.

It what follows we develop a few details concerning calculations with a discretized version of $H_0^{(m)}(\Omega)$ rather than $L_2(\Omega)$.

For $i \in 0,1,\ldots,m$, denote by $D^{(i)}$ a linear operator on K_0 which corresponds to a finite difference analogue of i-fold differentiation. Note that $D^{(i)*}D^{(i)}: K_0 \to K_0$. In analogy with $H_0^{(m)}(\Omega)$, define

$$||v||_S = [||v||^2 + ||D^{(1)}v||^2 + \ldots + ||D^{(m)}v||^2]^{\frac{1}{2}}, \; v \in K_0,$$

where the norms of the rhs of this expression are ℓ_2 norms. Define D so that if $v \in K_0$ then

$$Dv = (v, D^{(1)}v, \ldots, D^{(m)}v) .$$

Denote by $\nabla_S\alpha$ the gradient function of α calculated relative to $|| \; ||_S$.

<u>Theorem</u>. $(\nabla_S\alpha)(v) = (D^*D)^{-1}(\nabla\alpha)(v), \; v \in K_0$.

<u>Proof</u>: For $v, h \in K_0$,

$$\begin{aligned}\alpha'(v)h &= \langle L'(v)h, L(v)\rangle + \langle M'(v)h, M(v)\rangle \\ &= \langle h, L'(v)^*L(v) + M'(v)^*M(v)\rangle \\ &= \langle h, D^*D(D^*D)^{-1}[L'(v)^*L(v) + M'(v)^*M(v)]\rangle\end{aligned}$$

$$= \langle Dh, D(D^*D)^{-1}[L'(v)^*L(v) + M'(v)^*M(v)]\rangle$$

$$= \langle h, (D^*D)^{-1}[L'(v)^*L(v) + M'(v)^*M(v)]\rangle_S$$

and so the conclusion follows (see [5], p. 245).

For this new gradient we consider steepest descent as in (5) with $\nabla\alpha$ replaced by $\nabla_S\alpha$.

For m=1, D^*D is a finite dimensional approximation to $-\Delta + I$ restricted to functions satisfying homogeneous boundary conditions appropriate to the problem at hand. Hence in this case (and essentially in general) the net effect of multiplying $(\nabla\alpha)(v)$ by $(D^*D)^{-1}$ is to attenuate frequencies of higher order in $(\nabla\alpha)(v)$ when $(\nabla\alpha)(v)$ is expanded in an orthonormal sequence of eigenvectors of D^*D. These eigenvectors are numerical approximations, of course, to corresponding eigenfunctions of $-\Delta + I$ under appropriate boundary conditions.

For a simple linear case it was shown in [4] that

$$\frac{\alpha(v - \delta(\nabla_S\alpha)(v))}{\alpha(v)} \le 4/10, \quad v \in K_0, \ \alpha(v) \ne 0 ,$$

where δ is optimally chosen. This result was independent of mesh size. By contrast even for optimally chosen δ,

$$\sup_{\substack{v\in K_0 \\ \alpha(v)\ne 0}} \frac{\alpha(v - \delta(\nabla\alpha)(v))}{\alpha(v)} \to 1 \quad \text{as mesh } G \to 0.$$

Moreover, elements v are common which yield the maximum of this expression for a given grid. Computational evidence suggests that such phenomena persist into complicated cases such as the present one.

The above procedure has been coded for a system of eight nonlinear conservation equations in nine unknowns arising from magnetohydrodynamics [1]. An energy relation is reflected in a choice of S. The code has not yet been property evaluated but results seem promising nevertheless. Numerical results have been obtained for Burgers' equation with periodic boundary conditions:

$$u_t + uu_x = \eta u_{xx}$$

on $[0,T] \times [0,1]$ with $u(t,0) = u(t,1)$, $u_t(t,0) = u_t(t,1)$, $t \geq 0$.

For 1000 spatial grid points and $\eta = .01$, the method ran about 40 times faster than a corresponding explicit method. For $\eta = .001$, the two methods were about equal in CPU time. Four or five, on the average, steepest descent iterations were needed for each time step. Conjugate gradient enhancements gave no added efficiency since such a small number of steepest descent steps gave good accuracy.

The present steepest descent scheme has not been tested against Newton's iteration for solving $\nabla\alpha(v) = 0$, but it is expected that the two may be comparable for Burgers' equation. For systems of a number of equations and more than one space dimension the present scheme may have some important advantages compared with Newton's method in that $(D^*D)^{-1}$ is applied component-wise whereas the inverse in Newton's method is applied to all components at once. For 8 equations on p grid points, Newton's method requires solution of 8 p x p positive definite symmetric systems of a clearly understood type amenable to fast Poisson solvers.

REFERENCES.

1. D. Schnack and J. Killeen, Nonlinear, Two-Dimensional Magnetohydrodynamic Calculation, J. Comp. Physics 35 (1980), 110-145.

2. J.W. Neuberger, A type-Independent Method for Systems of Nonlinear Partial Differential Equations: Application to the Problem of Transonic Flow, Computers Math. Appl. 6 (1981), 67-78.

3. J.W. Neuberger, Steepest Descent for Systems of Nonlinear Partial Differential Equations, Oak Ridge National Laboratory, CSD/TM-161 (1981).

4. J.W. Neuberger, Steepest Descent for General Systems of Linear Differential Equations in Hilbert Space, Springer Lecture Notes (to appear).

5. J.M. Ortega and W.C. Rheinboldt, Iterative Solution of Nonlinear Equations in Several Variables, Academic Press (1970).

6. D.W. Peaceman, Fundamentals of Numerical Reservoir Simulation, Amsterdam - New York, Elsivier Publishing Co. (1977).

John W. Neuberger

North Texas State University

L E PAYNE

Some applications of 'best possible' maximum principles in elliptic boundary value problems

1. INTRODUCTION.

This paper presents a survey of some recent results on "best possible" maximum principles. This is a relatively new sub area of maximum principles - one that is still in the state of development. The paper is a write up of a talk given to an audience from a number of different disciplines, so highly technical details have been avoided and no attempt has been made to put the problems into their weakest possible settings.

The use of maximum principles in the study of boundary value problems for harmonic function dates back to C.F. Gauss (1838) and S. Earnshaw (1839); for equations more general than the Laplace equation they were first used by A. Paraf (1892). For an account of various contributions to the study of maximum principles over the years, we refer to the excellent source book of M.H. Protter and H.F. Weinberger (1967) [26].

The applications which we wish to discuss today have appeared since 1967 and are therefore not included in the text of Protter and Weinberger. However, many of them do appear in a recently published book by R. Sperb (1981) [30]. Sperb also deals with elliptic equations on manifolds and with parabolic equations, but since one often loses the "best possible" feature of the maximum principle in these cases we shall not discuss them in this series of lectures.

As we know, maximum principles have been used in the literature to study such qualitative questions as existence, uniqueness, regularity, stability, and orger of convergence of numerical schemes. They have also been used to study such quantitative questions as error bounds, limits of stability, decay rates, etc. For the qualitative questions discussed above, ordinary maximum principles (which abound in the literature) will usually suffice. However, for quantitative questions one often needs explicit bounds so it is helpful to have the type of "best possible" principles which I will discuss in my two lectures.

The idea of a best possible maximum principle is as follows : We are given a nonlinear partial differential equation in the form

$$F(x,u,Du,D^2u) = 0 \quad \text{in} \quad \Omega \subset R^N$$

and are interested in a solution of this problem which satisfies

$$B(x,u,Du) = 0 \quad \text{on} \quad \partial\Omega$$

We shall say that a function $\phi(u,q^2)$ of u and $q^2 \equiv |\text{grad } u|^2$ satisfies a "best possible" maximum principle if the following two properties hold:

1. $\phi(u,q^2)$ satisfies a maximum principle in Ω
2. $\exists$ some domain $\tilde{\Omega}$ (or a limiting domain)

such that $\phi \equiv$ constant in $\tilde{\Omega}$.

It is the second condition which suggests the term "best possible". Clearly a "best possible" maximum principle can be integrated over Ω or subdomains of Ω and the result will be isoperimetric.

Obviously maximum principle hold in many problems for which we cannot construct "best possible" maximum principles. Similarly "best possible" maximum principles can often be found for problems whose solutions do not satisfy ordinary maximum principles.

Before proceeding we recall what is meant by a maximum principle. To this end we will write down the two maximum principles attributed to E. Hopf [8,9] which apply to solutions of the inequality

$$Lu \equiv \sum_{i,j=1}^{N} a_{ij}(x) \frac{\partial^2 u}{\partial x_i \partial x_j} + \sum_{i=1}^{N} b_i(x) \frac{\partial u}{\partial x_i} \geq 0 \quad \text{in } \Omega$$

For our purposes we may assume that a_{ij} and b_i are continuous in $\overline{\Omega}$ and that the boundary Ω is a $C^{2+\alpha}$ surface for some $\alpha > 0$. Furthermore, it is assumed that the matrix a_{ij} is positive definite. Under these hypotheses the two principles may be stated as :

H.1 : If u assumes its maximum value M in Ω then $u \equiv M$ throughout $\overline{\Omega}$;

H.2 : If u assumes its maximum value M at a point P_1 on $\partial\Omega$ then either $u \equiv M$ in $\overline{\Omega}$ or $\frac{\partial u}{\partial n}(P_1) > 0$.

Here $\frac{\partial}{\partial n}$ is the normal derivative on $\partial\Omega$ directed outward from Ω.

Going back to our "best possible" maximum principle for $\phi(u,q^2)$ it will follow that ϕ satisfies a maximum principle if ϕ satisfies an inequality of the form $L\phi \geq 0$ in Ω. If we can then find a domain for which $\frac{\partial \phi}{\partial n} \leq 0$ at each point of $\partial\Omega$ than such a maximum principle will be "best possible".

The author first became interested in this area in response to two specific questions posed to him some years ago.

1. In the elastic torsion problem for a beam of fixed cross section can one rigorously determine the point on the boundary of the cross section at which the maximum stress occurs? This question was posed by Marvin Stippes.

2. In a simple reactor problem modeled by the Helmholtz equation can one find an isoperimetric bound for the average to peak ratio? This question which was asked by Ivar Stakgold, would give a bound for the efficiency of the reactor. Mathematically the first question involves the solution of

$$\Delta u + 2 = 0 \quad \text{in} \quad \Omega \subset R^2$$

$$u = 0 \quad \text{on} \quad \partial\Omega$$

and asks; at what point of $\partial\Omega$ does $|\text{grad } u|$ assume its maximum value ? Question 2 deals with the solution of

$$\Delta u + \lambda u = 0 \quad \text{in} \quad \Omega \subset R^3$$

$$u = 0 \quad \text{on} \quad \partial\Omega; \quad u > 0 \quad \text{in} \quad \Omega$$

and asks for a bound for the quantity

$$E = \frac{\int_\Omega u \, dx}{V u_M}$$

in terms of the geometry of Ω. Here V is the volume of Ω, and u_M is the maximum value of u in Ω.

In question 1, our "best possible" maximum principle fails to give a positive answer, but it does indicate that in general we can determine portions of the boundary on which the maximum stress cannot occur. We say more about this later. For question 2, a bound for E is obtained via the maximum principle, i.e., (see [24]).

$$E \leq (2 / \pi)G$$

where

$$G = \frac{[\lambda + K_0^2]^{1/2} + K_0}{\lambda^{1/2}}$$

and

$$K_0 = \max [0, \max_{\partial\Omega} (-K)],$$

K denoting the average curvature on $\partial\Omega$. This principle also gives a bound for the distance from the boundary of the point of maximum temperature (maximum u).

In a number of applications our principle will lead to an isoperimetric bound for the maximum value of the gradient of the solution of the problem in question. In various contexts this might be interpreted as a bound for the maximum stress, maximum velocity, maximum heat flux, etc. Clearly a principle which not only yields bounds for $\max_{\bar{\Omega}} |\text{grad } u|$ in terms of the geometry of Ω, but also gives some information on where this maximum value occurs and which at the same time is sharp, in the sense that the equality sign holds for some domain, should be very useful. There are in the literature a number of methods for obtaining bound for the maximum value of the gradient particularly in the case of second order quasilinear elliptic equations ([10], [27], [29]). Some kind of monotonicity principle combined with an ordinary maximum principle is often used. For instance, in the torsion problem Fu and Wheeler [5] recently made use the ordinary maximum principle to compute bounds for the maximum stress.

Other information on the maximum value of the gradient which is independent of any governing equation is known. Before proceeding let us briefly mention one or two results in this direction (see [16]).

A. <u>Let $u(x)$ be defined in a bounded region</u> $\Omega \subset R^2$ <u>and suppose</u> $\frac{\partial u}{\partial n}$ <u>vanishes on a C^2 portion</u> Σ <u>of</u> $\partial\Omega$ <u>which at each point has positive Gaussian curvature, if</u> $u \in C^2(D \cup \Sigma)$ <u>then</u> $|\text{grad } u|^2$ <u>can take its maximum value on</u> Σ <u>iff</u> $|\text{grade } u| \equiv 0$ <u>in</u> Ω.

B. <u>Let $u(x)$ be defined on a bounded region</u> $\Omega \subset R^N$ <u>and suppose u = constant and $\Delta u - \alpha(x)u = 0$ on a C^2 portion</u> Σ <u>of</u> $\partial\Omega$, <u>at each point of which the mean curvature K satisfies the following condition</u>

$$(N - 1)K - \alpha > 0$$

If $u \in C^2(D \cup \Sigma)$ then $|\text{grad } u|^2$ can take its maximum value on Σ iff $|\text{grad } u| \equiv 0$ in D; i.e., $u \equiv$ constant in Ω.

Application of A.

In the problem of two dimensional compressible subsonic flow (uniform at infinity) past an obstacle we know that the maximum speed occurs on the boundary of the obstacle. Theorem A(for $N = 2$) implies that this maximum speed cannot occur on a non convex portion of the boundary [14].

Application of B.

For an elastic plate problem the deflection w satisfies

$$\Delta^2 w = p(x) \quad \text{in} \quad \Omega \subset R^2$$

$$w = 0 \quad \text{on} \quad \partial\Omega$$

$$\Delta w - \sigma k \frac{\partial w}{\partial n} = 0 \quad \text{on} \quad \partial\Omega, \quad 0 < \sigma \leq 1/2.$$

Result B states that if Ω is convex than $|\text{grad } W|$ cannot take its maximum value on $\partial\Omega$ no matter what the form of the loading $p(x)$.

2. SOME SIMPLE EXAMPLES.

Consider again the torsion problem

$$\begin{aligned} \Delta u &= -2 \quad \text{in} \quad \Omega C R^2 \\ u &= 0 \quad \text{on} \quad \partial\Omega. \end{aligned} \tag{2.1}$$

We seek a "best possible" principle involving u and $|\text{grad } u|^2$, restricting our attention for the moment to convex regions Ω. The optimal region, i.e. the one for which $\phi(u,q^2)$ is to be constant throughout Ω, is likely to be either the most convex region (the interior of a circle) or the least convex region (a long thin ellipse as the ratio of lengths of major to minor axis tend to ∞). Since we can solve the problem in each case we take as possible candidates for ϕ,

$$\phi_1 = q^2 + 2u \quad \text{(constant for a circle)}$$

$$\phi_2 = q^2 + 4u \quad \text{(constant for a strip).}$$

If one of these satisfies a "best possible" maximum principle (in the usual sense) then clearly the second cannot.

We look first at ϕ_1. It is easily checked that

$$\Delta\phi_1 \geq 0 \quad \text{in} \quad \Omega$$

which in turn implies that

$$\phi_1 \equiv q^2 + 2u \leq \max_{\partial\Omega} q^2 \equiv q_0^2 \tag{2.2}$$

since $\phi_1 \equiv$ constant if $\partial\Omega$ is a circle, then ϕ_1 does satisfy a "best possible" maximum principle. Note that no convexity requirement was used.

Let us look now at ϕ_2 and observe that

$$\phi_{2,k} = 2u,_i u,_{ik} + 4u,_k,$$

and

$$\Delta\phi_2 = 2u,_{ik}\, u,_{ik} - 8$$

$$\geq \frac{1}{2}\, q_2^{[\phi_{2,k}-4u,_k][\phi_2,_k-4u,_k]} - 8 \quad .$$

where we have made use of Schwarz's inequality and the summation convention. This simplifies to

$$\Delta\phi_2 \geq q^{-2}\{\phi_{2,k}\phi_{2,k} - \Delta u,_k\phi_{2,k}\}$$

or

$$\Delta\phi_2 + w_k\phi_{2,k} \geq 0.$$

Note that w_k is singular at critical points of u, i.e. at points where $q = 0$. Thus by Hopf 1 the maximum value of ϕ_2 occurs either on $\partial\Omega$ or at a critical point of u. Recall that Hopf's second principle states if ϕ_2 takes its maximum value at a point P_1 on $\partial\Omega$ then either $\phi_2 \equiv$ constant in Ω or $\frac{\partial\phi_2}{\partial n} > 0$ at P_1. But at any point on $\partial\Omega$

$$\frac{\partial\phi_2}{\partial n} = 2\frac{\partial u}{\partial n}\frac{\partial^2 u}{\partial n^2} + 4\frac{\partial u}{\partial n}.$$

Writing the differential equation in normal coordinates on the boundary we have for $N = 2$,

$$\frac{\partial^2 u}{\partial n^2} + \frac{\partial^2 u}{\partial s^2} + K\frac{\partial u}{\partial n} + 2 = 0.$$

Thus on $\partial\Omega$

$$\frac{\partial\phi_2}{\partial n} = -2Kq^2,$$

and if Ω is strictly convex, then ϕ_2 can take its maximum value on $\partial\Omega$ only at a point where $q^2 = 0$. But under the continuity assumtions we have made q cannot vanish on $\partial\Omega$. (In any case this would not yield a max for ϕ_2). On the other hade if $K \geq 0$ on $\partial\Omega$, but K actually vanishes at a point on $\partial\Omega$, then Φ_2 can vanish at such a point only of $\phi_2 \equiv$ constant in Ω. Thus for convex region we have

i) The maximum value of ϕ_2 cannot occur on $\partial\Omega$ if Ω is strictly convex.

ii) The maximum value of ϕ_2 must occur at the interior critical point where u assumes its max value, if Ω is convex. In either case then we have

$$\phi_2 \equiv q^2 + 4u \leq 4u_M \tag{2.4}$$

where u_M is the maximum value of u in Ω. Again this is a type of best possible maximum principle since the equality sign is achieved in the limit for a long thin ellipse. We know from other considerations that if Ω is convex there is a single interior critical point of u [15]. We now indicate a few consequences of these two principles.

From (H.2) we know that at the point P_1 on $\partial\Omega$ at which ϕ_1 assumes its maximum value the normal derivative of ϕ_1 will be nonnegative. In fact at P_1

$$\frac{\partial\phi_1}{\partial n} = 2[-K(P_1)q_0 + 1]q_0 \geq 0 \tag{2.5}$$

with the equality sign holding iff $\phi_1 \equiv$ const. in $\overline{\Omega}$. Thus

$$K(P_1)q_0 \leq 1 \tag{2.6}$$

with equality iff $\partial\Omega$ is a circle. This result may be used in two different ways to yield

$$1. \quad q_0 \leq \frac{1}{K(P_1)} \leq \frac{1}{K_{min}} \tag{2.7}$$

and

$$2. \quad 2K(P_1)A = K(P_1)\left|\oint_{\partial\Omega} \frac{\partial u}{\partial n}\, ds\right| \leq K(P_1)q_0 L \leq L$$

Here K_{min} is the minimum value of K on $\partial\Omega$, A is the area of Ω and L the perimeter. The first result gives an upper bound for q_0 while the second states that the maximum value of q must occur at a point p on $\partial\Omega$ at which

$$K(P) \leq L/2A$$

If Ω is convex it is possible to show that ϕ_1 takes both its maximum and minimum values on $\partial\Omega$ (see [20]).

We now discuss a few consequences of (2.4). Let 0 be the interior point of Ω at which u takes its maximum value u_M, and let Q be the nearest point on $\partial\Omega$. Then integrating along a ray from 0 to Q (recall that Ω is assumed convex) we find

$$\int_0^{u_M} \frac{d\eta}{\sqrt{u_M - \eta}} \leq 2\delta .$$

where δ is the distance from 0 to Q. Evaluating the integral we conclue

$$u_M \leq \delta^2 \leq d^2 \tag{2.8}$$

where d is the radius of the largest inscribed circle in Ω.

Evaluating (2.4) at the point on the boundary at which q takes its maximu value q_0 we have

$$q_0^2 \leq 4u_M \leq 4d^2 .$$

Let us observe also that

$$2A = -\oint_{\partial\Omega} \frac{\partial u}{\partial n} ds \leq q_0 L \leq 2\delta L$$

which state that the point at which u assumes its maximum value must be at a distance of at least A/L from the boundary. A number of other consequences of (2.2) and (2.4) have been derived in the literature. A result which is somewhat sharper than (2.6) has been found by Fu and Wheeler [5] and results for nonconvex regions have also appeared. We indicate later how some of these inequalities can be sharpened.

The clamped membrane problem was introduced earlier, but for convenience we state it again. We are interested in the solution v of

$$\Delta v + \lambda v = 0 \quad \text{in} \quad \Omega \subset R^N$$

$$v = 0 \quad \text{on} \quad \partial\Omega$$

$$v > 0 \quad \text{in} \quad \Omega.$$

No ϕ_1 for this problem has as yet been constructed; however, it is possible for convex Ω to find a ϕ_2 given by

$$\phi_2 \equiv p^2 + \lambda v^2 \leq \lambda v_M^2$$

where $p^2 \equiv |\text{grad } v|^2$. Again this is a "best possible" result. From this we derive (letting V denote the measure of Ω)

i) $\lambda \geq \pi^2/4d^2$ (see 1)

ii) $\lambda \geq \pi/2\sqrt{\lambda}$

iii) $E \equiv \dfrac{\int_\Omega u dx}{V u_M} \leq 2/\pi$ (see [24])

Note that i) gives a lower bound for λ, while ii) may be interpreted as indicating that the maximum value of u occurs at a point in Ω which is at distance at least $\pi/2\sqrt{\lambda}$ from the boundary. It is worth remarking that all of these results hold for non convex regions provided the average curvature is non negative at each point of $\partial\Omega$.

3. A GENERAL RESULT.

We turn now to the more general result of Philippin and Payne [22], a result which applies to classical solutions of the nonlinear problem

$$\sum_{i,j=1}^{N} \frac{\partial}{\partial x_i} [g(u,q^2) \frac{\partial u}{\partial x_i}] + h(u,q^2) = 0, \quad \text{in} \quad \Omega \subset R^N \tag{3.1}$$

$$u = 0 \quad \text{on} \quad \partial\Omega.$$

We could actually consider the more general boundary condition

$$B(u, \frac{\partial u}{\partial n}) = 0$$

with appropriate assumptions on $\frac{\partial B}{\partial u}$ and $\frac{\partial B}{\partial u_n}$, but the above will suffice as an illustration. We assume that g is a C^1 function of its arguments and that h is continuous in its arguments. One can relax these hypotheses also, but we do not go into this type of generalization. In order to assure that the equation is strongly elliptic we make the assumption that

$$G(u,q^2) = g(u,q^2) + 2q^2 \frac{\partial g(u,q^2)}{\partial q^2} > 0 \tag{3.2}$$

on solutions. Our aim is again to find a $\phi(u,q^2)$ which satisfies a "best possible" maximum principle. It turns out that we are able to do a first integral in the case $N = 1$; so guided by this fact we choose as a candidate for a function ϕ_2 any solution of the first order partial differential equation

$$2 H(u,q^2) \frac{\partial \phi_2}{\partial q^2} = G(u,q^2) \frac{\partial \phi_2}{\partial u} \tag{3.3}$$

where

$$H(u,q^2) = h(u,q^2) + q^2 \frac{\partial g(u,q^2)}{\partial u}$$

which satisfies the further requirement that

$$\frac{\partial \phi_2}{\partial q^2} > 0 \qquad (3.4)$$

on solutions. This latter condition imposes some restrictions on the form of h, but in many cases of interest (4.4) can be easily satisfied.

The following two theorems were derived in [22]:

<u>Theorem 1:</u> Let u be a classical solution of (3.1) in Ω. Then any ϕ_2 which satisfies (3.3) and (3.4) must take its maximum value either on $\partial\Omega$ or at a point in Ω at which $|\text{grad } u| = 0$.

<u>Theorem 2:</u> If the average curvature K is nonnegative at every point of $\partial\Omega$ then ϕ_2 cannot take its maximum value on .

These results are again "best possible" in that $\phi_2 \equiv$ constant in the limit for the infinite slab. We postpone for the moment the question of determining a ϕ_1.

The proof of Theorem 1 is very messy so we do not go into it here. The proof of Theorem 2 involves showing that if $K > 0$, then $\frac{\partial \phi_2}{\partial n} \leq 0$ on . This contradicts Hopf's second principle if ϕ_2 is assumed to take its maximum value on .

In the special case in which

$$g = g(q^2), \quad h = f(u)\,\rho(q^2)$$

the equation for ϕ_2 separates and we find that a choice for ϕ_2 is

$$\phi_2 = \int_0^{q^2} \frac{G(s)}{\rho(s)}\, ds + 2\int_0^u f(\eta)\, d\eta$$

A number of interesting examples for $N = 2$ come quickly to mind

a) Surface of constant mean curvature : $g = \dfrac{1}{(1+q^2)^{1/2}}$, f = constant; ρ = constant.

b) Torsional creep : g arbitrary, f = constant; ρ = constant.

c) Meniscus problem (Capillary tube) : $g = \dfrac{1}{(1+q^2)^{1/2}}$, $f = -u$, ρ = constant.

d) Extensible film : $g = \dfrac{1}{(1+q^2)^{1/2}}$, $\rho = \dfrac{1}{(1+q^2)^{1/2}}$; f = constant.

e) Geometrical Problems:

Of course, various special problems in two dimensional nonlinear continuum mechanics also fall into this special category of problems.

Let us briefly sketch how one would hope to use the theorems we have established.

If our equation has a positive solution then our theorems may imply that

$$\phi_2(u,q^2) \le \phi_2(u_M,0)$$

In view of the fact that $\dfrac{d\phi_2}{dq^2} > 0$ we might hope to solve the inequality for

$$q^2 \le F(u,u_M) \tag{3.5}$$

Integrating the inequality along a ray from the interior point P_0 at which $u = u_M$ to the nearest boundary point we have

$$\int_0^{u_M} \frac{d\eta}{\sqrt{F(\eta,u_M)}} \le \delta(P_0, P_1) \le d.$$

In many cases this inequality can be solved to give

$$u_M \leq \sigma(d).$$

This result may then be inserted back into (3.5) to give

$$\max_{\partial\Omega} q^2 \leq F(0,\sigma(d)) \tag{3.6}$$

Let us go back for the moment to the question of whether it is possible to find a "best possible" maximum principle for the solution of our problem (3.1) in which the optimal region is the interior of the N sphere. This is not always possible but provided

$$2gH \frac{\partial H}{\partial q^2} - NG(g \frac{\partial H}{\partial u} - h \frac{\partial g}{\partial u}) \geq 0$$

(automatically satisfied if $g = g(q^2)$ and h = constant), it can be shown that any ϕ_1 which satisfies

$$2H(u,q^2) \frac{\partial \phi_1}{\partial q^2} = NG(u,q^2) \frac{\partial \phi_1}{\partial u}, \quad \frac{d\phi_1}{dq^2} > 0.$$

must take its maximum value on $\partial\Omega$. For special choices of h and g such a ϕ_1 will be constant throughout Ω if Ω is the N ball. Here there is no convexity requirement on Ω and $\partial\Omega$ need not be a $C^{2+\alpha}$ surface. Note that H.2. implies that

$$\frac{q_0 g(0,q_0^2)}{H(0,q_0^2)} \leq \frac{1}{NK_{min}}$$

We wish to emphasize again that it is often easy for such examples to find maximum principles involving grad u which are not "best possible", but we are not considering such inequalities in this lecture.

We now make a few remarks on the special case, $g = g(q^2)$, $h \equiv \beta$, where β is a positive constant. Then provided the equation is strongly elliptic $(G > 0)$ it follows that

i) $q = |\text{grad } u|$ takes its maximum value q_0 at a point P on $\partial\Omega$.

Furthermore if $\partial\Omega$ has positive average curvature K then

ii) $q_0 g(q_0^2) \leq \beta(NK_{min})^{-1}$

iii) $K(P) \leq \beta \dfrac{\text{measure } \partial\Omega}{N \text{ measure } \Omega}$

iv) various special results for special forms of g, e.g. a number of isoperimetic inequalities for minimal surfaces.

In continuum mechanics contexts the quantity on the left of (ii) is proportional to the maximum stress on $\partial\Omega$. The interesting feature of (iii) is that it is completely independent of the form of g. It states that maximum value of the stress cannot occur at a point on $\partial\Omega$ at which (iii) is violated.

In the capillary tube interpretations a number of inequalities follow from the principles for ϕ_2 and ϕ_1-inequalities which in the short tube interpretation give bounds for the height of the bottom of the meniscus, the volume of fluid in the tube and bounds for the slope of the free surface at the boundary. In the long tube setting they yield bounds for the maximum rise of the fluid at the tube boundary.

In the inextensible film setting ϕ_1 is inapplicable but ϕ_2 takes the form

$$\phi_2 = \log(1 + q^2) + 2Cu.$$

The principle state that

$$\log(1 + q^2) \le 2C(u_M - u) .$$

An integration to the boundary yields (provided $cd < \pi/2$)

$$u_M \le \frac{1}{c} \log \sec(cd)$$

and

$$\max_{\partial\Omega} q \le \tan(cd).$$

IV. GEOMETRICAL AND PLASMA PHYSICS INTERPRETATIONS:

Suppose we are interested in finding the surface $z = u(x_1, x_2)$, $(x_1, x_2) \in \Omega$ such taht $u = 0$ on $\partial\Omega$ and at points in Ω

$$K = f(z, n_z)$$

The equation of the surface would be of the form

$$\sum_{j=1}^{2} \left(\frac{u_{,j}}{\sqrt{1+q^2}} \right)_{,j} + \tilde{P}(u,q^2) = 0.$$

In particular if we ask for the surface such that at each point

$$K = CN_z^3$$

our equation becomes

$$\sum_{j=1}^{2} \left(\frac{u_{,j}}{\sqrt{1+q^2}} \right)_{,j} + \frac{C}{(1 + q^2)^{3/2}} = 0$$

In this case

$$\phi_2 = q^2 + 2Cu$$

the same function which appeared as ϕ_2 in the torsion problem. We note in fact that in the case in which $g = g(q^2)$, $h = f(u)\rho(q^2)$ and in addition

$$G(s) = \rho(s)$$

it follows that ϕ_2 takes the simple form

$$\phi_2 = q^2 + 2 \int_0^u f(\eta)\, d\eta.$$

We now mention a class of problems from plasma physics to which these maximum principles have been applied. As a mathematical model for a certain plasma confinement problem Grad, Hu and Stevens [6] have proposed the following boundary value problem for the magnetic potential u:

$$\Delta u + \frac{dp}{du} = 0 \quad \text{in } \Omega \subset R^2$$

$$u = \gamma \quad \text{on } \partial\Omega$$

$$-\int_{\partial\Omega} \frac{\partial u}{\partial n}\, ds = I.$$

the constant γ is unprescribed but I is specified. Everything looks relatively straight forward until we discover the form of p. In order to describe p we must indicate what kind of solution we are looking for. We seek a function u which has the property that if the solution surface $Z = u(x,y)$ is plotted over Ω it will have the shape of hill with γ either positive or negative. This is to be realized by a function p of the form

$$p = p(u, a(u), \frac{du}{da})$$

Where $a(\tilde{u})$ denotes the measure of the set of points for which $u > \tilde{u}$. This leads of course to a very nonlocal type of equation and almost nothing of a general nature is known about questions of existence, regularity, etc. However, for special functions p some of these questions have been answered. Let me remark that if we know that a C^1 solution of a particular problem of this type exists then our "best possible" maximum principle tells us at once that the function

$$\phi_2(u) = |\text{grad } u|^2 + 2p\left(u, a, \frac{du}{da}\right)$$

takes its maximum value either on $\partial\Omega$ or at an interior point at which grad $u = 0$. If Ω is convex the maximum cannot occur on $\partial\Omega$. Thus if in fact p were of the form

$$p = p(u,a)$$

and we could show that if S is the set of points in Ω at which $|\text{grad } u| = 0$ the following holds:

$$\max_{x \in S} p(u, a) = p(u_M, 0),$$

then we would have

$$|\text{grad } u|^2 \le 2\{p(u_M, 0) - p(u, a)\}$$

The "best possible" maximum principle has been used to study three special examples.

a) $p = \begin{cases} \frac{1}{2}\alpha u^2, & u \ge 0 \\ 0, & u < 0 \end{cases}$, Stakgold (1981), [32]

b) $p = u \int_0^u g(a(\eta))\, d\eta$, Mossino (1981), [13]

c) $p = \begin{cases} \frac{1}{\alpha}\lambda u^{\alpha}, & u \ge 0 \\ 0, & u < 0 \end{cases}$, Bandle & Sperb, [2]

In all cases the principle was used to obtain estimates for γ and for $\max_{\Omega} u$.

A ϕ_1 can be found only in very special cases.

V. COMPARISON THEOREMS

We now wish to show how "best possible" maximum principles may be used to obtain sharp inequalities relating solutions of nonlinear elliptic problems with those of simple linear ones and/or analogous one dimensional ones. As an example suppose u is a positive solution of

$$\begin{aligned} &\Delta u + f(u) = 0 \quad \text{in} \quad \Omega \subset R^N \\ &u = 0 \quad \text{on} \quad \partial\Omega , \end{aligned} \tag{5.1}$$

where $f(u)$ satisfies .

$$f(u) \geq 0, f'(u) \geq 0 \quad \text{for} \quad u > 0 \ .$$

We wish to compare solutions of (5.1) with those of

$$\begin{aligned} &\Delta\psi + 2 = 0 \quad \text{in} \quad \Omega \\ &\psi = 0 \quad \text{on} \quad \partial\Omega \end{aligned} \tag{5.2}$$

or to positive solutions W of

$$\begin{aligned} &\Delta W + \lambda_1 W = 0 \quad \text{in} \quad \Omega \\ &W = 0 \quad \text{on} \quad \partial\Omega . \end{aligned} \tag{5.3}$$

(We assume from this point that Ω is convex). To this end we define two function

$$x_1 = \left| \int_0^{u_M} \frac{d\eta}{\sqrt{F(u_M)-F(\eta)}} \right|^2 - \left| \int_u^{u_M} \frac{d\eta}{\sqrt{F(u_M)-F(\eta)}} \right|^2$$

and

$$x_2 = \cos \left| \frac{\sqrt{\lambda_1}}{2} \int_u^{u_M} \frac{d\eta}{\sqrt{F(u_M)-F(\eta)}} \right|$$

where

$$F(u) = \int_0^u f(\eta)d\eta \ .$$

By use of the "best possible" maximum principle one can show that

$$\Delta\chi_1 \geq -4 \ ,$$

and since $\chi_1 = 0$ on $\partial\Omega$ it follows immediately that

$$\chi_1 \leq 2\psi \quad \text{in } \Omega$$

$$|\text{grad } \chi_1| \leq |\text{grad } \psi| \quad \text{on} \quad \partial\Omega .$$

The first inequality implies that

$$\int_0^{u_M} \frac{d\eta}{F(u_M)-F(\eta)} \leq 2\psi_M \ .$$

On the other hand one can show that $\Delta\chi_2 + \lambda_1\lambda_2 \leq 0$ in Ω. This in turn implies that

$$\lambda_1 \int_0^{u_M} \left[\frac{d\eta}{\sqrt{F(u_M) - F(\eta)}} \right]^2 \geq \pi^2/2$$

In particular choosing $u = \psi$ we conslude that

$$\lambda_1 \psi_M \geq \pi^2/4$$

All of these inequalities are sharp in that the equality sign holds in the limit for an infinite slab. A number of more involved alternative theorems can actually be derived (see [17]).

Another use of the "best possible" maximum principle permits us to compare solutions of the torsional creep equation

$$\sum_{i=1}^{2} \frac{\partial}{\partial x_1} [g(q^2) \frac{\partial u}{\partial x_1}] + 2k_1 = 0 \quad \text{in} \quad \Omega \subset R^2,$$

$$u = 0 \quad \text{on} \quad \partial\Omega$$

with those of

$$[g(p^2)v']' + 2k_1 = 0 \qquad 0 \le x \le \sqrt{\psi_M}$$

$$v'(0) = 0, \qquad v(\sqrt{\psi_M}) = 0$$

where X is the solution of the torsion problem and Ω is assumed to be convex. If $g(s)$ satisfies

i) $g'(s) < 0, \quad s > 0$

ii) $G(s) = g(s) + 2sg'(s) > 0, \quad s > 0$

iii) $g'(s)[2sg^2(s))']^{-1}$ nonincreasing in s for $s > 0$

it follows that

$$u \le v(\sqrt{\psi_M} - \psi) \quad \text{in} \quad \Omega$$

$$|\text{grad } u| \le \frac{v'(\sqrt{\psi_M})}{2\sqrt{\psi_M}} |\text{grad } \psi| \quad \text{on} \quad \partial\Omega. \tag{5.4}$$

The proof of this result uses the best possible maximum principle for ψ as well as a result of Makar-Limanov which states that if Ω is convex then all of the level curves of ψ are convex [11]. Again inequalities (5.4) sharps.

In the torsional creep problem the quantity

$$S = \int_\Omega u \, dx$$

is propertional to the rigidity of the beam. By integrating the first of (5.4) over Ω we find that for g's satisfying i), ii) and iii)

$$S \le [\psi_M]^{-1} A \int_0^{\sqrt{\psi_M}} v(\sigma) d\sigma.$$

Inequalities (5.4) may also be used to prove that

$$Q_0 = \max_\Omega qg(q^2) \le k_1 \sqrt{\psi_M} ,$$

again a bound completely independent of the form of g.

It would be possible to derive additional results for solutions of

$$\Delta u + f(u) = 0 \text{ in } \Omega$$

$$u = 0 \quad \text{on } \partial\Omega$$

provided u satisfied the Makar-Limanov property. This property is known to hold for $f(u)$ = constant [11] and for $f(u) = \lambda_1 u$, [3]. Further results are contained in a paper of Caffarelli and Spruck [4].

It should be pointed out that the Makar-Limanov function used in establishing the convexity of level curves is the function

$$\Phi = 2|\text{grad } \psi|^2 + \sum_{i,j=1}^{2} \psi,_i \psi,_j \psi,_{ij} + \psi[4 - \sum_{i,j=1}^{2} \psi,_{ij} \psi,_{ij}]$$

It too satisfies a "best possible" minimum principle since Φ satisfies a minimum principle and $\Phi \equiv$ constant if is the interior of a circle.

The important property of Φ is that at interior points of Ω at which the curvature of the level curve of ψ is negative, Φ must also be negative. But Φ takes its minimum value on $\partial\Omega$. From this fact we conclude that if Ω is convex Φ must be nonnegative in Ω, and hence that the level curves of ψ are convex curves.

On the other hand the minimum principle permits us to find a lower bound for the quantity $K|\text{grad}\,\psi|$ which can be used to go back and sharpen some of the results obtained earlier for the torsion problem. In fact all of the results which we derived earlier by the use of ϕ_2 (where the optimal domain was the infinite strip) can now be sharpened to be optimal both for the strip and for the circle.

VI. CONCLUDING REMARKS.

By making use of relatively simple examples we have tried to describe the potential usefulness of a special class of maximum principles. We have dealt almost exclusively with Dirichlet boundary conditions and convex regions. More general boundary conditions can be easily dealt with as can nonconvex regions. It turns out, however, that one often loses the "best possible" feature unless the geometry of the domain is suitably restricted. For other uses of the techniques described herein see [18], [19], [21], [23], [25], [28].

REFERENCES.

1. C. Bandle, On isoperimetric gradient bounds for Poisson problems and problems of Torsional Creep, ZAMP, 30, (1979) pp. 713-715.

2. C. Bandle and R. Sperb, Qualitative behavior and bounds in a nonlinear plasma problem, SIAM J. Math. Anal. (to appear).

3. H.J. Brascamp and E.H. Lieb, Some inequalities for Gaussian measures and the long range order of the one dimensional plasma, Funct. Integ. Int. Conf. London (1974) A.A. Arthurs editor, Oxford (1975) pp. 1-14.

4. L.A. Caffarelli and J. Spruck, Convexity properties of polutions to some classical variational problems, (to appear).

5. L.S. Fu and L. Wheeler, Stress bounds for bars in torsion, J. Elasticity, 3 (1973) pp. 1-13.

6. H. Grad, P.N. Hu and D.C. Stevens, Adiabatic evolution of plasma equilibrium, Proc. Nat. Acad. Sci. U.S., 72, (1975), pp. 3789-3793.

7. J. Hersch, Sur la frequence fondamentale d'une membrane vibrante; Evaluation par defaut et principe de maximum, ZAMM, 11, (1960) pp. 387-413.

8. E. Hopf, Elementare Bemerkung über die Lösungen partieller Differentialgleichungen zwirter Ordung vom elliptischen Typus, Ber Sitzungsber. Preuss. Akad. Wiss., 19 (1927) pp. 147-152.

9. E. Hopf, A remark on elliptic differential equations of the second order, Proc. Amer. Math. Soc., 3, (1952), pp. 791-793.

10. O.A. Ladyzhenskaya and N.N. Uraltseva, Equations aux derivees partielles de type elliptique, Durod, Paris (1968).

11. L.G. Makar-Limanov, Solution of Dirichlet's problem for the equation $\Delta u = -1$ in a convex region, Math. Notes Acad. Sci. USSR, 9, (1971), pp. 52-53.

12. C. Miranda, Formule di maggiorazione e teorema di esistenza per le funzioni biarmoniche di due variabli, Giorn. Math. Battaglini, 78 (1948-49), pp. 97-118.

13. J. Mossino, A priori estimates for a model of grad Mercier type in plasma confinement, Applicable Analysis (to appear).

14. L.E. Payne, A priori bounds in problems of steady subsonic flows, Inequalities, Academic Press (1967), pp. 241-254.

15. L.E. Payne, On two conjectures in the fixed membrane eigenvalue problem, ZAMP, Vol. 24, (1973), pp. 721-729.

16. L.E. Payne, On explicit norm inequalities and their applications in partial differential equations, Constructive Approaches to Mathematical Models, Academic Press (1979), pp. 279-292.

17. L.E. Payne, Bounds for solutions of a class of quasilinear elliptic boundary value problems in terms of the torsion function. Proc. Royal Soc., Edinburgh 88A (1981), pp. 251-265.

18. L.E. Payne and G.A. Philippin, On some maximum principles involving harmonic functions and their derivatives, SIAM J. Math. Anal. 10 (1979), pp. 96-104.

19. L.E. Payne and G.A. Philippin, Some applications of the maximum principle in the problem of torsion creep, SIAM J. Appl. Math. 33 (1977).

20. L.E. Payne and G.A. Philippin, Some remarks on the problems of elastic torsion and of torsional creep, Some Aspects of Mech. of Continua, Sen Memorial Volume (1977) pp. 32-40.

21. L.E. Payne and G.A. Philippin, Some maximum principles for nonlinear elliptic equations in divergence form with applications to capillary surfaces and to surfaces of constant mean curvature, Nonlinear Analysis, 3 (1979), pp. 193-211.

22. L.E. Payne and G.A. Philippin, On maximum principles for a class of nonlinear second order elliptic equations, J. Diff. Eqtns. 37 (1980) pp. 39-48.

23. L.E. Payne, R. Sperb and I. Stakgold, On Hopf type maximum principles for convex domains, Nonlinear Analysis, 1, (1977), pp. 547-599.

24. L.E. Payne and I. Stakgold, On the mean value of the fundamental mode in the fixed membrane problem, Appl. Anal., Vol. 3 (1973), pp. 295-303.

25. L.E. Payne and I. Stakgold, Nonlinear problems in nuclear reactor analysis, Lecture Notes in Mathematics #322 Springer Verlag, (1972) pp. 298-307.

26. M.H. Protter and H.F. Weinberger, Maximum principles in differential equations, Prentice-Hall, Inc. (1967).

27. M.H. Protter and H.F. Weinberger, A maximum principle and gradient bounds for linear elliptic equations, Indiana Univ. Math. J. 23, (1973-74) pp. 239-249.

28. P.W. Schaefer and R.P. Sperb, Maximum principles and bounds in some inhomogeneous boundary value problems, SIAM J. Math Anal., 8, (1977), pp. 871-878.

29. J.B. Serrin, Gradient estimates for solutions of nonlinear elliptic and parabolic equations, Contrib. to Nonlinear Fctnl. Analysis, Univ. of Wisconsin Press, (1971), pp. 565-601.

30. R.P. Sperb, Maximum principles and their applications, Math. in Science and Engineering, Vol. 157 Academic Press (1981).

31. R.P. Sperb, Extension of two theorems of Payne to some nonlinear Dirichlet problems, ZAMP, 26, (1975), pp. 721-726.

32. I. Stakgold, Gradient bounds for plasma confinement, Report #44A, Univ. of Delaware, (1979).

L.E. Payne
Cornell University

J SMOLLER*

Two lectures on nonlinear parabolic systems

1. INTRODUCTION

We consider two systems of nonlinear parabolic equations in a single space variable. The first system arises in population dynamics; specifically, it is a predator-prey system where diffusion and spatial effects are taken into account. It is of the form

$$u_t = u_{xx} + f(u) - uv \ , \ v_t = dv_{xx} + v[-v + m(u-\gamma)], \ |X|<L, \ t>0, \qquad (1)$$

together with homogeneous Dirichlet boundary conditions

$$v(\pm L,t) = u(\pm L,t) = 0 \ . \qquad (2)$$

Here u is the prey density and v the predator density, so we only consider solutions (u,v) with $v \geq 0$, $u \geq 0$. The nonlinear function f is given by $f(u) = au(1 - u)$, where $a > 0$, and, $d \geq 0$, $0 < \gamma < 1$, $m > 0$. The problem is to find all the stationary solutions of (1) and to describe their stability properties. The quantities d and L are considered as bifurcation parameters; the number of stationary solutions changes as these parameters vary.

The second system is a model system for isentropic gas dynamics with viscosity:

$$v_t - u_x = \varepsilon v_{xx} \ , \ u_t + p(v)_x = \mu u_{xx} \ , \ x \in \mathbb{R} \ , \ t > 0 \ . \qquad (3)$$

*Research supported in part by the N.S.F.

We consider the Cauchy problem for (2); ie, we assign initial data

$$v(x,0) = v_o(x) \ , \ u(x,0) = u_o(x) \ , \ x \in \mathbb{R} \ . \tag{4}$$

In this model, v is the specific volume (i.e., the reciprocal of the density ρ ; $v = \rho^{-1}$) , u is the velocity, and $p(v)$ denotes the pressure. The hypotheses on p are the usual ones; namely

$$p' < 0 \ , \ p'' > 0 \ . \tag{5}$$

The problem here is to prove the existence of a solution of (3)-(4) under mild hypotheses on the data functions (v_o,u_o) .

2. THE PREDATOR-PREY SYSTEM

It is easy to check that the system (1), (2) admits arbitrarily large invariant rectangles of the form $0 \le u \le U$, $0 \le v \le V$, and thus the problem (1)-(2) has a unique solution for any preassigned (smooth) bounded initial data; see [1]. Moreover, one can also see that all solutions tend to the fixed rectangle

$$0 \le u \le 1 \ , \ 0 \le v \le m(1 - \gamma) \tag{6}$$

as $t \to \infty$. This implies, in particular, that all steady-state solutions of (1) , (2) , that is, all solutions of the system on $|x| < L$,

$$u'' + f(u) - uv = 0 \ , \ dv'' + v[-v + m(u - \gamma)] = 0 \ , \tag{7}$$

with boundary data

$$(u,v) \ (\pm L) = 0 \ , \tag{8}$$

must lie in the region (6) .

We now turn to the problem of determining the solutions of (7), (8). First, it is not too hard to show that there is a number $L_f > 0$ such that if $0 < L < L_f$, then $(0,0)$ is the only solution of (7), (8) , and in this case, all solutions of the time dependent problem tend to $(0,0)$ as $t \to \infty$, uniformly on $|x| \le L$.

Next, if $L > L_f$, the trivial solution bifurcates into itself and a solution of the form $(\bar{u},0)$, where $\bar{u}$ satisfies

$$u' + f(u) = 0 \ , \ |x| < L \ , \ u(\pm L) = 0 \ . \tag{9}$$

We can show that there is a number $L_\gamma > L_f$ such that if $L_f < L < L_\gamma$, then $(\bar{u},0)$ is a global attractor of all solutions of (1), (2) (having non-negative initial data), as $t \to +\infty$, with the exception, of course, of the trivial solution $(0,0)$. This theorem is proved by a comparison method whereby we use as comparsion functions, solutions of a nonlinear scalar equation. As L crosses L_γ , the solution $(\bar{u},0)$ undergoes a bifurcation and becomes unstable.

At this point, we must consider two cases; $d = 0$ and $d > 0$. We first assume $d = 0$, and consider the equation

$$u'' + h(u) = 0 \ , \ |x| < L \ , \ u(\pm L) = 0$$

where

$$h(u) = \begin{cases} f(u) \ , & , \text{ if } u \le \gamma \\ f(u) - mu(u-\gamma) & , \text{ if } u > \gamma \ . \end{cases}$$

If $L > L_\gamma$, (10) admits a non constant solution, $\tilde{u}$, and we set

$$\tilde{v}(x) = \begin{cases} 0 & , \text{ if } \tilde{u}(x) < \gamma \\ m(\tilde{u}(x)-\gamma) & , \text{ if } \tilde{u}(x) \ge \gamma \ . \end{cases}$$

The function $(\tilde{u},\tilde{v})$ is a solution of (7), (8) in the case $d = 0$; it bifurcates out of $(\bar{u},0)$. The function $\tilde{v}(x)$ is interesting in that it has compact support in the interior of the interval $|x| < L$. Thus the solution $(\tilde{u},\tilde{v})$ corresponds to a situation in which the predator survives in the interior of the spatial domain; its "dead zone" is near the boundary, and has thickness which is bounded away from zero, independently of L .

If we linearize equations (1), (2) about $(\bar{u},0)$, we find that when $L = L_\gamma$, zero comes into the spectrum of the linearized operator, and for $L > L_\gamma$, this spectrum contains an interval of the form $0 \leq \lambda \leq \varepsilon$, for some $\varepsilon > 0$. This leads one to suspect that there are many other bifurcations as L exceeds L_γ . This contention is correct. In fact, we can show that in this case the situation becomes highly degenerate, and there appears continuum of stationary solutions which bifurcate out of $(\bar{u},0)$.

In the case when $d > 0$, we can show that there exists a monotone curve $L = L(d)$, with $L' > 0$, starting at the point $(0,L_\gamma)$ in the L-d plane, with the property that above this curve $(\bar{u},0)$ is a global attractor and below it, $(\bar{u},0)$ is unstable. Using some standard bifurcation techniques, [7], we can show that $(\bar{u},0)$ bifurcates across this curve, into itself and a stable solution, (u_1,v_1) . It would be of great interest to investigate the stability and bifurcation properties of (u_1,v_1) for fixed L , as $d \searrow 0$.

We can obtain similar results in the case when $f(u)$ is a cubic polynomial, only here the bifurcation diagrams are more complicated than in the quadratic case.

3. THE VISCOSITY EQUATIONS

The interest in the equations (3) stems from the fact that they can be considered as models of the compressible Navier-Stokes equations; their solutions, for small ε and μ should be near to those of the isentropic gas dynamics. The initial-value problem (3), (4) has been studied in the case $\varepsilon = \mu$ by Kanel' [4], and more recently, by Matsumura and Nishida [5], for the full compressible Navier-Stokes equations in three space dimensions. These authors have used the so-called "energy" method to obtain the existence of a global solution defined in $t > 0$. But in order for their methods to work, it is necessary to assume that the data functions lie in some Sobolev space H^s with $s > 1$. This means that the data must be "small" at infinity and excludes some natural choices. Thus it excludes "Riemann problem" data, and it does not allow one to study the stability of those travelling waves which approximate shock waves.

We shall study the system (3) by the method of finite differences. Thus we choose mesh parameters Δt, $\Delta x > 0$ and we divide the upper half plane $t \geq 0$ be the lines $x = n\Delta x$, $n \in \mathbb{Z}$, and $t = k\Delta t$, $k \in \mathbb{Z}_+$. If $f = f(x,t)$ we write $f_n^k = f(n\Delta x, k\Delta t)$. In these terms, we consider the following difference approximation to (3):

$$\begin{aligned} \frac{v_n^{K+1} - v_n^K}{\Delta t} - \frac{u_{n+1}^k - u_{n-1}^k}{2\Delta x} &= \varepsilon \frac{v_{n+1}^k - 2v_n^k + v_{n-1}^k}{(\Delta x)^2} \\ \frac{u_n^{K+1} - u_n^k}{\Delta t} + \frac{P(v_{N+1}^k) - P(v_{n-1}^k)}{2\Delta x} &= \mu \frac{u_{n+1}^k - 2u_n^k + u_{n-1}^k}{(\Delta x)^2} \end{aligned} \tag{11}$$

Notice that if $\varepsilon = \mu = (\Delta x)^2/2\Delta t$, these equations become the Lax-Friedrichs equations.

We assume that the initial data is of bounded variation, so that we have the estimate

$$\sum_{n\in\mathbb{Z}} |v_n^0 - v_{n-1}^0| + |u_n^0 - u_{n-1}^0| \leq c_0 .$$

If we set $U_n^k = u_n^k - u_{n-1}^k$, and $P_n^k = P(v_n^k) - P(v_{n-1}^k)$, then we obtain

$$U_n^{k+1} = \gamma(U_{n+1}^k + \beta U_n^k + U_{n-1}^k) - \frac{\tau}{2}(P_{n+1}^k - P_{n-1}^k) , \tag{12}$$

where $\gamma = \mu\Delta\tau/(\Delta x)^2$, and $\tau = \Delta t/\Delta x$. Introducing the translation operators $SU_n^k = U_{n+1}^k$, $S^{-1}U_n^k = U_{n-1}^k$, $SP_n^k = P_{n+1}^k$, $S^{-1}P_n^k = P_{n-1}^k$, we can write (12) as

$$U_n^{k+1} = \gamma(S+\beta+S^{-1})U_n^k - \frac{\tau}{2}(S - S^{-1})P_n^k .$$

If we iterate this formula we obtain

$$U_n^k = \sum_{|m|\le k} \alpha_m^k U_{n+m}^{0} - \tau \sum_{\ell=1}^{k} \sum_{0\le m\le\ell} A_m^\ell(P_{n+m}^{k-\ell} - P_{n-m}^{k-\ell}) ,$$

where

$$\begin{cases} \alpha_m^\ell = \gamma^\ell \sum_{j=0}^{\ell} \binom{\ell}{j}\beta^{\ell-j}\binom{j}{\frac{1}{2}(j-m)} , & |m| \le \ell \\ A_m^\ell = \frac{1}{2}\gamma^{\ell-1} \sum_{j=0}^{\ell-1} \binom{\ell-1}{j}\beta^{\ell-j-1} \frac{m}{j+1}\binom{j+1}{\frac{1}{2}(j+1-m)}, & m \le \ell+1 . \end{cases} \tag{13}$$

One sees that the α_m^ℓ and A_m^ℓ are linear combinations of random walk coefficients with unrestricted and absorbing boundaries, respectively. Using these facts, we can prove that

$$\sum_{m=0}^{\ell} A_m^\ell \le c/\sqrt{\ell} , \tag{14}$$

where c is independent of ℓ .

Now if we assume that the difference approximants for $u\Delta\tau \le T$ all lie in a region of the form $v \ge c(T)$, where $c(T)$ is independent of the mesh parameters Δt , Δx , then we can uniformly bound the total variation of the approximants, where the bounds are independent of the mesh parameters. Having this estimate, it is easy to prove that the difference approximants are L^1-Lipschitz continuous with respect to t. These facts imply that the difference approximants converge to a (classical) solution of (3), (4).

As a consequence of these results, we have the following theorem.

Theorem. Consider the equations (3), (4) with $\varepsilon = \mu > 0$, and assume that u_0 and v_0 have bounded total variation, and that $v_0(x) \geq \delta > 0$ for all $x \in \mathbb{R}$. There is a $\delta' > 0$ such that if

$$\max_{v \geq \delta'} \sqrt{-P'(v)} \leq \Delta x/\Delta t \quad \text{and} \quad \varepsilon \leq \Delta x^2/2\Delta t \quad ,$$

then the entire set of difference approximations converges to a classical solution of (3), (4) in $t > 0$. This solution is unique and is of bounded variation on each line $t > 0$. The boundary data is assumed in the classical sense at points of continuity and in general, in the sense that

$$\lim_{t \searrow 0} \int_{\mathbb{R}} \left\{ \phi(x,t) \begin{pmatrix} v(x,t) \\ u(x,t) \end{pmatrix} - \phi(x,0) \begin{pmatrix} v_0(x) \\ u_0(x) \end{pmatrix} \right\} dx = 0$$

for every continuous function ϕ which vanishes for $|x|$ sufficiently large.

The fact that the difference approximations all lie in a region of the form $v \geq \delta' > 0$, is a consequence of a theorem of David Hoff, [3].

Finally, we remark that we can show, directly from the difference approximants, that our solution satisfies the representations

$$u(x,t) = \int_{\mathbb{R}} \frac{1}{\sqrt{4\pi\mu t}} e^{-y^2/4\mu t} u_0(x+y)dy$$

$$- \int_0^t ds \int_{\mathbb{R}} \frac{1}{\sqrt{4\pi\mu(t-s)}} \frac{y}{2\mu(t-s)} e^{-y^2/4\mu(t-s)} p(v(x+y,t))dy \ ,$$

$$v(x,t) = \int_{\mathbb{R}} \frac{1}{\sqrt{4\pi\varepsilon t}} e^{-y^2/4\varepsilon t} v_o(x+y)dy$$

$$+ \int_0^t ds \int_{\mathbb{R}} \frac{1}{\sqrt{4\pi\varepsilon(t-s)}} \frac{y}{2\varepsilon(t-s)} e^{-y^2/4\varepsilon(t-s)} v(x+y),t)dy \; .$$

Notes. The discussion in §2 is taken from the paper of E. Conway, R. Gardner and myself, [2]. Complete proofs, as well as more precise statements are contained in that paper. The material in §3 is taken from the paper of T. Nishida and myself, [6]. In this paper we give complete proofs of all of the estimates, as well as the derivation of the formulas (13). The important inequality (14) is obtained from Stirlings approximation to $n!$, together with the use of Bernstein polynomials.

REFERENCES.

1. N. Chueh, C. Conley and J. Smoller, Positively invariant regions for systems of nonlinear diffusion equations, Ind. U. Math. J., 26, (1977), 373-392.

2. E. Conway, R. Gardner, and J. Smoller, Stability and bifurcation of steady-state solutions for predator-prey equations, Adv. Appl. Math., 3, (1982), 288-339.

3. D. Hoff, A finite difference scheme for a system of two conservation laws with artificial viscosity, Math. Comp., 33, (1979), 1171-1193.

4. Ya. Kanel', One some systems of quasilinear parabolic equations, Zh. Vychislit., Matem. i Matem Fiziki, 6, (1966), 446-477; eng. transl. in USSR Comp. Math. and Math. Phys., 6, (1966), 74-88.

5. A. Matsumura, and T. Nishida, The initial-value problem for the equations of motion for compressible viscous, and heat conductive fluids, Proc. Japan Acad., 55, Ser. A, (1979), 337-342.

6. T. Nishida, and J. Smoller, A class of convergent finite difference schemes for certain nonlinear parabolic systems, Comm. Pure Appl. Math., to appear.

7. M. Crandall, and P. Rabinowitz, Bifurcation, perturbation of simple eigenvalues and linearized stability, Arch. Rat. Mech. Anal., 52, (1973), 161-181.

Joel Smoller

The University of Michigan

H F WEINBERGER

Long-time behaviour of a class of biological models

We wish to extend results obtained for continuous models in population genetics and population ecology to cover a larger class of models in which the space and time variables are discrete. This generalization is desirable for a variety of reasons: In the first place, it is generally not practical to measure more than finitely many aggregate quantities. Secondly deterministic models by necessity rely on the assumption that the random events found in nature are averaged out. Finally computations for continuous models are generally carried out by discrete approximation. Moreover continuous models often appear as approximations of discrete ones. (cf. [2], [38], [39]) Therefore it is useful to know that all models manifest the same qualitative long term behavior. We shall limit our discussion to models which are deterministic and which only involve a scalar as an independent variable. Both of these restrictions are serious ones. The work discussed here appears with proofs in [51].

The most frequently used model for the spread of a mutant or new population in a homogeneous environment is the Fisher equation

$$\frac{\partial u}{\partial t} = D\,\Delta u + f(u) \tag{1.1}$$

This one dimensional equation was introduced by R.A. Fisher [15] to model the diffusion of an advantageous form (allele) of a single gene in a population of diploid individuals. It is assumed that there are only two allelic forms A and a of the gene. The function u(t,x) is the gene frequency, that is, the ratio of A alleles to the total number of A and a alleles at time t in the population near point x. Fisher found that if $f(u) = u(1-u)$, there is a constant c^* with the property that whenever $|c| \geq c^*$, equation (1.1) has a traveling wave solution of the form $u(t,x) = W(x-ct)$ and whenever $|c| < c^*$ no such traveling wave exists. He also conjectured that the minimal wave speed c^* is the speed at which the newly arrived advantageous allele advances into the established population. Kolomogoroff, Petrowsky, and Piscounoff [31] established this for the case where $u(x,0)$ is zero for $x < 0$ and 1 for $x > 0$ provided that $f(0) = f(1) = 0$ and f is concave. The Fisher equation has also been used to study the spread of a population when u is interpreted as a population density.

We shall consider discrete time models of the form

$$u_{n+1} = Q[u_n] \tag{1.2}$$

where $u_n(x)$ represents the gene fraction or population density at time n at the point x of the habitat and Q is an operator on a certain set of functions on the habitat. The habitat may be of any dimension. If the model is discrete, x ranges over a finite set of niches or census tracts. Our model is deterministic in the sense that u_{n+1} is determined by u_n. We continue to assume that u_n is scalar-valued and that the habitat is homogenous.

A few biologically reasonable hypotheses about Q permit us to derive results of Fisher and of Kolomogroff, Petrowsky, and Piscounoff type for this class of models. To be more precise, we can show that for each unit vector ξ there is a wave speed $c^*(\xi)$ with the property that asymptotically a new mutant or population confined to a bounded set spreads like a solution of a wave equation whose plane wave solutions with normals ξ

have speeds $c^*(\xi)$. Under additional conditions we further establish that $c^*(\xi)$ is a threshold value for the existence of traveling wave solutions.

Equation (1.1) implies (1.2) when $Q[v]$ is defined to be the solution at some generation time τ of the initial value problem for (1.1) with initial values $v(x)$. Therefore our results generalize the one-dimensional results of Fisher [15], Kolmogoroff, Petrowsky, and Piscounoff [31], Kanel [25], [26], Kametaka [24], and Aronson and Weinberger [4], as well as the multi-dimensional results of Aronson and Weinberger [5] for the rotationally symmetric model (1.1).

Translational invariance (homogeneity) does not imply rotational invariance (isotropy). Because prevailing winds and other phenomena can produce preferred directions, the analogue of (1.1) without rotational symmetry is also of interest. In this case the operator $D\Delta$ in (1.1) is replaced by an elliptic operator of the form

$$\Sigma a_{ij} \frac{\partial^2}{\partial x_i \partial x_j} + \Sigma b_i \frac{\partial}{\partial x_i}$$

This problem can again be put into the form (1.2) by defining $Q[v]$ to be the solution at time τ of the initial value problem with initial values $v(x)$. We can, in fact, allow for seasonal variation by permitting the coefficients and f to be periodic functions of t of period τ .

Since the migrational behavior can be affected by the genotype [41] or by the population density [2], [3], [19], it is also of interest to consider the problem in which the coefficients a_{ij} and b_i in the above operator depend upon u . This problem can again be converted to the form (1.2) and treated by our methods.

Our results also generalize work by Diekmann [10], [11], Diekmann and

Kaper [12], Thieme [48], [49], and the author [53], [54] on one-dimensional and rotationally symmetric discrete-time models of the form (1.2) where Q is a nonlinear integral operator.

We have not been able to generalize the more delicate results on convergence in shape obtained by Kolmogoroff, Petrowsky, and Piscounoff [31], Kanel [25], [26], Fife and McLeod [13], Bramson [7], and Uchiyama [50], [51], which show that for the one-dimensional Fisher equation the width of the transition layer between the high and low parts of a spreading pulse remains bounded and which approximate the space-time path of this layer.

The results presented here can also be applied to certain models for the spread of epidemics [1], [10], [11], [28], [29], [37], [47], [56].

Some examples of biological models are given in § 2. A precise statement of the hypotheses on the recursion operator Q is in § 3. In § 4 we show that the rather small set (3.1) of hypotheses leads to a wave speed $c^*(\xi)$ for disturbances which depend only on the single variable $x \cdot \xi$, where ξ is any unit vector.

We state our results in § 5. Essentially these are that, in an asymptotic sense, an initially confined disturbance is propagated along the ray cone which corresponds to the wave speed $c^*(\xi)$, and that there are plane wave solution of the form $W(x \cdot \xi - nc)$ exactly when $c \geq c^*(\xi)$. The remainder of the paper is devoted to proofs ot these theorems together with some examples and counterexamples.

Our results show that a rather large class which contains both continuous and discrete models is robust with respect to qualitative asymptotic behavior. This robustness means, of course, that a very crude model can give correct predictions about the asymptotic behavior of an ecological system. The other side of the coin of robustness, of course, is that a model which is found to predict such behavior correctly may be far from being a good model for predicting other phenomena.

2. SOME MODELS IN POPULATION GENETICS AND POPULATION GROWTH.

To obtain an example of a recursion of the form (1.2), we consider the so-called stepping stone model [27], [30], [35], [40], [45] in population genetics. We classify the individuals of a certain diploid population according to their genotypes with respect to a single gene locus, which occurs in two allelic (that is, variant) forms, which we label A and a. There are then three genotypes: the homozygotes AA and aa, and the heterozygote Aa.

The habitat is either naturally or arbitrarily divided into discrete regions or niches. The population which lives in any one niche during the nonmigratory part of a life cycle is called a deme. We suppose that the generations are nonoverlapping, and that members of different demes do not mix or interact except during a brief period of migration at the end of the life cycle. Migration moves the individuals to new niches, so that it changes the memberships of the various demes. Just after migration the individuals form gametes and die. The individuals are assumed to be monoecious (that is, hermaphroditic). The gametes mate at random to form the new generation of diploids.

The ratio of the number of alleles of type A to the total number of alleles of types A and a in a certain deme is called its gene fraction. Let $u_n(i)$ denote the gene fraction in the newly born individuals of the nth generation in the ith deme.

The Hardy-Weinberg law states that the ratios of the numbers of newly born individuals of genotypes AA, aA, and aa, respectively, are: $u_n(i)^2$, $2u_n(i)(1 - u_n(i))$, $(1 - u_n(i))^2$. These individuals undergo various hazards during the growth cycle, and we assume that their abilities to survive these hazards depend only on their genotypes with respect to the gene under consideration. If the fitnesses (that is, survival rates to the migration stage) of the three genotypes have the ratios $1 + s_i : 1 : 1 + \sigma_i$, then the ratios of the numbers of survivors at the time of migration are $(1 + \sigma_i)u_n(i)^2 : 2u_n(i)(1 - u_n(i)) : (1 + s_i)(1 - u_n(i))^2$. We suppose that the total number of individuals in the ith deme who survive to migrate is a fixed carrying capacity p_i, which does not depend upon the genotypic makeup of the population. We further suppose that a fraction ℓ_{ji} of the population of each genotype in the ith deme migrates to become part of deme j. (Of course the fraction ℓ_{ii} remains in the ith deme.) Under our assumptions the gene fraction $u_{n+1}(j)$ born in the jth deme in the next generation is equal to the gene fraction in the jth deme after migration. It is given by the formula

$$u_{n+1}(j) = \sum_i m_{ji} g_i(u_n(i)) , \tag{2.1}$$

where

$$g_i(u) = \frac{2(1+s_i)u^2 + 2u(1-u)}{2[(1+s_i)u^2 + 2u(1-u) + (1+\sigma_i)(1-u)^2]} \tag{2.2}$$

is the gene fraction of the ith deme just before the migration and

$$m_{ji} = \frac{\ell_{ji}p_i}{\sum_k \ell_{jk}p_k} \tag{2.3}$$

is the fraction of those individuals who are in the jth deme after the migration who came from the ith niche.

Thus the function $\{u_n(i): i = 1, 2, \ldots\}$ satisfies a recursion of the form (1.2) with

$$Q[u](j) = \sum_i m_{ji} g_i(u(i)) \ . \qquad (2.4)$$

In the present work we will only deal with a homogeneous habitat. By this we mean that all the niches are identical, that they are obtained by applying the elements of a group of translations to any one niche, and that the migration function ℓ_{ij} depends only on the translation which takes the ith niche into the jth niche.

Suppose, for instance, that the organisms are living in a plane R^2. This plane is divided into the squares $\{(x,y) | (k-\frac{1}{2})h \le x < (k+\frac{1}{2})h \ , \ (l-\frac{1}{2})h \le y < (l+\frac{1}{2})h \ ; \ k,l = 0, \pm 1, \pm 2, \ldots\}$ of side h, which are the niches. We label each niche by the coordinates of its center, which are integral multiples of h. Then we can think of the habitat as the set of points of R^2 whose coordinates are integral multiples of h. We shall usually write these centers as vectors x. $\mathcal{H}$ can be obtained from any one niche by applying all translations by two-vectors whose components are multiples of h.

The fact that $\mathcal{H}$ is homogeneous means that the fitnesses s and σ and the adult population p are the same for all the niches, and the migration function ℓ_{ij} depends only on the vector difference $x_i - x_j$ between the centers of the niches. For the moment we also assume that s, σ, and p do not depend on u, so that they are constants. Since

$$\sum_k \ell_{jk} = \sum_k \ell(x_j - x_k) = \sum_i \ell(x_i) = \sum_i \ell_{i0} = 1$$

because all the individuals in niche x_0 must go somewhere, (2.3) becomes

$$m_{ij} = \ell_{ij} = m(x_i - x_j) \ .$$

Thus the operator Q in the recursion (1.2) is

$$Q[u](x) = \sum_{y \in \mathcal{H}} m(x-y) g(u(y)) , \tag{2.5}$$

where $\sum_{x \in \mathcal{H}} m(x) = 1$ and

$$g(u) = \frac{su^2 + u}{1 + su^2 + (1-u)^2} . \tag{2.6}$$

We see that

$$g(u) - u = \frac{u(1-u)\ su-\sigma(1-u)}{1 + su^2 + \sigma(1-u)^2} . \tag{2.7}$$

The definition of y shows that we must only consider functions $u(x)$ such that $0 \leq u \leq 1$. It is easily seen that g increases from 0 to 1 as u increases from 0 to 1. Therefore $Q[u]$ again has values on the interval $[0,1]$.

We see from (2.7) that there are three possible behaviors of the function g:

(i) If $s > 0 > \sigma$, the AA homozygotes are the most fit and the aa homozygotes are the least fit, and

$$g(u) > u \qquad \text{for} \quad 0 < u < 1 . \tag{2.8}$$

This is called the <u>heterozygote intermediate</u> case. (Note that if $s < 0 < \sigma$, then $g(u) < u$. We can reduce this to the above case by replacing the variable u by $1 - u$, which amounts to interchanging the labels A and a .)

(ii) If s and σ are both negative, the heterozygote is more fit than either of the homozygotes, and we reger to this as the <u>heterozygote superior</u> case. (2.7) shows that this case is characterized by the fact that

$$g(u) \begin{cases} > u & \text{for } 0 < u < \pi_1 , \\ < u & \text{for } \pi_1 < u < 1 , \end{cases} \tag{2.9}$$

where

$$\pi_1 = \frac{\sigma}{s + \sigma} . \tag{2.10}$$

(iii) When s and σ are positive, we speak of the heterozygote inferior case. Here

$$g(u) \begin{cases} < u & \text{for } 0 < u < \pi_0 , \\ > u & \text{for } \pi_0 < u < 1 , \end{cases} \tag{2.11}$$

where π_0 is given by the same formula (2.10) as π_1.

For the sake of simplicity, we shall confine our attention to the heterozygote intermediate case (2.8) here. For other cases and the general models see [57].

The above model can be modified in various other ways. The niches may consist of parallelograms, hexagons, or other regions instead of squares. We can consider the limiting case of very small regions, so that $u_n(x)$ becomes a function of the continuous variable x and the formula (2.5) for Q is replaced by

$$Q[u](x) = \int_{R^2} m(x-y)g(u(y))\, dy . \tag{2.12}$$

If $m(x)$ depends only on the Euclidean distance $|x|$, the habitat is isotropic (rotationally symmetric). This case was treated in previous work [11], [12], [48], [49], [53], [54]. A discrete habitat cannot be rotationally symmetric[1], and therefore we need to treat nonisotropic operators Q here. Because of such factors as prevailing winds and the position of the sun, the migration may well be nonisotropic even in a homegeneous isotropic habitat.

Fisher's equation (1.1) can be obtained from the recursion (1.2) with Q defined by (2.5) by taking a suitable limit in which both the size of the niches and the length of the time step approach zero.

The above model involves various assumptions which may or may not be satisfied. One can think of the recursion (1.2) itself as the model. It says that the gene fraction of each deme of one generation is uniquely determined by the set of all gene frequencies of the preceding generation. If the niches are not too small relative to the maximum distance of migration, $Q[u](x)$ only depends on the values of u at a few nearby niches, and the operator Q can be determined experimentally by varying these values and watching the outcome.

The major assumption inherent in the class of models (1.2) is that the number of individuals born in each deme is so large that the numbers of individuals of the three genotypes who survive to adulthood do not depend upon the sizes of the initial populations but only on their gene fraction. It is this assumption which allows us to work with a scalar-valued rather than a vector-valued dependent variable.

One must, of course, also assume that all the external influences on growth, selection, and migration remain the same for all generations.

Recursions of the form (1.2) can serve as models for many other biological processes. We may for example consider the population growth of a single migrating species with synchronized nonoverlapping generations. Let the habitat again consist of the plane R^2 divided into squares centered at (ih,jh), and let $u_n(x)$ represent the size of the population which is in the square centered at x at a certain part of the life cycle of the nth generation. If one assumes that as far as interaction with the species under consideration is concerned, the rest of the world does not change in time, one obtains a model of the form (1.2) where Q may be measured experimentally.

If $u_n(x)$ is the number of individuals in the square centered at x just after migration, these individuals produce $g(u_n)$ new individuals and die, where $g(u)$ is a function whose graph is called the reproduction curve [44]. A fraction $m(y-x)$ of these new individuals migrates to y. Thus one obtains the model (2.5). One can again obtain space limits such as (2.12) by letting the size of the squares approach zero and partial differential equations like the Fisher equation (1.1) by letting both the square size and the length of the life cycle approach zero.

In the above models we have made the assumption of determinism. That is, we assume that u_{n+1} is completely determined by the function u_n, the assignment being given by the operator Q. In nature things are, of course, not that simple, and the map from u_n to u_{n+1} is more likely to be a Markov process. That is, given the function u_n, one has a probability distribution for the function u_{n+1}.

It may, however, happen that, if u_n is known, one can say with probability one that $u_{n+1}(x)$ lies in a certain interval $[v^-(x), v^+(x)]$ for each value of x. The function $v^-(x)$ and $v^+(x)$ are determined by u_n and hence are given by applying two operators Q and Q' to u_n. Thus, instead of the recursion (1.2), we are led to the "interval statement"

$$Q[u_n] \le u_{n+1} \le Q'[u_n] .$$

We shall show in §4 that if the operators Q and Q' have some of the properties mentioned in this section, the theorems of §6 will give information about the behavior of u_n for large n .

3. FORMULATION OF THE PROBLEM.

In this section we shall formulate a mathematical model which contains many of the models discussed in the preceding section as special cases.

Since we wish to treat spaces of one, two, and three dimensions, we shall work in a general Euclidean space R^N of N dimensions. We shall often identify a point of R^N with the vector from the origin to this point, so that we can speak of the vector sum of two points and of the point αx where α is a scalar.

Definition. A habitat $\mathcal{H}$ is defined to be a set of points in R^N with the following property: If x and y are in $\mathcal{H}$, then $x + y$ and $x - y$ are also in $\mathcal{H}$. This property implies that the origin 0 is in $\mathcal{H}$. (In fact, $\mathcal{H}$ is a group under addition.)

In order to avoid trivialities we shall assume that N is actually the dimension of $\mathcal{H}$. That is, we suppose that there is no unit vector which is orthogonal to all of $\mathcal{H}$. We also suppose that $N \geq 1$ so that $\mathcal{H}$ contains a point other then the origin. Since it contains all multiples of this point, $\mathcal{H}$ is unbounded.

In our applications the dependent variable u has only nonnegative values. If u is a gene fraction, its values lie on the interval $[0,1]$. In the case 1 population growth we normalize the carrying capacity to be 1 and assume that this value is never exceeded.

Definition. B is the set of continuous functions defined on with values on the interval $[0,1]$.

For any y in $\mathcal{H}$ the translation operator

$$T_y u(x) = u(x-y)$$

is defined.

Our basic evolution law is the recursion

$$u_{n+1} = Q[u_n] ,$$

where u_n and u_{n+1} are elements of B and Q is a given operator on B. The properties of the model are the properties of Q.

The homogeneity of the habitat is equivalent to the assumption that Q commutes with any translation. That is, $Q[T_y[u] = T_y[Q[u]]$ for all u in B.

A constant function α is clearly translation invariant. That is, $T_y\alpha = \alpha$ for all y. Consequently, $T_y Q[\alpha] = Q[\alpha]$ for all y, which means that $Q[\alpha]$ is again a constant. This simply states that in a homogeneous habitat the effects of migration cancel when u is constant. Thus, the properties of the model in the absence of migration can be found by looking at what Q does to constant functions.

Because $\sum_{x \in \mathcal{H}} m(x) = 1$ in the model (2.5), we see that $Q[\alpha] = g(\alpha)$ there. We shall therefore require the function $Q[\alpha]$ to have properties like those of the function g.

We shall also assume that increasing u everywhere in one generation increases u everywhere in the next one. More precisely, if $u(x) \geq v(x)$ for all x, then $Q[u](x) \geq Q[v](x)$ for all x. In the xase of the model (2.5), this means that the function $g(u)$ is nondecreasing. This is certainly true of the function (2.6) when $1 + s$ and $1 + \sigma$ are positive.

In the model (2.5) for population growth this assumption implies that the reproduction curve is nondecreasing. This excludes the case of a humped reproduction curve, which even in the absence of migration leads to oscillations [42], [52] and sometimes to chaos [23], [32], [36].

In addition to these assumptions we shall require $Q[u]$ to behave continuously with respect to changes in u. We summarize our basic hypotheses:

(i) $Q[0] \in B$.

(ii) $Q[T_y[u]] = T_y[Q[u]]$ for all $u \in B$, $y \in \mathcal{H}$.

(iii) $Q[\sigma] > \sigma$ for $\sigma \in (0,1)$,

$$Q[0] = 0, \tag{3.1}$$

$Q[1] = 1$.

(iv) $u \leq v$ implies that $Q[u] \leq Q[v]$.

(v) $u_n \to u$ as $n \to \infty$ uniformly on each bounded subset of $\mathcal{H}$ implies that $Q[u_n](x) \to Q[u](x)$ for each $x \in \mathcal{H}$.

The operator Q will be understood to satisfy these conditions throughout the remainder of this work.

The hypothesis (iii) coresponds to the heterozygote intermediate case. Other cases are also treated in [57], but we wish to keep the discussion simple.

As we shall see, these hypotheses suffice to define wave speeds and to determine the large-time behavior of solutions u_n which for $n = 0$ are less then 1 and vanish outside a bounded set.

In order to establish the existence of traveling waves (Theorem 6.6) we shall need the following compactness property:

Every sequence v_n of functions in B with $v_n \leq 1$ has a subsequence v_{n_i} such that the sequence $Q[v_{n_i}]$ converges uniformly on every bounded subset of $\mathcal{H}$. (3.7)

We are given an initial function u_0 in B and we shall be interested in predicting the behavior for large n of the sequence of functions $u_n(x)$ which is determined by the recursion

$$u_{n+1} = Q[u_n] .$$

We now state some propositions which are used to establish our results. The first of these is a comparison principle.

<u>Proposition 3.1</u>. (comparison principle) Let R be an operator from B to B which is order preserving in the sense that

$$v \geq w \Rightarrow R[v] \geq R[w] . \tag{3.9}$$

If the sequence v_n satisfied the inequalities

$$v_{n+1} \geq R[v_n] \tag{3.10}$$

while the sequence satisfies

$$w_{n+1} \leq R[w_n] , \tag{3.11}$$

and if $v_0 \geq w_0$, then $v_n \geq w_n$ for all n .

This proposition is proved by induction: if $v_n \geq w_n$, then $v_{n+1} \geq R[v_n] \geq R[w_n] \geq w_{n+1}$. If we let $v_n = w_{n+1}$, we obtain

<u>Proposition 3.2</u>. Let R have the order-preserving property (4.1), and suppose that $R[w_0] \geq w_0$. If the sequence w_n is defined by the recursion

$$w_{n+1} = R[w_n] ,$$

then $w_{n+1} \geq w_n$ for all n .

4. THE WAVE SPEED.

In this section we shall define a wave speed $c^*(\xi)$ corresponding to any operator Q which satisfies the hypotheses (3.1). The name wave speed will be justified by the theorems which are stated in the next section. For the moment, $c^*(\xi)$ will be defined as a scalar-valued function of unit vectors ξ. We can think of $c^*(\xi)$ as the speed of plane waves whose normal is in the direction ξ.

In order to define c^* we begin by choosing a function $\phi(s)$ of one real variable with the properties

$$\begin{aligned} &(i)\ \phi \text{ is continuous and nonincreasing} \\ &(ii)\ \phi(-\infty) \in (0,1)\ , \\ &(iii)\ \phi(s) = 0 \quad \text{for } s \geq 0\ . \end{aligned} \tag{4.1}$$

For any real number c and any unit vector ξ we define the operator

$$R_{c,\xi}[a](s) \equiv \max \{\phi(s), Q[a(x\cdot\xi + s + c)](0)\} \tag{4.2}$$

on continuous functions of one real variable $a(s)$ with $0 \leq a \leq 1$. Here the maximum just means the larger of two numbers for each s. The function $a(x\cdot\xi + s + c)$ is to be regarded as a function of x in $\mathcal{H}$ with ξ, s, and c fixed.

We shall suppose that Q has the properties (3.1). Then $R_{c,\xi}$ still has the property

$$R_{c,\xi}[\alpha] > \alpha \quad \text{for } \alpha \in (0,1) \tag{4.3}$$

and the order preserving property (3.1.iv).

We now define a sequence $a_n(c,\xi;s)$ by the recursion

$$a_{n+1} = R_{c,\xi}[a_n]\ , \quad a_0 = \phi\ . \tag{4.4}$$

We begin by stating some simple properties of this sequence.

<u>Lemma 4.1</u>. The sequence $a_n(c,\xi;s)$ is nondecreasing in n, nonincreasing in s and c, and continuous in c, ξ, and s.

We remark that since Q may only be defined on continuous functions, the continuity of the a_n as functions of s is needed in order that the recursion (5.4) really define the sequence a_n.

We now consider the limits of a_n at $s = \pm\infty$.

<u>Lemma 4.2</u>. Define the sequence of constants α_n by the recursions

$$\alpha_{n+1} = Q[\alpha_n] , \qquad \alpha_0 = \sigma(-\infty) . \tag{4.5}$$

Then α_n increases to 1 as $n \to \infty$, and for all n, c, and ξ

$$a_n(c,\xi;-\infty) = \alpha_n , \quad a_n(c,\xi;+\infty) = 0 \tag{4.6}$$

It follows that $a(c,\xi;-\infty) = 1$. The value $a(c,\xi;+\infty)$ may or may not be 1. Since a is nonincreasing in s, $a(c,\xi:+\infty) = 1$ if and only if $a = 1$. We give a criterion to determine whether or not this is the case.

<u>Lemma 4.3</u>. The value $a(c,\xi:+7) = 1$ if and only if there is an n such that

$$a_n(c,\xi;0) > \phi(-\infty) . \tag{4.7}$$

Since $a_1(c,\xi;0) > \phi(-\infty)$ when c is sufficiently negative, we see that $a(c,\xi;+\infty) = 1$ for such c. We now define

$$c^*(\xi) \equiv \sup\{c|a(c,\xi;+\infty) = 1 . \tag{4.8}$$

If $a\ (c,\xi;+\infty) = 1$ for all c, we set $c^*(\xi) = +\infty$.

Since $a(c,\xi;s)$ is the limit of a nondecreasing family of continuous functions, it is a lower semicontinuous function of c, ξ, and s. Lemma 5.3 shows that $c(c,\xi;+\infty) = 1$ if and only if $a(c,\xi;0) > \phi(-\infty)$, and the lower semicontinuity shows that the set of (c,ξ) where this is valid is open. This fact has two consequences:

Proposition 4.1. $a(c,\xi;+\infty) = 1$ if and only if $c < c^*(\xi)$, and $c^*(\xi)$ is a lower semicontinuous function of ξ.

The sequence a_n depends upon the choice of the function $\phi(s)$. Consequently $c^*(\xi)$ appears to depend upon this choice. The following lemma shows that this is not the case.

Lemma 4.4. Let $\hat{\phi}(s)$ be any continuous nonincreasing function of s with the properties

$$\hat{\phi}(-\infty) \in (0,1) .$$
$$\hat{\phi}(s) = 0 \quad \text{for} \quad s \geq 0 \tag{4.9}$$

Define the sequence $\hat{a}_n(c,\xi;s)$ by the recursion

$$\hat{a}_{n+1}(s) = \max\{\hat{\phi}(s),\ Q[\hat{a}_n(x\cdot\xi + s + c)](0)\} ,$$
$$\hat{a}_n(s) = \hat{\phi}(s) , \tag{4.10}$$

so that $\hat{a}_n$ increases to a nonincreasing limit function $\hat{a}(c,\xi;s)$. Then $\hat{a}(c,\xi;+\infty) = a(c,\xi;+\infty)$.

Lemma 5.4 shows that the definition (5.9) gives the same value of $c^*(\xi)$ when a is replaced by $\hat{a}$, so that c^* does not depend on the choice of the function ϕ.

It is important to know the behavior of a as $s \to +\infty$ when $c \geq c^*(\xi)$.

Proposition 4.2.

$$\lim_{s \to +\infty} a(c,\xi;s) = 0$$

uniformly in c and ξ on every set of the form $\{(c,\xi) | c \geq c^*(\xi), c^*(\xi) \leq C, |\xi| = 1\}$ where C is a constant.

While each $a_n(c,\xi;s)$ is continuous in c, ξ, and s, Lemma 4.3 shows that the limit function $a(c,\xi;s)$ is discontinuous in c at $c = c^*(\xi)$. Moreover, a may also be discontinuous in s. Consider, for example the operator

$$Q[u](s) = \frac{1}{3} u(s+1)[2 - u(s+1)] + \frac{2}{3} u(s-1)[2 - u(s-1)]$$

on $\mathcal{H} = R^1$.

It is obvious from the definition that

$$a(c,1;s) = 0 \quad \text{for } c \geq 1 , \ s \geq 0 .$$

Moreover, because

$$Q[u](s+1) \geq \frac{2}{3} u(s)[2 - u(s)] > u(s) \quad \text{when } u(s) < \frac{1}{2} ,$$

we find that if $\phi(s) > 0$ for $s < 0$,

$$a(1,1;s) \geq \frac{1}{2} \quad \text{for } s < 0 .$$

Thus the function $a(1,1;s)$ has a jump at $s = 0$.

It is easily seen that $a(c,1;s) \equiv 1$ for $c < 1$, so that $c^*(1) = 1$ in this example.

Of course, since $a(c,\xi;s)$ is lower semicontinuous and nonincreasing in c and in s, it is continuous from the right in c and in s.

The following three properties of the wave speed give an intuitive feeling for this function.

<u>Proposition 4.3</u>. If $c^*(\xi)$ is the wave speed corresponding to the operator Q, then the wave speed corresponding to the operator $T_y Q$ is $c^*(\xi) + y \cdot \xi$.

Proposition 4.4. If Q has the property

$$u(x) = 0 \quad \text{for} \quad |x| < b \quad \text{implies} \quad Q[u](0) = 0$$

then

$$c^*(\xi) \le b \qquad \text{for all} \quad \xi .$$

Proposition 4.5. Let Q_1 and Q_2 be two operators with the properties (3.1). Let $c_1^*(\xi)$ be the wave speed corresponding to Q_1. If

$$Q_1[u] \le Q_2[u]$$

for all continuous functions u on B with values on $[0,1]$ then

$$c_1^*(\xi) \le c_2^*(\xi)$$

for all ξ.

5. STATEMENT OF THE THEOREMS. We shall state our results in this section.

For any set V of vectors in R^N we define

$$nV = \{v_1 + v_2 + \ldots + v_n | v_j \in V \quad \text{for} \quad j = 1, \ldots, n\} . \qquad (5.1)$$

It is easily seen that if V is convex, then

$$nV = \{nv | v \in V\}$$

We denote the unit sphere $\{\xi \in R^N | |\xi| = 1\}$ by S^{N-1}.

We shall show that, in an asymptotic sense, $c^*(\xi)$ is a propagation speed for arbitrary initial disturbances.

We define the convex set

$$\mathscr{S} = \{x \quad R^N | x\cdot\xi \quad c^*(\xi) v \xi \in S^{N-1}\} . \qquad (5.2)$$

If $c^*(\xi)$ were a true propagation speed, so that $u = 0$ for $x \cdot \xi \le 0$ implied that $Q[u] = 0$ for $x \cdot \xi \le c^*(\xi)$, but not that $Q[u] = 0$ for $x \cdot \xi \le c$ when $c < c^*(\xi)$, then it would follow that, if the initial disturbance u_0 is concentrated at the origin, the support of $u_1 = Q[u_0]$ is $\mathcal{S}$. More generally it would follow that if u_n is defined by the recursion $u_{n+1} = Q[u_n]$, the support of u_n is $n\Phi + \mathrm{supp}(u_0)$. (Note that $n\mathcal{S} = \{x | x \cdot \xi \le nc^*(\xi)\ \xi \in X^{N-1}\}$.)

Our first two theorems say that this property is approximately true when n is large. The first theorem says that if the support of u_0 is bounded, then at time n very little disturbance lies for beyond $n\mathcal{S}$, while the second theorem states that the disturbance at time n fills most of $n\mathcal{S}$.

In all the theorems the sequence u_n of functions in B is a solution of the recursion $u_{n+1} = Q[u_n]$, and Q satisfies the hypotheses (3.1).

<u>Theorem 5.1</u>. Suppose that the set $\mathcal{S}$ defined by (6.2) is bounded and not empty. Let $\mathcal{S}'$ be any open set which contains $\mathcal{S}$. Suppose that $u_0 = 0$ outside a bounded set. If $u_0 < 1$, then

$$\lim_{n\to\infty} \max_{x \notin n\mathcal{S}'} u_n(x) = 0 . \tag{5.3}$$

If $\mathcal{S}$ is empty and c^* is bounded, (5.3) holds when the maximum is over the whole space.

<u>Theorem 5.2</u>. Suppose that the interior of $\mathcal{S}$ is not empty and let $\mathcal{S}''$ be any closed bounded subset of the interior of $\mathcal{S}$. For any $\sigma > 0$ there exist a radius r_σ with the property that if $u_0(x) \ge \sigma$ on a ball of radius r_σ and if $u_{n+1} = Q[u_n]$, then

$$\lim_{n\to\infty} \min_{x \in n\mathcal{S}''} u_n(x) = 1 .$$

The next two theorems show how to estimate the wave speed in some cases.

Theorem 5.3. If $m(x,dx)$ is a bounded nonnegative measure on $\mathcal{H}$ with the property that for all continuous u with $0 \le u \le 1$

$$Q[u](x) \le \int u(x-y)m(y,dy) \ , \tag{5.4}$$

then

$$c^*(\xi) \le \inf_{\mu>0} \frac{1}{\mu} \log \int e^{\mu x\cdot\xi} m(x,dx) \ . \tag{5.5}$$

Theorem 5.4. Suppose that $\ell(x,dx)$ is a bounded nonnegative neasure on with the properties that

$$\int \ell(x,dx) > 1$$

and that there is a positive ε such that for all continuous u with $0 \le u \le \varepsilon$

$$Q[u](x) \ge \int u(x-y)\ell(y,dy) \ .$$

Then $\mathcal{S}$ is not empty and

$$c^*(\xi) \ge \inf_{\mu>0} \frac{1}{\mu} \log \int e^{\mu x\cdot\xi} \ell(x,dx) \ , \tag{5.6}$$

where the right-hand side is $+\infty$ if the integral on the right diverges for all positive μ .

If the measure ℓ is not concentrated on any hyperplane $x\cdot\xi$ = constant , then $\mathcal{S}$ has interior points, and the radius r_σ in Theorem 5.2 can be chosen so that it does not depend upon σ .

We note the following immediate corollary of Theorems 5.3 and 5.4.

Corollary. If the nonnegative bounded measure $m(x,dx)$ satisfies (5.6) when $0 \le u \le 1$ and if for every positive σ there is a positive ε such that

$$Q[u](x) \ge (1 - \delta) \int u(x-y)m(y,dy) \quad \text{when} \quad 0 \le u \le \varepsilon , \tag{5.7}$$

then

$$c^*(\xi) = \inf_{\mu \to 0} \frac{1}{\mu} \int e^{\mu x \cdot \xi} m(x,dx) \tag{5.8}$$

for all unit vectors ξ .

Theorem 5.5. (Hairtrigger effect) Let Q and ℓ satisfy the hypotheses of Theorem 5.4 and suppose that the support of ℓ contains a set W of vectors in $\mathcal{H}$ with the property that any bounded subset of $\mathcal{H}$ is contained in a translate of the set $n W$ for some integer n . Then if u_0 is not identically zero on W and $\mathcal{S}''$ is as in Theorem 5.2,

$$\lim_{n\to\infty} \min_{x \in n\mathcal{S}''} u_n(x) = 1 \tag{5.9}$$

This theorem states that r_σ in Theorem 5.2 may be chosen to be arbitrarily small, regardless of the value of σ .

The last theorem asserts the existence of a traveling wave of speed c with normal ξ whenever $c \ge c^*(\xi)$. We now need the property (3.7).

Theorem 5.6. Suppose that, in addition to (3.1), Q has the compactness property (3.7). Then if $c \ge c^*(\xi)$, there is a nonincreasing function $W(s)$ which is defined for all s of the form $x \cdot \xi - nc$ with x in $\mathcal{H}$ and n an integer such that the sequence $u_n(x) = W(x \cdot \xi - nc)$ satisfies the recursion (1.2), $W(-\infty) = 1$, and $(W + \infty) = 0$.

REFERENCES

1. D.G. Aronson, The Asymptotic Speed of Propagation of a Simple Epidemic, Nonlinear Diffusion, (W.E. Fitzgibbon and H.F. Walker, eds.) Research Notes in Mathematics 14, Pitman, London, 1977, pp. 1-23.
2. D.G. Aronson, Density Dependent Interaction Diffusion Systems, Dynamics and Modeling of Reactive Systems. Academic Press, New York.
3. D.G. Aronson and L.A. Peletier, Large Time Behavior of Solutions of the Porous Medium Equation in Bounded Domains, J. Differential Equations.
4. D.G. Aronson and H.F. Weinberger, Nonlinear Diffusion in Population Genetics, Combustion, and Nerve Propagation, Partial Differential Equations and Related Topics, (J. Goldstein, ed.), Lecture Notes in Mathematics 446, Springer, New York, 1975, pp. 5-49.
5. D.G. Aronson and H.F. Weinberger, Multidimensional Nonlinear Diffusion Arising in Population Genetics, Adv. in Math., 30 (1978), pp. 33-76.
6. T. Bonnesen and W. Fenchel, Theorie der Konvexen Korper, Ergeb.d. Math. u. ihrer Grenzgeb., 3, Chelsea, New York, 1948.
7. M. Bramson, Maximal Displacement of Branching Brownian motion, Comm. Pure Appl. Math., 31 (1978), pp. 531-581.
8. C. Conley, An Application of Wazewski's Method to a Nonlinear Boundary Value Problem Which Arises in Population Genetics, J. Math. Biol., 2 (1975), pp. 241-249.

9. R. Courant, Differential and Integral Calculus, vol. I., Interscience New York, 1936.

10. O. Diekmann, Thresholds and Travelling Waves for the Geographical Spread of Infection, J. Math. Biol., 6 (1978), pp. 109-130.

11. O. Diekmann, Run for Your Life, A Note on the Asymptotic Speed of Propagation of an Epidemic, J. Differential Equations, 33 (1979), pp. 58-73.

12. O. Diekmann and H.G. Kaper, On the Bounded Solutions of a Nonlinear Convolution Equation, J. Nonlin. Analysis, 2 (1978), pp. 721-737.

13. P.C. Fife and J.B. McLeod, The Approach of Solutions of Nonlinear Diffusion Equations to Travelling Wave Solutions, A.M.S. Bull., 81 (1975), pp. 1076-1078; Arch. Rational Mech. Anal., 65 (1977), pp. 335-361.

14. P.C. Fife and L.A. Peletier, Nonlinear Diffusion in Population Genetics, Arch. Rational Mech. Anal., 64 (1977), pp. 93-109.

15. R.A. Fisher, The Advance of Advantageous Genes, Ann. Eugenics, 7 (1937), pp. 355-369.

16. R.A. Fisher, Gene Frequencies in a Cline Determined by Selection and Diffusion, Biometrics, 6 (1950), pp. 353-361.

17. W. Fleming, A Selection Migration Model in Population Genetics, J. Math. Biol., 2 (1975), pp. 219-233.

18. H. Fujita, On the Blowing Up of Solutions of the Cauchy Problem for $u_1 = \Delta u + u^{1+\alpha}$, J. Fac. Sci. Univ. Tokyo (1), 13 (1966), pp. 109-124.

19. M.E. Gurtin and R.C. MacCamy, On the Diffusion of Biological Populations, Math. Biosci., 33 (1977), pp. 35-49.

20. K.P. Hadeler and F. Rothe, Travelling Fronts in Nonlinear Diffusion Equations, J. Math. Biol., 2 (1975), pp. 251-263.

21. J.B.S. Haldane, The Theory of a Cline, J. Genetics, 48 (1948), pp. 277-284.

22. J.M. Hammersley, Postulates for Subadditive Processes, Ann. Probab., 2 (1974), pp. 652-680.

23. F. Hoppenstead and J.M. Hyman, Periodic Solutions of a Logistic Difference Equation, SIAM J. Appl. Math, 32 (1977), pp. 73-81.

24. Y. Kametaka, On the Nonlinear Diffusion Equations of Kolmogorov Petrovsky Piskunov Type, Osaka J. Math., 13 (1976), pp. 11-66.

25. Ja.I. Kanel, Stabilization of Solutions of the Cauchy Problem for Equations Encountered in Combustion Theory, Mat. Sbornik (N.S.) 101, 59 (1962) supplement, pp. 245-288.

26. Ja.I. Kanel, On the Stability of Solutions of the Equations of Combustion Theory for Finite Initial Functions, Mat. Sbornik (N.S.) 107, 65 (1964), pp. 398-413.

27. S. Karlin, Population Subdivision and Selection Migration Interaction, in Population. Genetics and Ecology, S. Karlin and E. Nevo, eds., Academic Press, New York, 1976, pp. 617-657.

28. D.G.Kendall, Mathematical Models of the Spread of Infection, in Mathematics and Computer Science in Biology and Medicine, H.M.S.O., London, 1965, pp. 213-225.

29. W.D. Kermack and G. McKendrick, A Contribution to the Mathematical Theory of Epidemics, Proc. Royal Soc. A, 115 (1927), pp. 700-721.

30. M. Kimura, Stepping-stone Model of Population, Ann. Report National Inst. of Genetics of Japan, 3 (1953), pp. 62-63.

31. A. Kolmogoroff, I. Petrovsky and N. Piscounoff, Etude de l'equations de la Diffusion Avec Croissance de la Quantite de Matiere et Son Application a un Probleme Biologique, Bull Univ. Moscow, Ser. Internal., Sec. A, 1 #6, pp. 1-25.

32. T.Y. Li and J.A. Yorke, Period Three Implies Chaos, Amer. Math. Monthly (1975), pp. 985-992.

33. D. Ludwig, D.G. Aronson and H.F. Weinberger, Spatial Patterning of the Spruce Budworm, J. Math. Biol., 8 (1979), pp. 217-258.

34. D. Ludwig, D.D. Jones and C.S. Holling, Qualitative Analysis of Insect Outbreak Systems: The Spruce Budworm and the Forest, J. Anim. Ecol., 47 (1978), pp. 315-332.

35. G. Malecot, Quelques Schemas Probabilistes sur la Variabilite des Populations Naturelles, Ann. Univ. Lyon Sci. A, 13 (1950), pp. 37-60.

36. R.M. May and G.F. Oster, Bifurcations and Dynamic Complexity in Simple Ecological Models, Amer. Naturalist, 110 (1976), pp. 573-599.

37. D. Mollison, Possible Velocities for a Simple Epidemic, Adv. Appl. Prob., 4 (1972), pp. 233-257.

38. T. Nagylaki, Conditions for the existence of Clines, Genetics, 80 (1975), pp. 595-615.

39. T. Nagylaki, A Diffusion Model for Geographically Structured Populations, J. Math. Biol., 6 (1978), pp. 375-382.

40. T. Nagylaki, The Geographical Structure of Populations, Studies in Mathematics, vol. 16: Studies in Mathematical Biology, Part II, S.A. Levin, ed., Math. Assoc. of America, Washington, 1978, pp. 588-623.

41. T. Nagylaki and M. Moody, Diffusion Model for Genotype Dependent Migration, Proc. Nat. Acad. of Sci. U.S.A., 77 (1980), pp. 4842-4846.

42. A.J. Nicholson, An Outline of the Dynamics of Animal Populations, Austr. J. Aool., 2 (1954), pp. 9-65.

43. M.H. Protter and H.F. Weinberger, Maximum Principles in Differential Equations, Prentice Hall, Englewood Cliffs, N.J., 1967.

44. W.E. Ricker, Stock and Recruitment, L. Fish. Res. Bd. Can., 11 (1954), pp. 559-623.

45. S. Sawyer, Results for the Stepping Stone Model for Migration in Population Genetics, Ann. Probab., 4 (1976), pp. 699-728.

46. M. Slatkin, Gene Flow and Selection in a Cline, Genetics, 75 (1973), pp. 733-756.

47. H.R. Thieme, A Model for the Spatial Spread of an Epidemic, J. Math. Biol., 4 (1977), pp. 337-351.

48. H.R. Thieme, Asymptotic Estimates of the Solutions of Nonlinear Integral Equations and Asymptotic Speeds for the Spread of Populations, J. Reine Angew. Math., 306 (1979), pp. 94-121.

49. H.R. Thieme, Density-dependent Regulation of Spatially Distributed Populations and Their Asymptotic Speeds of Spread, J. Math. Biol., 8 (1979), pp. 173-178.

50. K. Uchiyama, The Behavior of Solutions of the Equation of Kolomogorov-Petrovsky-Piskunov, Proc. Japan. Acad. Ser. A, 53 (1977), pp. 225-228.

51. K. Uchiyama, The Behavior of Solution of Some Nonlinear diffusion Equations for Large Time. J. Math. Kyoto U., 18 (1978), pp. 453-508.

52. S. Utida, Studies on Experimental Population of the Azuki Bean Weevil, Mem. Coll. of Agricult., Kyoto Imp. Univ., 48 (1941), pp. 1-30.

53. H.F. Weinberger, Asymptotic Behavior of a Model in Population Genetics, in Nonlinear Partial Differential Equations and Applications, J. Chadam, ed., Lecture Notes in Math. , 648, Springer, New York, 1978, pp. 47-98.

54. H.F. Weinberger, Asymptotic Behavior of a Class of Discrete Time Models in Population Genetics in Applied Nonlinear Analysis, V. Lakshmikanthan, ed., Academic Press, New York, 1979, pp. 407-422.

55. H.F. Weinberger, Genetic Wave Propagation, Convex Sets, and Semi-Infinite Programming, in Constructive Approaches to Mathematical Models, C.V. Coffman and G.J. Fix, eds., Academic Press, New York, 1979, pp. 293-317.

56. H.F. Weinberger, Some Deterministic Models for the Spread of Genetic and Other Alterations, Biological Growth and Spread---Mathematical Theories and Applications, W. Jager, H. Rost, and P. Tautu, eds., Springer Lecture Notes in Biomath 38, Springer, Berlin-Heidelberg-New York, 1980, pp. 320-349.

57. H.F. Weinberger, Long Time Behavior of a Class of Biological Models, S.I.A.M. J. on Math. Anal. 13 (1982), pp. 353-396.

H. F. Weinberger
University of Minnesota

L WHEELER

Applications of the maximum principle to the minimization of stress concentration in elastic solids subject to antiplane shear deformation

1. INTRODUCTION

An important practical problem in the mechanics of elastic solids is the analysis of stress concentration. The objective is to find the maximum stress under given conditions of loading and geometry and determine how the maximum can be reduced by variations in these conditions.

The task of reducing stress concentration is normally approached by a trial method in which the shape of the solid is remodelled iteratively until the stress is brought to within acceptable levels. This process suggests a variational problem in which we seek a shape that minimizes the stress concentration. This variational problem is unconventional in that the extremal is sought in the supremum norm. Despite this drawback, a degree of success has been obtained indirectly by reducing the problem to a free-boundary problem. The notion behind this reduction is natural. One seeks a shape that renders the stress uniform, an ideal which has been exploited in [1-4]. None of these works offer a rigorous proof that the free-boundary problem solves the optimization problem. They do, however, indicate that the free-boundary problem is sensible from the standpoint of existence of solutions.

In [5-7], the maximum principle for subharmonic functions was used to establish mathematically conditions under which the free-boundary problem yields the solution to the problem of minimum stress concentrations. The purpose of the present paper is to describe the optimization problem of stress concentration for states of antiplane shear deformation and to find conditions under which it can be rigorously solved by the free-boundary problem. An interesting aspect of the free-boundary problem in the present case is that it is equivalent to the two-dimensional free-streamline problem of classical hydrodynamics, which is the best understood of all free-boundary problems.

2. THE TRACTION BOUNDARY-VALUE PROBLEM FOR ANTI-PLANE SHEAR DEFORMATION.

Let x_i $(i = 1, 2, 3)$ denote rectangular Cartesian coordinates, and let D denote the plane domain formed by the intersection of the x_1-x_2 plane with a cylinder R whose generators extend in the x_3-direction to $x_3 = \pm\infty$. Occupying R is an elastic medium whose response is governed by the linearized theory for homogeneous and isotropic solids. For anti-plane shearing deformation parallel to the generators of R, the displacement components obey

$$u_1 = u_2 = 0 \text{ on } R, \quad u_3 = u_3(x_1, x_2) \text{ on } R , \tag{2.1}$$

and the governing equations reduce to

$$\nabla^2 u_3 = 0 \text{ on } D . \tag{2.2}$$

The stress components of interest are related to u_3 through

$$\tau_{31} = \frac{\partial u}{\partial x_1} , \quad \tau_{32} = \frac{\partial u}{\partial x_2} \text{ on } D , \tag{2.3}$$

where

$$u = \mu u_3 , \tag{2.4}$$

μ being the shear modulus, and the stress magnitude τ is given by

$$\tau = (\tau_{31}^2 + \tau_{32}^2)^{\frac{1}{2}} = |\nabla u| . \tag{2.5}$$

We assume that on the boundary of R the tractions are prescribed. Because of (2.1), these reduce to a single component t which is in the x_3-direction, but does not vary with x_3. There follows

$$\frac{\partial u}{\partial n} = t \quad \text{on} \quad \partial D , \tag{2.6}$$

where ∂D stands for the boundary of D. Thus, u is the solution of the two-dimensional Neumann problem

$$\nabla^2 u = 0 \quad \text{on} \quad D, \quad \frac{\partial u}{\partial n} = t \quad \text{on} \quad \partial D . \tag{2.7}$$

The stress function u is determined to within a constant by (2.7), and the stresses are uniquely determined from (2.3). Clearly, the stress does not depend upon the elastic constants.

We introduce the function v which is conjugate to u. Thus, we assume the existence of a function v related to u through the Cauchy-Riemann relations

$$\frac{\partial u}{\partial x_1} = \frac{\partial v}{\partial x_2} , \quad \frac{\partial u}{\partial x_2} = -\frac{\partial v}{\partial x_1} . \tag{2.8}$$

For D simply connected and bounded, we may take v to be the solution of the Dirichlet problem

$$\nabla^2 v = 0 \quad \text{on} \quad D, \quad v = f(s) \quad \text{on} \quad \partial D , \tag{2.9}$$

where

$$f(s) = \int_0^s t(\sigma)\,d\sigma , \tag{2.10}$$

in which s denotes arc length.

3. THE OPTIMIZATION PROBLEM.

The boundary of D is partitioned into two complementary parts consisting of a fixed part and an arc C which is variable and traction free. The idea is to give the loading on the fixed part and seek a form for C which minimizes the stress concentration. More specifically, by the optimization problem of minimum stress concentration, we mean the problem of finding within a class S of competing curves C, one which minimizes the quantity

$$\tau_M\{C\} = \sup_{C} \tau. \tag{3.1}$$

Thus, we seek an arc $C^* \epsilon S$ for which τ^* defined by

$$\tau^* = \sup_{C^*} \tau \tag{3.2}$$

satisfies

$$\tau^* = \inf_{C \epsilon S} \tau_M\{C\}. \tag{3.3}$$

4. REDUCTION OF THE OPTIMIZATION PROBLEM TO A CLASSICAL FREE-BOUNDARY PROBLEM.

It is natural to ask whether the optimization problem can be solved by finding an arc C in S along which τ is constant. In this case, the problem gives way to a free-boundary problem of the type familiar in the hydrodynamics of jets and cavities. Our purpose in the present section is to describe conditions under which the validity of this reduction can be established mathematically.

The idea behind the reduction to classical free-boundary problems stems from the Min-Max Principle of Garabedian and Spencer [8]. According to this principle, the free streamline connecting two points has the property that among all connecting streamlines, it has the least maximum speed. The proofs given below are based upon methods used in [9].

Theorem 1: Let S be such that for every C in S, the corresponding domain D consists of the intersection of the half space $x_2 > 0$ with the complement of a closed region which is star-shaped with respect to a point $x_2 = -a$ $(a \geq 0)$ of the x_2-axis (see Figure 1). Assume $\nabla^2 v = 0$ on D, $v = 0$ on ∂D, $v = x_2 + o(1)$ as $x_1^2 + x_2^2 \to \infty$, and assume there exists a curve C_0 in S such that

$$\tau = \tau_0 \quad \text{on} \quad C_0 ,$$

where τ_0 is a constant. Then

$$\inf_{C \in S} \tau_M\{C\} = \tau_0 .$$

Proof. Consider a curve $C \in S$, and let D denote the corresponding domain. Suppose first that $D_0 \subset D$. Since v is non-negative on D and $v_0 = 0$ on ∂D_0, the function

$$\bar{v} = v - v_0$$

satisfies

$$\bar{v} \geq 0 \quad \text{on} \quad \partial D_0 .$$

Therefore, and since

$$\nabla^2 \bar{v} = 0 \quad \text{on} \quad D_0 \quad \text{and} \quad \bar{v} = o(1) \quad \text{as} \quad x_1^2 + x_2^2 \to \infty,$$

it follows that

$$\bar{v} \geq 0 \quad \text{on} \quad D_0 . \tag{4.1}$$

The curves C and C_0 have common endpoints, at which they are tangent. Therefore, and by (4.1),

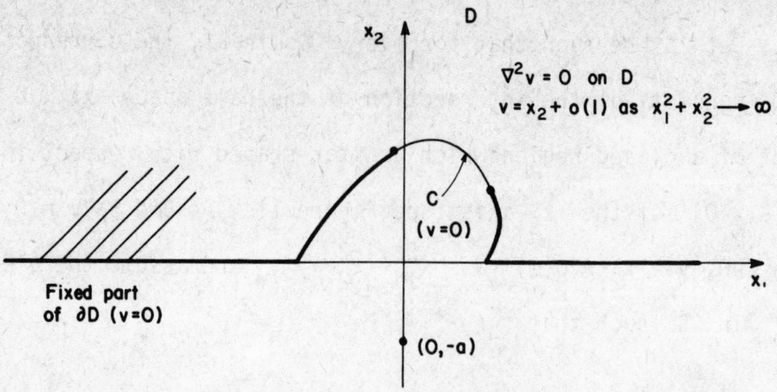

Figure 1. Optimum notch bottom

$$\frac{\partial v}{\partial n} \leq \frac{\partial v_0}{\partial n} \tag{4.2}$$

at the end points. At points of C_0,

$$\frac{\partial v_0}{\partial n} = -\tau_0 . \tag{4.3}$$

and at points of C,

$$\frac{\partial v}{\partial n} = -\tau .$$

Therefore, and by (4.2), we see that

$$\tau \geq \tau_0$$

at the endpoints, whence

$$\tau_M\{C\} \geq \tau_0 .$$

This takes care of the case $D_0 \subset D$.

Assume now that D_0 is not contained in D, and perform a contraction of D into a new domain $\hat{D}$, through

$$x = (x_1, x_2) \to (kx_1, kx_2) = (\hat{x}_1, \hat{x}_2) = \hat{x}$$

such that $D_0 \subset \hat{D}$ and C_0, $\hat{C}$ have a point $x^* = (x_1^*, x_2^*)$ in common. At x^*, the normals to C_0 and $\hat{C}$ coincide.

Define

$$\hat{v}(\hat{x}) = kv(\hat{x}k^{-1}) \quad \text{for all} \quad \hat{x} \in \hat{D} . \tag{4.4}$$

then

$$\hat{v},_{\alpha}(\hat{x}) = v,_{\alpha}(x) , \tag{4.5}$$

and

$$\tau(x) = |\nabla\hat{v}(kx)| \quad \text{for all} \quad x \in D . \tag{4.6}$$

Set

$$\bar{v} = \hat{v} - v_0 \quad \text{on} \quad D_0 . \tag{4.7}$$

Then

$$\nabla^2\bar{v} = 0 \quad \text{on} \quad D_0 , \tag{4.8}$$

$$\bar{v} \geq 0 \quad \text{on} \quad \partial D_0 , \tag{4.9}$$

and

$$\bar{v} \to 0 \quad \text{as} \quad x_1^2 + x_2^2 \to \infty . \tag{4.10}$$

Consequently, the maximum principle implies

$$\bar{v} \geq 0 \quad \text{on} \quad D_0 . \tag{4.11}$$

Therefore, and since

$$\bar{v}(x^*) = 0 , \tag{4.12}$$

there follows

$$\frac{\partial\bar{v}}{\partial n} \leq 0 \quad \text{at} \quad x^* , \tag{4.13}$$

so that

$$\frac{\partial \hat{v}}{\partial n} \le \frac{\partial v_0}{\partial n} \quad \text{at} \quad x^* . \tag{4.14}$$

Since

$$\left.\frac{\partial \hat{v}}{\partial n}\right|_{x^*} = -\tau(k^{-1}x^*) \quad \text{and} \quad \left.\frac{\partial v_0}{\partial n}\right|_{x^*} = -\tau_0 , \tag{4.15}$$

this gives

$$\tau_0 \le \tau(k^{-1}x^*) .$$

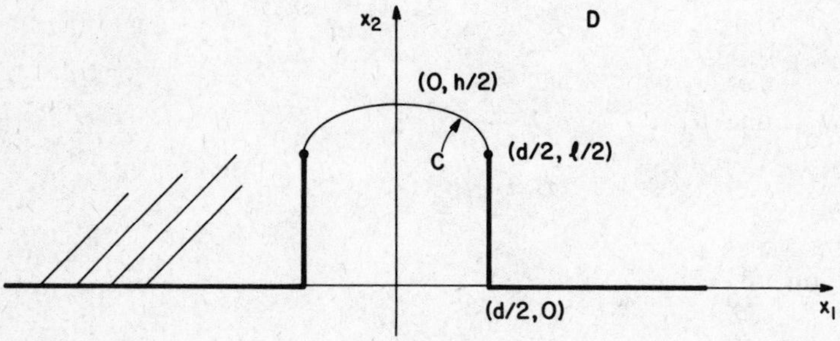

Figure 2

Accordingly, and because $k^{-1}x^*$ is a point of C, the desired conclusion now follows.

The foregoing theorem includes as a special case the domain depicted in Figure 2. Because of the theorem, we are able to obtain results for this case from long-established results for the RIABOUCHINSKI cavity problem (see, for example [9]) of hydrodynamics.

The following theorem concerns a class of bounded domains whose boundary (see Figure 3) is made up of two lines

$$L_1 : x_1 = a, \quad L_2 : x_1 = b, \quad \text{where} \quad a < b \tag{4.16}$$

and two parametric curves

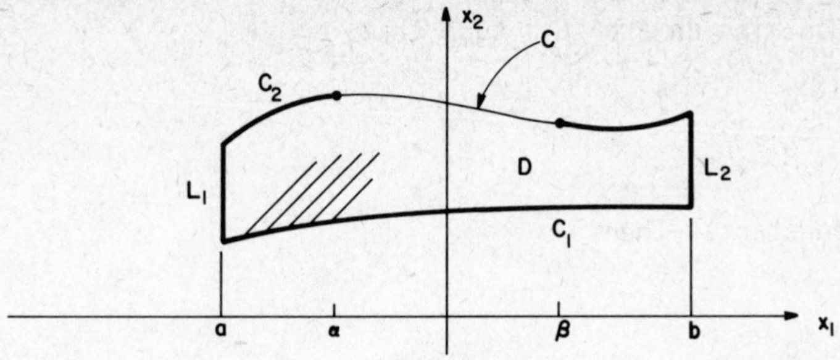

Figure 3

$$C_1 : x_2 = f(x_1), \quad C_2 : x_2 = g(x_1) \quad \text{for} \quad a < x_1 < b , \tag{4.17}$$

where

$$f(x_1) < g(x_1) \quad \text{for} \quad a < x_1 < b . \tag{4.18}$$

The variable points of ∂D, which make up the elements of S, comprise a sub-arc of C_2,

$$C : x_2 = g(x_1), \quad \alpha < x_1 < \beta \quad , \tag{4.19}$$

where $\alpha > a$ and $\beta < b$. While a, b, and f are completely fixed, the arc C_2 is fixed only on the subarcs

$$x_2 = g(x_1) \quad \text{for} \quad a < x_1 < \alpha \quad \text{and} \quad \beta < x_1 < b . \tag{4.20}$$

Theorem 2: Let S, a, α, β, b, f and g be as described in the preceding paragraph. Assume that

$$\nabla^2 v = 0 \quad \text{on} \quad D , \tag{4.21}$$

$$v = 0 \quad \text{on} \quad C_1 , \tag{4.22}$$

$$v = 1 \quad \text{on} \quad C_2 , \tag{4.23}$$

$$v(a,x_2) \quad \text{and} \quad v(b,x_2) \quad \text{are non-decreasing in} \quad x_2 , \tag{4.24}$$

and suppose S contains an arc C_0 such that

$$\tau = \tau_0 \quad \text{on} \quad C_0 \, , \tag{4.25}$$

where τ_0 is a constant. Then

$$\inf_{C \in S} \tau_M\{C\} = \tau_0 \, . \tag{4.26}$$

<u>Proof</u>. Let $C \in S$. If $D_0 \subset D$, proceed as in the proof of Theorem 1. Assume that D_0 is not contained in D.

Let $\hat{D}$ denote the domain obtained from D by performing a translation in the negative x_2-direction such that $\hat{C}$ (the corresponding translation of C) has in common with C_0 only points of tangency, and let x^* denote such a point. Let $\bar{D}$ stand for the component of $\hat{D} \cap D_0$ which contains x^* in its boundary.

Set

$$\bar{v} = \hat{v} - v_0 \quad \text{on} \quad \bar{D} \, . \tag{4.27}$$

Then

$$\nabla^2 \bar{v} = 0 \quad \text{on} \quad \bar{D} \tag{4.28}$$

and

$$\bar{v}(x^*) = 0 \, . \tag{4.29}$$

Now, points of the boundary $\partial\bar{D}$ lie within $\hat{C}$, L_1, L_2, C_1. Accordingly, and by (4.22)-(4.24), there follows

$$\bar{v} \geq 0 \quad \text{on} \quad \partial\bar{D} \, . \tag{4.30}$$

Thus, the maximum principle furnishes

$$\bar{v} \geq 0 \quad \text{on} \quad D \, . \tag{4.31}$$

Therefore, (4.29) yields

$$\frac{\partial \bar{v}}{\partial n} \leq 0 \quad \text{at} \quad x^* . \tag{4.32}$$

But

$$\frac{\partial v}{\partial n} = -\tau \quad \text{and} \quad \frac{\partial v_0}{\partial n} = -\tau_0 \quad \text{at} \quad x^* , \tag{4.33}$$

so

$$\tau \geq \tau_0 \quad \text{at} \quad x^* . \tag{4.34}$$

The proof is now complete.

Let us now consider the class of domains described by Figure 4. Here, the boundary consists of two curves C_1 and C_2 which are star-shaped with respect to a point 0 and two radial segments R_1 and R_2 through 0.

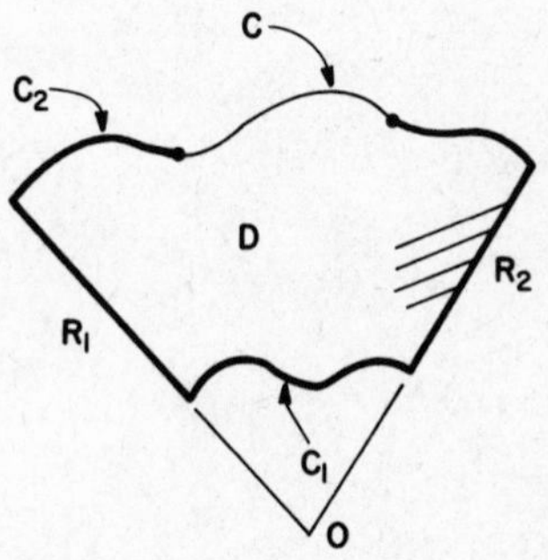

Figure 4

<u>Theorem 3</u>: Let S and the corresponding domains D be as in the preceding paragraph. Assume that

$$\nabla^2 v = 0 \quad \text{on} \quad D , \tag{4.35}$$

$$v = 0 \quad \text{on} \quad C_1 , \tag{4.36}$$

$$v = 1 \quad \text{on} \quad C_2 , \tag{4.37}$$

and that v does not decrease on R_1 or R_2 with distance from 0. Suppose that S contains an arc C_0 such that

$$\tau = \tau_0 \quad \text{on} \quad C_0 , \tag{4.38}$$

where τ_0 is a constant. Then

$$\inf_{C \in S} \tau_M\{C\} = \tau_0 . \tag{4.39}$$

<u>Proof</u>. Let $C \in S$. The case $D_0 \subset D$ can be handled as it was in Theorem 1, so assume D_0 is not contained in D.

Let $\hat{D}$ stand for the domain found from D by performing a contraction $\hat{x} = kx$ such that $\hat{C}$ intersects C_0 only at points of tangency. Let x^* stand for such a point, and let $\bar{D}$ stand for the component of $\hat{D} \cap D_0$ that contains x^* in its boundary.

Set

$$\hat{v}(\hat{x}) = kv(\hat{x}k^{-1}) \quad \text{for all} \quad \hat{x} \in \hat{D} , \tag{4.40}$$

and note that

$$\tau(x) = |\nabla\hat{v}(kx)| \quad \text{for all} \quad x \in D . \tag{4.41}$$

Let

$$\bar{v} = \hat{v} - v_0 \quad \text{on} \quad \bar{D} . \tag{4.42}$$

Then

$$\nabla^2\bar{v} = 0 \quad \text{on} \quad \bar{D} , \tag{4.43}$$

and

$$\bar{v} \geq 0 \quad \text{on} \quad \partial\bar{D} , \tag{4.44}$$

so

$$\bar{v} \geq 0 \quad \text{on} \quad \bar{D} \tag{4.45}$$

by the maximum principle.

Accordingly, and because

$$\bar{v}(x^*) = 0 , \tag{4.46}$$

there follows

$$\frac{\partial \bar{v}}{\partial n} \leq 0 \quad \text{at} \quad x^* . \tag{4.47}$$

But

$$\left.\frac{\partial \hat{v}}{\partial n}\right|_{x^*} = -\tau(k^{-1}x^*) \quad \text{and} \quad \left.\frac{\partial v_o}{\partial n}\right|_{x^*} = -\tau_o , \tag{4.48}$$

so (4.42) and (4.47) yield

$$\tau_o \leq \tau(k^{-1}x^*) ,$$

and the conclusion follows immediately. This completes the proof.

REFERENCES.

1. A.J. Durelli and W.M. Murry, Stress distribution around an elliptical discontinuity in any two-dimensional, uniform, and axial system of combined stress, Exper. Stress Analysis Proc., Vol. 1, No. 1, 19-31 (1943).
2. H. Neuber, Der zugbeanspruchte Flachstab mit optimalem Querschnittübergang, Forsch. Ingenieurwesen, 35, 29-30 (1969).
3. H. Neuber, Zur Optimierung der Spannungskonzentration, Continuum Mechanics and Related Problems of Analysis, Nauka, Moscow, 375-380 (1972).

4. G.P. Cherepanov, Inverse problems of the plan theory of elasticity, P.M.M., 38, No. 6 (1974).

5. L.T. Wheeler, On the role of constant stress surfaces in the problem of minimizing elastic stress concentration, Int. J. Solids and Structures, Vol. 12, 779-789 (1976).

6. L.T. Wheeler, On optimum profiles for the minimization of elastic stress concentration, ZAMM, 58, T235-T236 (1978).

7. L.T. Wheeler and I.A. Kunin, On voids of minimum stress concentration, Int. J. Solids and Structures, Vol. 18, 85-89 (1982).

8. P. Garabdeian and D.C. Spencer, Extremal methods in cavitational flow, J. Rat. Mech. Analysis, 1, 359-409 (1952).

9. D. Gilbarg, Jets and cavities, Encyclopedia of Physics, S. Flugge, Ed., Springer, New York: 1960, Vol. IX/III.

Lewis Wheeler

University of Houston